Strategic Management of Marine Ecosystems

RENEWALS 458-4574

DATE DUE

NATO Science Series

A Series presenting the results of scientific meetings supported under the NATO Science Programme.

The Series is published by IOS Press, Amsterdam, and Springer (formerly Kluwer Academic Publishers) in conjunction with the NATO Public Diplomacy Division.

Sub-Series

I. **Life and Behavioural Sciences**   IOS Press
II. **Mathematics, Physics and Chemistry**   Springer (formerly Kluwer Academic Publishers)
III. **Computer and Systems Science**   IOS Press
IV. **Earth and Environmental Sciences**   Springer (formerly Kluwer Academic Publishers)

The NATO Science Series continues the series of books published formerly as the NATO ASI Series.

The NATO Science Programme offers support for collaboration in civil science between scientists of countries of the Euro-Atlantic Partnership Council. The types of scientific meeting generally supported are "Advanced Study Institutes" and "Advanced Research Workshops", and the NATO Science Series collects together the results of these meetings. The meetings are co-organized by scientists from NATO countries and scientists from NATO's Partner countries — countries of the CIS and Central and Eastern Europe.

Advanced Study Institutes are high-level tutorial courses offering in-depth study of latest advances in a field.
Advanced Research Workshops are expert meetings aimed at critical assessment of a field, and identification of directions for future action.

As a consequence of the restructuring of the NATO Science Programme in 1999, the NATO Science Series was re-organized to the four sub-series noted above. Please consult the following web sites for information on previous volumes published in the Series.

http://www.nato.int/science
http://www.springeronline.com
http://www.iospress.nl

Series IV: Earth and Environmental Series – Vol. 50

Strategic Management of Marine Ecosystems

edited by

Eugene Levner
Holon Academic Institute of Technology,
Holon, Israel

Igor Linkov
Cambridge Environmental Inc.,
Cambridge, MA, U.S.A.

and

Jean-Marie Proth
INRIA, Metz, France

Proceedings of the NATO Advanced Study Institute on
Strategic Management of Marine Ecosystems
Nice, France
1–11 October 2003

A C.I.P. Catalogue record for this book is available from the Library of Congress.

ISBN 1-4020-3158-0 (PB)
ISBN 1-4020-3157-2 (HB)
ISBN 1-4020-3198-X (e-book)

Published by Springer,
P.O. Box 17, 3300 AA Dordrecht, The Netherlands.

Sold and distributed in North, Central and South America
by Springer,
101 Philip Drive, Norwell, MA 02061, U.S.A.

In all other countries, sold and distributed
by Springer,
P.O. Box 322, 3300 AH Dordrecht, The Netherlands.

Printed on acid-free paper

All Rights Reserved
© 2005 Springer
No part of this work may be reproduced, stored in a retrieval system, or transmitted in any form or by any means, electronic, mechanical, photocopying, microfilming, recording or otherwise, without written permission from the Publisher, with the exception of any material supplied specifically for the purpose of being entered and executed on a computer system, for exclusive use by the purchaser of the work.

Printed in the Netherlands.

Table of contents

Preface 5

Acknowledgments 7

Chapter 1. Disturbance of Marine Ecosystems: Problems and Solutions 9

Meta-analysis of the radioactive pollution of the ocean 11
Alexey V. Yablokov

Marine protected areas: a tool for coastal areas management 29
C. F. Boudouresque, G. Cadiou, L. Le Diréac'h

Damage control in the coastal zone: improving water quality by harvesting aquaculture-derived nutrients 53
Dror L. Angel, Timor Katz, Noa Eden, Ehud Spanier, Kenny D. Black

A modular strategy for recovery and management of biomass yields in large marine ecosystems 65
Kenneth Sherman

Express methods of nondestructive control for physiological state of algae 81
T.V.Parshikova

Chapter 2. Modeling Approaches and Mathematical Foundations of Environmental Management 93

Strategic management of ecological systems: a supply chain perspective 95
E. Levner, J.-M. Proth

Modelling the environmental impacts of marine aquaculture 109
William Silvert

Addressing uncertainty in marine ecosystems modelling 127
Lyne Morissette

Environmental games and queue models 143
Charles S. Tapiero

Computational complexity of modeling ecosystems 159
Vladimir Naidenko, Inna Bouriako and Jean-Marie Proth

Chapter 3. Policy/Stakeholder Process in Marine Ecosystem Management 167

Performance metrics for oil spill response, recovery, and restoration: a critical review and agenda for research 169
T.P. Seager, I. Linkov, C. Cooper

The challenges to safety in the east mediterranean: mathematical modeling and risk management of marine ecosystems 179
K. Atoyev

Strategic management of marine ecosystems using whole-ecosystem simulation modelling: the 'back to the future' policy approach 199
Tony J. Pitcher, Cameron H. Ainsworth, Eny A. Buchary, Wai Lung Cheung, Robyn Forrest, Nigel Haggan, Hector Lozano, Telmo Morato and Lyne Morissette

Chapter 4. Management of Contaminated Sediments: Example of Integrated Management Approach 259

Towards using comparative risk assessment to manage contaminated sediments 261
T. Bridges, G. Kiker, J. Cura, D. Apul, I. Linkov

Multi-criteria decision analysis: a framework for managing contaminated sediments 271
I. Linkov, S. Sahay, G. Kiker, T. Bridges, T.P. Seager

Barriers to adoption of novel environmental technologies: contaminated sediments 299
T.P. Seager, K.H. Gardner

Index of authors 313

Preface

The demand for advanced management methods and tools for marine ecosystems is increasing worldwide. Today, many marine ecosystems are significantly affected by disastrous pollution from industrial, agricultural, municipal, transportational, and other anthropogenic sources. The issues of environmental integrity are especially acute in the Mediterranean and Red Sea basins, the cradle of modern civilization. The drying of the Dead Sea is one of the most vivid examples of environmental disintegration with severe negative consequences on the ecology, industry, and wildlife in the area. Strategic management and coordination of international remedial and restoration efforts is required to improve environmental conditions of marine ecosystems in the Middle East as well as in other areas.

The NATO Advanced Study Institute (ASI) held in Nice in October 2003 was designed to: (1) provide a discussion forum for the latest developments in the field of environmentally-conscious strategic management of marine environments, and (2) integrate expertise of ecologists, biologists, economists, and managers from European, American, Canadian, Russian, and Israeli organizations in developing a framework for strategic management of marine ecosystems.

The ASI addressed the following issues:
- Key environmental management problems in exploited marine ecosystems;
- Measuring and monitoring of municipal, industrial, and agricultural effluents;
- Global contamination of seawaters and required remedial efforts;
- Supply Chain Management approach for strategic coastal zones management and planning;
- Development of environmentally friendly technologies for coastal zone development;
- Modeling for sustainable aquaculture; and
- Social, political, and economic challenges in marine ecosystem management.

Papers presented in this book were submitted by the ASI lecturers and participants. In addition, several papers were invited from the leading scientists in the field. The organization of the book reflects discussions during the meeting. The papers in the first chapter review and summarize problems related to marine ecosystems. They provide the background and examples of environmental challenges and potential solutions. The second chapter provides modeling and mathematical foundations for specific environmental management methods and tools useful for marine ecosystem management. These methods provide a means for coordinating technological, economical, and ecological contradicting demands and offer an exciting prospect for efficient utilization of environmental resources. For example, Strategic Supply Chain management methodology permits detailed characterization of the functional and structural aspects of ecosystems, assesses the impact of human activity on biological systems, and evaluates practical consequences stemming from the activity. The third chapter presents several papers dealing with integration of political and stakeholder priorities with environmental modeling. A key paper by Pitcher and his colleagues introduces an integrative approach to the strategic management of marine ecosystems with policies based on restoration ecology, and an understanding of marine ecosystem processes in the light of findings from terrestrial ecology. The critical issues include whether past ecosystems make viable policy goals, and whether desirable goals may be reached from today's ecosystem. The final chapter provides another integrated approach for marine ecosystem management that is based on comparative risk assessment and multi-criteria decision analysis. Three papers presented in the chapter illustrate the theoretical foundation of these methods and review applications for a wide range of issues related to sediment management – from highly technical issues (such as selection of optimal technology) to political (assessing value judgment for policy decision makers and stakeholders).

An important objective of the ASI was to identify specific initiatives that could be developed by those in attendance and their broader network of institutions to enhance the progress of environmental risk assessment in developing countries. Consistent with this goal, this book presents the interpretation and perception of issues related to strategic management of marine ecosystems by individual scientists, while also illustrating a wide variety of environmental problems in developing countries.

Eugene Levner, Igor Linkov, and Jean-Marie Proth
August 2004.

Acknowledgments

Editors would like to thank the many authors who have contributed significantly to the quality of these proceedings.
The work presented in this book could not have been started if it were not for the generous support provided by the Scientific and Environmental Affairs Division of the North Atlantic Treaty Organization (NATO). In particular, we would like to thank Doctor Alain H. Jubier for his valuable help.
The staff of INRIA (Institut National de Recherche en Informatique et en Automatique – The French National Institute of Research in Computer Science and Automatic) has been extremely helpful in the organization of the ASI. Missis Christel Wiemert who had the difficult task to assemble the final manuscript and adjust the presentation of most of the papers, have been extremely helpful.
Finally, ASI Directors would like to thank Doctor Igor Linkov for joining them in preparing this book.

Chapter 1

Disturbance of Marine Ecosystems: Problems and Solutions

META-ANALYSIS OF THE RADIOACTIVE POLLUTION OF THE OCEAN

Alexey V. YABLOKOV
*Center for Russian Environmental Polic, Russia, 119991, Moscow,
Vavilova Street, 26, e-mail: yablokov@ecopolicy.ru; www.atomsafe.ru*

Abstract
Humans have been altering the marine environment for millennia. Up till now, five critical environmental issues have affected the oceans: over-fishing, chemical pollution and eutrophication, habitat destruction, invasion of exotic species and global climate change. However, one of the major threats the oceans may face in the twenty-first century is radioactive pollution over the second half of the twentieth century.

1. Introduction

The following may be listed among the main anthropogenic sources of radioactive pollution of the ocean:
- Dumping of solid (SRW) and liquid radio-wastes (LRW);
- Pollution from underwater N-explosions;
- Radioactive pollution from land (including river run-off and land-based activities);
- Radioactive fallout from the atmosphere;
- Radioactive pollution originating from accidents (lost N-warheads and radio-emission from thermo-electric generators, sunken craft and ships, falling satellites with radioactive materials, etc.);
- Discharge from ships with N-reactors.

In spite of intensive studying [1,2,3,4], we are still far from having established a really comprehensive inventory of all the anthropogenic radioactive sources of the Ocean. This is mostly due to the fact that much of this data is connected with military activities and remains classified. In India – possibly one of the world's most marine polluted country – for example, the Nuclear Energy Act prohibits the release of information related to nuclear facilities [5]. It looks like only the Russian Federation after the collapse of the USSR have published a more or less complete inventory of radioactive pollution of adjacent seas [6,7]. These circumstances call for some meta-analysis, which will include official as well as unofficial data, for a general ecological understanding of the situation as far as radionuclides pollution of the ocean is concerned.

2. Radioactive dumping

Beginning in the late 40s and up till 1983, at least 13 countries with a nuclear industry (Belgium, Britain, Germany, Japan, Italy, France, the Netherlands, New Zealand, South Korea, Sweden, Switzerland and the United States) dumped their SRW and LRW into deep parts of ocean (more than 4 km deep). All these countries (excluding the USSR) officially reported dumping up to 1,2 million Ci in radioactive materials [7], including (at time of dumping): USSR – 1 037 kCi; Great Britain - 948 kCi; Switzerland - 119 kCi; USA – 95 kCi; Belgium - 57 kCi; France - 9.6 kCi; the Netherlands – 9.1 kCi; Japan – 0.4 kCi; Sweden – 88 Ci; New Zealand – 28; Germany – 5 Ci; Italy - 5 Ci [2]The largest single radioactive object ever dumped into the Ocean over that period of time was USS *Seawolf*'s sodium-cooled reactor (up to 33 000 Ci) which was scuttled 3000 m deep off the Delaware coast (Maryland) in 1954 [8].

It's possible to use the USSR' dumping activity as the most well known case study [6,7,9]. In 1959, 600 m^3 of low-level liquid waste (LWR) was discharged in the White Sea (20 mCi) and in 1960, the *Lenin* discharged 100 m^3 of LRW (200 mCi) near Gogland Island in the Gulf of Finland. The total activity of LRW dumping is 24 kCi (903 TBq, including 87 TBq for ^{239,240,241}Pu): Baltic Sea - 0.2 Ci (0.0007 TBq); White Sea - 100 Ci (3.7 TBq); Barents Sea - 12153 Ci (450 TBq); Kara Sea - 8500 Ci (315 Tbq). A total of at least 12 335 Ci (456 TBq) of LRW was dumped by the USSR between 1966 and 1991 in the Sea of Japan and near the southeastern coast of the Kamchatka Peninsula.

Low- and intermediate-level SRW dumped over 65 operations between 1967 and 1991 in the White, Barents and Kara seas was enclosed into more than 11 000 metal containers, barges, lighters, and tankers (a total of 17 craft). The total activity of sunken intermediate and low-level SRW, was over 15.5 kCi (574 TBq) in the Kara Sea and 40 Ci (1.5 TBq) in the Barents Sea. The total activity of intermediate and low-level SRW (6868 sunken containers, 38 sunken ships, and over 100 other individual sunken large objects) dumped by the USSR in the Sea of Japan and other areas of the Pacific was 6 851 Ci (254 TBq). SRW comprised mainly contaminated film coverings, tools, personal protective devices, uniforms, fittings, pipelines, activity filter boxes, pumps, steam generators, and various objects contaminated during ship repair work.

Among all RW dumping by the USSR in the ocean, the greatest ecological hazard is presented by objects with SNF (Table 1).

In total, during the 70s and the-80s the former USSR dumped at least 10 reactors with SNF, two shielding assemblies, and 14 reactors without SNF from N-subs and 3 – from icebreakers in the North Atlantic, Arctic and Pacific oceans. The maximum activity of solid radio-wastes that entered seas adjacent to Russia may have been in excess of 2,5 million Ci (at the time of disposal). The main radionuclides were ^{134,137}Cs, ^{90}Sr, ^{239,240}Pu, ^{63}Ni, ^{60}Co.

After more precise calculations and accounts [10,11,12] it was revealed that the activity of the icebreakers' shielded assembly was 3.5 times higher (not 5600 but 19500 TBq), and the radioactivity of the N-subs' reactors dumped in the Kara Sea was 4.8 times lower (not 2.25 but 0.46 million Ci). At the same time it was discovered that data about the dumping of the two N-subs' reactors in the Sea of Japan. was missing from the White Book Therefore the total activity of the N-reactors which were dumped in the Sea of Japan, was not 1.7 but 396 TBq (940 kCi). According to the new estimate the USSR dumped not up to 2.5 million Ci, but 1.64 million. Because of natural decaying, by the year 2000 there are 0.7 millions Ci in the North Pacific's dumping sites, and 107 kCi (4 PBq) – in the Kara Sea.

TABLE 1. Objects with Spent Nuclear Fuel Dumped by the USSR in the oceans [6,7]

Object	Place, Year	Depth, Meters	Max. activity kCi*	Radionuclides
Compartments of NS's with 4 reactors, 3 containing SNF	Abrosimov Inlet, Kara Sea, 1965	20	1200	Fission products
Shielding assembly of reactor from icebreaker *Lenin* with residual	Tsivolka Inlet, Kara Sea, 1967	49	100	^{137}Cs (50 kCi), ^{90}Sr (50 kCi), ^{238}Pu, ^{241}Am, ^{244}Cm(2 kCi)
Reactor from NS	Novaya Zemlya' Depression, Kara Sea, 1972	300	800	Fission products
NS with two reactors	Stepovoy Inlet, Kara Sea, 1981	50	200	Fission products
Two NS reactors	Sea of Japan, 1978	3000	0,046	Fission products
Core plate from the Reactor of NS No. 714	East of Kamchatka, Pacific, 1989	2500	70	Fission products
Total: 8 reactors, two shield assemblies	Arctic and Pacific	20 – 3000	Up to 2500	86% fission products, 12% activation products, 2% actinides

*- At the time of dumping

In the Arctic Ocean the reactors were dumped mainly in the shallow fjords of Novaya Zemlya at a depth ranging from 12 to 135 m and in the Novaya Zemlya Trough at depths of up to 380 m [6]. Before dumping, the reactor compartments with SNF were filled with a hardening furfurol-based mixture. This filling was supposed to prevent the SNF from being in contact with seawater for up to 500 years. The shield assembly with SNF from the icebreaker was additionally placed in a reinforced concrete container and a metal shell. Between 1992 and 2000 some studies were carried out around several dumping places in the Kara Sea. It was revealed that in the Stepovoy Inlet, Abrosimov Inlet and Tsivolka Inlet, leakage from dumped objects reached worrisome levels of radioactivity – by ^{137}Cs up to 109 kBq/kg dry sediment, ^{90}Sr – 3,8 kBq/kg, ^{60}Co – 3,2 kBq/kg, ^{239,240}Pu – 18 Bq/kg [4]. There is up to 31 Bq/m^3 in the surrounding water, which is six times more than at the surface [13]. As some calculation show, maximum possible release in the Abrosimov Bay may reach 1 TBq per year of ^{137}Cs [14]. Similar investigations carried out in 2003 along the coast of Russia's far eastern Maritime Territory and Sakhalin island revealed elevated concentrations of Cs-137 in two locations at depths of 3,000 m [15]. It means, that in spite of the absence of any immediate danger, it is obvious that the situation is slowly getting out of control. A special study of the radiological impact of this dumping [16] revealed that it may be dangerous to stay on some Novaya Zemlya beaches adjacent to the fjords used as dumping sites.

Official records from the radioactive dumping into other parts of the ocean are spotty. The figure of 47 dumping sites in the North Atlantic and Pacific [17] is far from the truth. The U.S. officially reported on 30 dumping sites in the Atlantic and Pacific (1946 – 1970). Three of them near the Farallon Isl., off San Francisco Bay, at depths of 90, 900 and 1800 meters, totaling 52 530 55-gallon drums, with a total activity of some 14,7 kCi (540 TBq) [18]. The aircraft carrier *Independence* used as target in the Bikini Atoll's U.S. bomb tests is believed to have been sunk here [19]. 33 998 containers were dumped by the U.S. between 1951 and 1967 in the West Atlantic, with a total radioactivity of up to 77,5 kCi at the time of dumping. Total official U.S. dumping included 52 530 containers with 14,7 kCi [18].

It was not unusual in U.S. and Soviet dumping practice to shoot radio-waste containers with guns when they would not sink. Now after 40-30 years many containers can become corroded and cracked and can disintegrate as, for example, in the Hurd Deep, off the Channel Isl. in UK territorial water (official European dumping site between 1950 and 1963). The UK dumped 50 570 containers with 44,1 kCi beta- and 3,3 kCi alfa-emitters here between 1950 and 1967 [20].

In spite of a special international agreement strongly prohibiting the dumping since 1983 this practice has been continuing up till now in some places, using some loops in the legislation. For example, every year up to 200 tons of low-grade radio-wastes generated by the oil industry are pumped out into the Northern Sea from *Scotoil's* purification plant in Aberdeen, Great Britain [21].

3. Radioactive pollution due to military accidents

There were about 90 publicly reported military accidents involving nuclear weapons (59 American, 25 Soviet/Russian, four French and one British). Most of them involved N-submarines, but also involved planes, missiles, nuclear-waste storage facilities and surface ships [22,23,24,18,25]. There are four Russian and two U.S. N-submarines (with more than half-dozen reactors and nearly 50 nuclear warheads) already at the bottom of the Ocean. Among them:

- 1963. U.S. NS *Thresher*, Western Atlantic. In 1990 ^{60}Co in sediments were detected near the NS.
- 1968. U.S. NS *Scorpion,* 400 miles southwest of the Azores Isl., Atlantic, with two N-warheads on board. In 1990 ^{60}Co in sediments were detected near the NS.
- 1968. Soviet NS K-129, near Hawaiian Islands, Pacific, more than 6000 m deep, with five N-warheads (in 1974 U.S. *Glomar Explorer* retrieved two N-warheads during operation *Jennifer*).
- 1970. Soviet NS K-8 (*November),* Bay of Biscay, 4680 m deep, with two N- reactors with a total activity of 250 kCi and 10 N-warheads. 1986. Soviet NS K-219 (*Yankee*) with two reactors (activity of 250 kCi) and 32

(50?) N-warheads, 600 miles northeast of Bermuda, 5500 meters deep. The warheads (total activity about 7 kCi) were scattered on the sea floor and have surely been leaking ^{239}Pu.

- 1989. Soviet NS K-278 (*Komsomoletz*) with one reactor and two N-torpedoes (total activity 150 kCi including 2,9 PBq ^{90}Sr, 3,1 PBq ^{137}Cs and 25 TBq for actinides), Norwegian Sea, 1685 m deep. One important difference between this accident and others, including those involving US N-subs, is the threat of accelerated release of radionuclides into the marine environment. The reason is that the *Komsomolets* has a titanium pressure hull. The rate of corrosion is increased a thousandfold when titanium reacts with the ship's steel components in seawater. In 1993 ^{137}Cs concentration near the NS was five times higher (10-30 Bq/m^3) than on surface [4].
- 2002. Russian NS K-141 (*Kursk*) with two reactors (activity up 150 kCi), Barents Sea, 105 m deep (it was later raised).
- 2003. Russian NS K-159 with two reactors (activity about 500 kCi), Barents Sea, 240 m deep.
- In 1985, while NS K-431 (*Viktor*) was having its reactor refueled in Soviet Maritime Territory, Chazhma Bay, an uncontrolled spontaneous chain reaction occurred. A radioactive fallout occurred on the water surface for up to 30 km, and the total release of radioactive substances into the atmosphere was at about 2 000 kCi for short living gases and 5 000 kCi (185 PBq) for other fissions, mostly iodine isotopes, ^{60}Co, ^{54}Mn and other activation radionuclides [26]. A large part of the water area of Ussury Bay was radioactively contaminated. One hour after the explosion, the activity of short-living radionuclides in the seawater reached 2 Ci/l. The radioactivity of bottom sediments is mainly due to ^{60}Co and to ^{137}Cs.

At least five other accidents involving N-subs and ships resulted in the release of radionuclides into the ocean:

- 1966. Soviet NS, NS base in Poliarny, Kola Bay, Barents Sea;
- 1971. U.S. NS SSN-583 (*Dase*), Western Atlantic;
- 1986. Soviet NS K-175, Kamran' Bay, South China Sea;
- 1989. Soviet NS K-192 (Echo-2), Barents Sea;
- 1989. Soviet NS, Ara Bay, Barents Sea; released up to 2 kCi (74 TBq) LWR;
- 1993. Russian icebreaker *Arctic*, Kara Sea;
- 1997. Russian vessel *Imandra* (floating storage for SNF), Kola Bay, Barents Sea.

Sunken N-bombs are one of the serious sources of radioactive pollution. Although this information was always kept top-secret (and never officially reported), several (from many?) cases like this are known [27,28,29,25]:

- 1950. A U.S. B-36 dropped the N-bomb off the coast of British Columbia;
- 1952. U.S. C-124 "*Globemaster*" transport aircraft with three N-bombs and "nuclear capsule" jettisoned two of the bombs east of Rehobeth, Delaware, and Cape May, Wildwood, New Jersey. In spite of an intensive search, they are still there at the bottom of the ocean;1956. the U.S. Air Force lost a

bomber with two nuclear-weapon cores in their carrying cases over the Mediterranean Sea;
- 1958. A U.S. B-47 bomber collided with another jet near the U.S. Air Force's base on Tybee Island, Georgia; An H-bomb was jettisoned into water several miles off the mouth of the Savannah River in Wassaw Sound off Tybee Beach, Georgia. In spite of an intensive search it was never found.;
- 1959. A U.S. Navy P-5M aircraft carrying a nuclear depth charge (without fissile core) crashed into Puget Sound near Whidbey Island;
- 1965. U.S. A-4E bomber *Skyhawk* loaded with a N-bomb B-43 rolled off an elevator on aircraft carrier *Ticonderoga* (CVA-14) and fell into the sea several miles off the Ryakyu Islands, Japan;
- 1966. A U.S. B-52 aircraft carrying four multi-megaton N-bombs crashed into an Air Force KC-135 refueling tanker and dropped all its weapons near Palomares, Spain (two of the bombs off shore; two bombs ruptured, scattering radioactive particles over 100 km^2, The 3rd bomb landed instantly, and the 4th was lost 19 km off the coast (It was found after 870 days of intensive search, involving about 80 ships and thousands of servicemen);
- 1968. A U.S. B-52 aircraft with four N-bombs crashed 7 miles south of the Thule Air Force Base, Bylot Sound, Greenland; ^{239,240,241}Pu from the bombs spread over the ice (up to 11 TBq) and sank to the bottom (^{239,240}Pu concentration in sediments in 2000 had reached 7600 Bq/kg);
- 1977. A N-warhead was dropped into the ocean from Soviet NS K-171 (*Delta-1*) in the Western Pacific, near the Kamchatka Peninsula; it was successfully found and recovered.

Special assessment [30] revealed that 27 types of N-sub failure at sea or in base, 3 types of failure of other ships and 4 types of failures connected with storing and transporting nuclear weapons have radiological consequences.

At least in one case the ocean was polluted by a missile: in 1962 a nuclear test device atop a Thor rocket booster fell into the Pacific near Johnston Atoll.

4. Pollution from underwater N-explosions

Underwater and close-to-surface (above-water) N-explosions are a very serious source of radioactive contamination of the Ocean. There are several places where such pollution occurs:
- Bikini atoll, Marshall Islands (one underwater, 13 above-water U.S. N-explosions);
- Enewetak atoll, Marshall Islands (two underwater, 18 above-water U.S. N-explosions);
- Pacific Ocean (two underwater U.S. N-explosions);
- Christmas Island, Polynesia (several U.S. and British underwater and above-water N-explosions);

- South Atlantic, between 38^0 and 49^0 S.L. and between 8^0 and 11^0 W.L (three U.S. above-water N-explosions);
- Moruroa and Fangataufa atolls, Tuamotu Archipelago, French Polynesia (four above-water French N-explosions).
- Chemaya Bay, South-Eastern Barents Sea (three underwater and three above-water USSR N-explosions).

Now several thousand square kilometers of the bottom of the South Barents Sea are the most Pu-polluted place in the ocean – up to 15 kBq/kg in sediments [31]. There are also here ^{241}Am, ^{137}Cs, ^{90}Cr, ^{155}Eu and other radionuclides (Table 2).

TABLE 2. Distribution of Radionuclides (Bq/kg dry weight) in Surface Bottom Deposits of the Southeastern Barents Sea, 1992 [32]

	Chernaya Bay	Karskie Vorota strait	Pechora Gulf	Of Vaygach Island
^{241}Am	2622.0	0.0	0.0	0.0
^{137}Cs	1444.2	9.4	23.8	6.7
^{60}Co	618.0	0.0	0.0	0.0
^{155}Eu	344.4	0.0	0.0	0.0
^{212}Pb	298.5	25.1	66.2	24.1
^{212}Bi	260.9	0.0	22.8	24.3
^{226}Ra	206.4	62.2	0.0	0.0
^{214}Pb	170.0	30.1	45.8	27.7
^{224}Ra	139.9	7.6	22.4	26.9
^{214}Bo	131.7	26.5	39.4	22.6
^{208}Ti	89.6	8.8	19.2	8.6
^{228}Ac	33.5	18.7	51.3	16.6

The total amount of the radionuclides concentrated in the Chernaya Bay area is about 3×10^{12} Bq or 81 kCi [33]. Some of them could be diffused all over the Eastern part of the Barents Sea over the course of this Century with a potential negative impact on Norwegian and Russian fisheries and marine life. Through differences in the concentration of Pb, Cs, Pu, Co radionuclides in the sediment profiles, it is possible to calculate when the pollution started (depths of 10 to 15 cm), and when it maximized (3-4 cm depth [34]). Even 40 year after the N-explosions, levels in bottom sediments here were up to 200 mCi/kg ^{239,240}Pu, and in water – ^{137}Cs up to 200 Bq/m^3, ^{90}Sr up to 140 Bq/m^3 [4]. Chernaya Bay may now be one of the most radioactive polluted places in the Ocean, only comparable with the waters of the Enewetok atoll.

5. Radioactive fallout from the atmosphere.

Atmospheric deposition of radionuclides originating from more than 500 atmospheric N-tests and N-industry activities in the 1960s and 1970s were the main sources of radioactive pollution of the Ocean. Before 1980, the following was released into the atmosphere from N-tests [29,25]: ^{140}Ba – 732 Ebq; ^{131}J – 651 Ebq; ^{141}Ce – 254 Ebq; ^{3}H

– 240 Ebq; ^{103}Ru –238 Ebq; ^{95}Zr – 143 Ebq; ^{91}Y – 116 Ebq; ^{89}Sr – 91.4 Ebq; ^{144}Ce – 29.6 Ebq; ^{106}Ru – 11.8 Ebq; ^{137}Cs – 0.912 EBq, ^{90}Sr – 0.604 Ebq, $^{239, 240, 241}$Pu – (0.151 – 0.375) Ebq, etc, for an overall total of more than 2 510 Ebq. Although the USSR, the USA, France and Great Britain stopped atmospheric tests after 1963 (China – after 1983), the fallout of fission residuals (mostly ^{137}Cs, ^{90}Sr, ^{239,240}Pu, ^{131}J, ^{14}C, ^{3}H) with an activity of many millions Ci from the atmosphere to the Ocean will continue for many centuries to come.

The second a source of radioactive pollution of the ocean from the atmosphere after N-tests is accidental and regular discharges from Nuclear Power Plants. If N-test pollution is steadily declining with time, the level of NPP pollution due to annual emission is increasing. Some calculations [35] indicate that due to the activities of the 2000 NPPs (now – 440) that produce about 1000 GWt annually the individual effective dose will soon reach 1 mSv annually.

The Chernobyl catastrophe (the atmospheric discharge was about 50-250 MCi) immediately resulted in the serious radioactive pollution of the North Atlantic (Baltic, Norwegian, North and Irish Seas) and the Mediterranean (especially the Black Sea). ^{134}Cs concentrations in the Baltic's waters were up to 6 000 Bq/m^3, in dry alga *Fucus vesiculosus* they reached 4900 Bq/kg (^{131}J - up to 29 kBq/kg), ^{137}Cs and in dry plankton they came up to 2500 Bq/kg [4]. These characteristics are three times higher than pre-Chernobyl levels. In the Black Sea the highest ^{137}Cs radioactivity from Chernobyl was about 500 Bq/m^3, i.e. 30 times higher than pre-accident levels [36]. It must be noted that in the days and weeks following the catastrophe the levels of tens of other radionuclides were hundreds to thousands of times higher than the level of ^{137}Cs: this was bound to have an enormous negative impact on the marine environment.

The atmospheric radioactive fallout which was collected by pack and floating sea-ice can be transported over long distance, for instance from the Kara shelf to the waters off Greenland and Norway [4].

6. Radioactive pollution from land

The man-originated radioactive flow from land to the seas includes two main sources: river runoffs and direct discharges from land-based activities.

Radioactive river runoff to the Ocean mainly consists of atmospheric fallout and direct radioactive discharges into the rivers. Atmospheric fallout results from N-tests, Chernobyl and other radioactive catastrophes, and is also due to "permissible" everyday discharges into air from up to 440 commercial Nuclear Power Plants, from hundreds of scientific N-reactors and from several reprocessing facilities. As a result the soils of the Northern Hemisphere contain ^{137}Cs at a level of about 40 mCi/km^2, and ^{90}Sr at about 30 mCi/km^2, on average. These radionuclides collected by catchment areas are slowly moving to the ocean. Entries of ^{90}Sr and ^{137}Cs to the Barents Sea from river runoff between 1961 and 1989 were about 6 kCi (200 TBq). The total entry of ^{137}Cs and ^{90}Sr to the Barents Sea from the atmosphere with global fallout of the products of nuclear explosions over the same period is estimated at approximately 100 kCi (3700 TBq). Calculations for the Kara Sea give corresponding values of 33 kCi (1200 TBq) and 70 kCi (2600 TBq).

During the production of the USSR's nuclear arsenal (up to 40 000 N-warheads) three USSR uranium/plutonium plants, "MAYAK" in South Ural (Tobol – Irtysh – Ob' river system), Siberian Chemical Combine in Western Siberia (Tom' – Ob' river system), and Mining Chemical Combine in Eastern Siberia (Enysey river system) produced up to 2.71 billion Ci LRW. From these 1.61 billion Ci are stored in surface water bodies. Sooner or later, a considerable part of these radionuclides will go to the Arctic Ocean due to the Northern Asia Slope. According to expert estimate, the radioactive discharge to the Ob-Irtysh river system alone was up to 63 PBq [4]. There are concentrations of these "weaponry" radionuclides in the sediments of the Enisey and Ob' deltas: up to 100 Bq/kg by ^{137}Cs, and up to 16 Bq/m^3 by ^{90}Sr (in sediments). In the 1960s and 70s, the Kara Sea waters had ^{90}Sr up to 85 Bq/m^3, ^{137}Cs up to 40 Bq/m^3 and ^{239,240}Pu up to 16 mBq/m^3 [4]. It is reasonable to calculate that for their production of 30 000 N-warheads, the USA must have produced similar amounts of LRW (about 2 billion Ci). North American lakes Ontario and Erie were radioactively contaminated from the West Valley N-reprocessing plant in 1972. 400 millions m^3 from a liquid radioactive uranium mill tailing were sent to Church Rock, New Mexico after breaking the dam in 1979 [8]. Over their 40 years of producting fissile material for the U.S. nuclear arsenal, the Hanford Engineering Works discharged many billions of m^3 of radioactive water directly into the Columbia River. The same phenomenon occurred at the Savannah River production plant in the southeast of the USA, and at the Sequoiah Fuel Corp. uranium processing plant, which contaminated the Arkansas River. Through the rivers, a considerable part of these "military" radionuclides were transported to the ocean.

Mining is an additional source of inland radioactive water pollution. It never released any new manmade radionuclides, but introduced large amounts of natural uranium, radium (like in the Pechora River, North Ural, Russia), and thorium isotopes from the deep geological formations into surface ecosystems.

Direct discharges from land-based activities even caught the attention of the UN General Assembly, which in December 1996 adopted Resolution 51/189 about the Global Programme of Action for the Protection of the Marine Environment from Land-Based Activities.

France and Great Britain discharged large amounts of LRW to the ocean, from Sellafield in England (the nuclear fuel production and reprocessing facility) and from La Hague (France's similar plant in Normandy) through long pipes (in La Hague - 1700 m from the shore): 1.1 million Ci ^{137}Cs, 0.5 million Ci ^{241}Pu, 0.5 million Ci ^{3}H, about 0.3 million Ci ^{106}Ru, about 0.5 million Ci of other radionuclides [11]. The peak of the Sellafield beta-nuclides discharge was reached in 1975, when a total of more than 9000 TBq (including more than 5200 TBq by ^{137}Cs) were released. Discharges by alpha-emitters peaked in 1973– up to 180 TBq. Between 1966 and 1984, Sellafield discharged 20 821 TBq (566 kCi) ^{241}Pu and up to 631 TBq ^{239}Pu [37]. In 1991, a total of were 100 Еий for ^{238}Pu, 610 TBq for ^{239,240}Pu, and 945 TBq for ^{241}Am [25] were discharged. In 1980-1984 ^{137}Cs concentration in Scotland's coastal waters was up to 400 Bq/m^3 [4]. The Irish Sea sediments near Sellafield had ^{137}Cs up to 5.5 kBq/kg, ^{90}Sr – 2 kBq/kg, ^{239,240}Pu – 34.8 kBq/kg, ^{238}Pu – 9.6 kBq/kg, ^{241}Am – 2.2 kBq/kg [4,38,39]. Between 1968 and 1979 about 180 kg of Pu was discharged from Sellafield. In 1983 the GB Ministry of Environment closed off more than 20 km of the beach area near Sellafield due to dangerous levels of radioactive contamination [40]. In 1999 - 2000 Sellafield discharged about 130 GBq alpha-nuclides, and La Hague – 40 GBq [29]. In a single year, La Hague discharges five times more ^{129}O than was released during all N-tests worldwide [41].

Till now radioactive discharge from Sellafield and La Hague have polluted not only the Irish Sea and the Channel, but also Northern Seas like the Barents and even the Kara Sea and could have contributed about 200 kCi (7400 TBq) - up to 7% of the whole anthropogenic radionuclides budget here: up to 20% ^{137}Cs and 30% ^{90}Sr from Sellafield ended up in the Barents Sea [42]. Since the 1990s, detectable increases in the concentrations of $^{127, 129}$J, and 99Tc have been revealed in the North Sea near Norway [43,44,45]. In 2003 ^{99}Tc (up to 20 Bq/Kg) were found in smoked and fresh farmed salmon sold in Sweden's six leading supermarkets [46]. in 2001, The Nordic Council (Finland, Sweden, Denmark, Norway, Iceland) wrote to Britain over continuing radioactive emissions (especially ^{99}Tc and ^{125}Sb) from Sellafield. In 2001 Norway initiated a lawsuit against Britain, insisting that Sellafield's discharge release represents a serious threat to the Norwegian Fishing industry [47]. Official sources (review see: [4]) insist that Sellafield and La Hague discharged no more than 1.1 mln Ci ^{137}Cs, 0.5 mln Ci 241 Pu, 0.5 mln Ci ^{3}H, 0.3 mln Ci ^{106}Ru and more than 0.1 mln Ci other long-living radionuclides into the North Atlantic

There are three Indian re-processing plants use water from THE Arabian Sea as a secondary coolant, AND discharge radioactive wastes back into the Sea (*Trombay* in Bombay Harbour and *Tarapur*, about 80 km to the North West). These places are called the "radiation coast" due to high levels of radioactivity directly discharged into the sea by these reprocessing plants [48]. The third Indian reprocessing plant, *Kalpakkam*, is located on the Tamil Nadu shore, on the Bengal Bay coast. Due to state secrecy it is impossible to detect the real scale and radionuclide composition of the Indian Ocean's pollution, but it can reasonably be supposed that it may compare with the North Atlantic's pollution.

Numerous NPPs located on the shores of the the USA, Great Britain, South Africa, Japan etc, regularly discharge radionuclide. There are some examples of illegal discharges or LRW dumping from such NPPs. One of the latest happened in October 2002 on Scotland's North Sea coast from NPP *Torness* [49].

There are three coastal radioactive waste storage facilities for Russian N-sub SNF assemblies (Andreev Bay, on the Barents Sea; Vilyuchinsk, on the Okhotsk Sea and Bol'shoi Kamen', on the Sea of Japan). Beginning in the 60s, some nuclear fuel assemblies were just set right on the ground without even a roof [50,51]. The total activity of the SNF in the Kola Peninsula is about 10 000 kCi [52]. All such coastal N-installations are potential (and real) sources of serious local pollution. In 1996-1998 the sediment of the Kola Peninsula water contained: ^{137}Cs up to 115 Bq/kg; ^{60}Co – up to 74 Bq/kg, ^{239,240}Pu up to 9 Bq/kg [53]. The Kola peninsula N-icebreakers base annually discharges about 1.6×10^7 of ^{137}Cs, 7.6×10^7 of ^{90}Sr and even some Plutonium –1998 figures reached 70 mBq/m^3 ^{239,240}Pu, which is 5 to 6 times higher than local background concentration [53]. The amount of ^{152}Eu, $^{134, 137}$Cs, ^{60}Co in the algae *Laminaria digitata* and *Fucus vesiculosus* is several times higher than in other places along the Barents Sea shelf [53]. At least seven types of failures on coastal objects are connected to radioactive waste storing and conditioning [30].

7. Pollution from Radioisotope Thermal Generators

There are two types of thermal radioisotope generators: the more powerful are based on up to 90% ^{235}U enriched N-reactors (for satellites), and the others are Radioisotope Thermal (Thermoelectric) Generators (RTGs), which are used to supply power to lighthouses and meteorological posts and in deep sea acoustic beacon signal transmission, and also for satellites. RTGs have either a ^{90}Sr or a ^{238}Pu core. Since 1960, nine models of RTGs have been developed in the USSR (for a total number of up to 1500 RTGs with a total ^{90}Sr activity of up to 1.5 Ebq). The most common is the Beta-M type (230 Watts of power, 35 to 40 kCi of activity). The radioactivity of an RTG at a distance of 0.5 meters is up to 800 roentgens per hour.

There are at least six cases when RTGs were sources of radioactive sea pollution [54,22,55,56]:
- 1987. An RTG was lost at sea during transportation near the eastern coast of Sakhalin Island, in the Okhotsk Sea, with an activity of up to 750 kCi (27.8 PBq);
- 1997. An RTG was lost at sea during transportation near the eastern coast of Sakhalin Island, in the Okhotsk Sea, with an activity of about 35 kCi (1.3 PBq);
- 2003. The disintegrated core of an RTG (up to 40 kCi or 1.5 PBq) was found in the waters of the Finnish Bay, in the Baltic Sea;
- 2003. Two disintegrated RTGs were found underground on the coast of the Laptev Sea, Yakutia;
- 2003. RTG Beta-M type # 255 was found completely dismantled in Olenya Bay lighthouse # 414.1, the Kola Harbor; the radioisotope core was found in the water near the shore, 1.5 to 3 meters deep.
- 2003. RTG Beta-M type # 256, which powered lighthouse No 437 was found completely dismantled on Yuzhny Goryachinksy island, the Kola Harbor; its radioisotopes core could not be found.

Satellite RTGs usually have a plutonium core. The following are among the accidents involving satellites with RTGs falling into the ocean :
- 1964. U.S. satellite *Transit 5NB-3*, with SNAP-9A RTG, ^{238}Pu 16 kCi (629 TBq); burned up upon reentering the atmosphere over the West Indian Ocean north of Madagascar;
- 1968. U.S. *Nimbus B-1*, with two SNAP-19 RTGs, ^{238}Pu total up to 1265 TBq, in the Santa Barbara Channel, on the Californian shore (was recovered);
- 1970. Parts of U.S. station *Appolo-13* with SNAP-27 RTG, ^{238}Pu up to 1.63 PBq, re-entered the atmosphere over the South Pacific and fell into the ocean south the Fiji Islands, in the vicinity of the Tonga Trench;
- 1973. USSR satellite with RTG, into the Western Pacific north of Japan;
- 1983. Part of the USSR satellite *Cosmos-1402* with reactor core ^{235}U, ^{90}Sr and ^{137}Cs (up to 1 PBq) re-entered the atmosphere and fell into the South Atlantic, 1600 km East of Brazilia;
- 1996. Russian space station *Mars-96*, 18 RTGs ^{238}Pu total activity 174 Еий

(4.7 kCi), fell into the South Pacific between the Easter Isl. and the Chilean shore.

RTGs are not such an important source of radiation compared with many others, but they can be a reason for heavy local pollution in any part of the Ocean.

8. Sunken ships with radioactive materials

In 1996 six members of the U.S. Congress sent President Bill Clinton a letter expressing deep concern over the existing practice of shipping radioactive waste by sea [57]. There are at least three cases when commercial cargo ships transporting radioactive material sank:

- 1984. Cargo "Mont-Louise" (France), transporting 350 t uranium hexafluoride from France to the USSR (30 containers), sank after a collision with a ferry 15 km off the coast of Ostende (where the sea is 15 m deep). All the containers were successfully found and raised within the next 40 days.
- 1997. Cargo MSC "Karla" (Panama) transporting 5 cesium chloride sources (^{137}Cs up to 9 kCi or 330 TBq) from France to the USA ran into heavy sea, split in two and sank 70 nautical miles off the Azores islands where the sea is 3 000 m deep.
- August 2003. Cargo "Sealand Express" (US Ship Management) transporting 56 t uranium dioxide ("yellow cake") from South Africa to the US ran aground in Table Bay about 150 m off Sunset Beach in Milnerton, Cape Town [58]. The shipment was headed to a uranium processing plant in Newport News, Virginia. AngloGold Co., a subsidiary of London-based Nuclear Fuels Corporation (Nufcor) transport a thousand tons of uranium oxide (a by-product of gold mining) from South Africa each year.

Statistically, there is one serious accident during sea transportation for each 10 million ship/km [59]. The world's annual maritime traffic of radioactive materials is about 3 to 5 million ship/km. It means that one serious shipwreck involving radioactive material may occur every 3 to 4 years.

9. Discussion

ADD – dangerous consequences of previous irradiations.
Up to 10 000Bq/kg in plankton everywhere from underwater N-explosions/
And 2500 Bq/kg in the Finnish Bay several weeks after the Chernobyl catastrophe.

The above-mentioned facts are only a part of the real picture of radioactive marine pollution from anthropogenic sources.

Emission of radionuclides from underwater sources instrumentally detected in a radius of 10 to 70 km after 30 years, which represents several hundred meters to several km per year. Some data indicate that the steel corpus of the nuclear warhead has now disappeared and that ^{239}Pu is escaping into the Norway Sea. This process of plutonium escape could create a zone of contamination by ^{239}Pu corrosion products (which are both highly active and chemically toxic). 6.4 kilograms of ^{239}Pu from two N-warheads (total activity at 430 Ci) are enough to poison the local fishing grounds in the Norway Sea.

The common feeling [1,3,11,13,60,61,62,4] that the level of anthropogenic radioactive pollution of the world's ocean is not so serious is only justified according to modern average levels of ^{137}Cs concentration in surface water 2-6 Bq/m^3, ^{90}Sr concentration at 1 – 4 Bq/m^3, ^{239,240}Pu concentration at 5 – 30 mBq/m^3., ^{241}Am – 1-2 mBq/m^3. It is well known that the total radioactivity of naturally occurring radionuclides ^{40}K, ^{226}Ra, ^{232}Th and ^{210}Po in the ocean is many times higher than that of anthropogenic ones. But it would be a mistake to conclude that the level of anthropogenic radioactive pollution of the Ocean is "negligible". Natural background radiation is like a ship filled to the brim with water. The tiniest additional drop (hundreds of thousands of times smaller than the volume of water in the ship) can initiate an overflow. This analogy leads to the conclusion that even comparatively small additional amounts of manmade radionuclides in the ocean can have some negative consequences for ecosystems and humans. Through bioaccumulation, manmade radionuclides in the water can concentrate in marine animals and plants up to many thousands of times (Table 3).

TABLE 3. Maximal Concentration factors for radionuclides (concentration in tissues compared with concentration in water) in some Marine Organisms [63,11,25]

	Invertebrates	Fish
P-32		100 000
Zn-65	50 000	87 200
^{210}Po	50 000	26 500
Cs-137	200	17 580
Fe-55	30 000	3 000
^{241}Am	20 000	2 500
^{210}Pb	837	1600
Cm	30 000	3000
Pu	10 600	300
^{226}Ra	1000	100
^{3}H	10	10

From a biological point of view, it is impossible to carry out an exhaustive study of all the consequences, such as bio-concentration for all marine ecosystems, and it is impossible to study the specific radiotoxic impact of each radionuclide on all species. What is known about Tc influence on many thousands of marine species of the North Atlantic? But namely Tc has now become a more detectable pollutant originating from Sellafield. Outside the food-chain concentration, there are also numerous natural processes including horizontal and vertical water transportation, different sediment concentrations, resuspensions, etc. All these can result in manifold concentration of different radionuclides in some unpredictable places.

Another important reason for disillusion about the safety of existing ocean radioactive pollutions is a methodologically wrong conclusion on safety based on statistical average data. Like averaging temperature for hospital patients has nothing to do with each particular person's real health condition, the average concentration of radionuclides in the ocean has nothing to do with the real radiological situation in particular places. Dumping and lost radioactive objects, including nuclear reactors and N-warheads, can and do create thousandfold concentration in hundreds of places all over the ocean.

The third reason for serious concern about the ocean's radioactive pollution is the lack of real data and the continuing secrecy on the subject. The known cases of radioactive pollution seem to be just the tip of the iceberg. Who could predict 20 years ago that the USSR would secretly dump radioactive waste whose total activity reaches several million Ci in the Arctic seas? When Russia published all the data and called for all other countries to open their secret files [6,7] they never did. In the words of Ph. M. Klasky, director of the Bay Area Nuclear Waste Coalition (California, USA) : *"For years, we have asked ... to conduct a survey so that we know how much radioactive waste is being produced ... Without this information and oversight, abuses do occur"* [64].

10. Acknowledgements

I would like to express my deep appreciation for Prof. G.G. Policarpov (Sebastopol Biological Institute, Ukraine), who many years ago brought marine radioactive pollution to my attention, and Prof. Dennis Woodhead (EG CEFAS) whose strong criticism helped to improve this article. My sincere thanks to Prof. Eugene Levner, who intensively pressed me to prepare a lecture, based on this paper. I also thank Col.(Ret.) Alexander K. Nikitin (Bellona, Sankt-Peterburgh), Dr.Yury. S. Tsaturov (Roshydromet, Moscow) and Prof. Yury A. Israel (Global Ecology and Climate, Moscow) for consultations. I would like to thank my assistants in the Center for Russian Environmental Policy Rimma D. Philippova and Anna S. Egorova for their invaluable help.

11. References

1. IAEA-TECDOC-481 (1988) Inventory of Selected Radionuclides in the Oceans, IAEA Technical documents Series # 481, Vienna (http://www-pub.iaea.org/MTCD/publications/ResultsPage.asp).
2. IAEA-TECDOC-1105 (1999) Inventory of Radioactive Waste Disposal at Sea, IAEA Technical document # 1105, Vienna (http://www-pub.iaea.org/MTCD/publications/ResultsPage.asp).
3. IAEA-TECDOC-1330 (2003) Modeling of the Radiological Impact of Radioactive Waste Dumping in the Arctic Seas, IAEA Technical document # 1330, Vienna (http://www-pub.iaea.org/MTCD/publications/ResultsPage.asp)
4. Matishov, D.G. (2001) Anthropogenic Radionuclides in Marine Ecosystems, Doctor thesis, Institute of Limnology, Sank-Petersburg (in Russian).
5. Makhijany, A., Hu, H., Yih, K. (1995) Nuclear Wasteland. A global Guide to Nuclear Weapons Production and its Health and Environmental Effects, The MIT Press, Cambridge – London, XXV.
6. Yablokov, A.V., Karasev, V.K., Rumyantsev, B.M. et al. (1993) Facts and Problems Related to Radioactive Waste disposal in the Seas Adjacent to the Territory of the Russian Federation, Office of the President of the Russian Federation, Moscow (in Russian).
7. Yablokov, A.V. (2001) Radioactive Waste Disposal in Seas Adjacent to the Territory of the Russian Federation, Marine Pollution Bulletin, 43, # 1-6, 8-18.

8. Lutins, A. (2003) U.S. Nuclear Accidents (http://www.lutins.org/nukes.html).
9. IAEA-TECDOC-938 (1997) Predicted radionuclides release from marine reactors dumped in the Kara Sea, IAEA Technical documents Series # 938, Vienna, 69
10. Sivintsev, Yu.V., Kiknadze, O.E. (1998) Proc. Intern. Seminar Afterword to the "White Book", Nizny Novgorod, 19 – 21 January, 1998, "Lazurit" Intern. Center, 169 -175.
11. Vasiliev, F., Sivinsev, Yu. (2002) White Book – 2000, or radiological consequences of disposition of radioactive wastes into Arctic and Far Eastern seas, Bull. Nucl. Energy, #5, 37–41.
12. Vasiliev, A. (2003). White Book – 2000. Barrier of Safety, 8, 28 – 29 (in Russian).
13. Strand, P., Tsuturov, Y., Howard, B., McClelland, V., Bewers, M., Dahlgaard, H., Joensen, H.P., Nikitin, A., Palsson, S., Rissanen, K. (2002) Radioactive contamination in the Arctic, 5th Int. Conf. on Environmetal Radioactivity in the Arctic and Antarctic, St. Petersburg, Russia 16-20 June 2002, 1-5.
14. Povinec, P., Osvath, I., Baxter, M. (1995) Marine scientists on the Arctic Seas Documenting the radiological record, IAEA Bulletin, 37, # 2 (http://www.iaea.org/worldatom/Periodicals/Bulletin/ Bull372/povoinec.html).
15. Russia … (2003) Russia checking on Nuclear Waste Dumped in Sea of Japan, Agence France Presse, August 12 (Russian Environmental Digest, 11 - 17 August 2003, # 5).
16. IASAP (1997) Assessment of the Impact of Radioactive Waste Dumping in the Arctic Seas, Report of the International Arctic Seas Assessment Project.
17. Fry, R. M.,Control of the dumping of radioactive wastes in the oceans (http://f40.iaea.org/search97cgi/s97-cgi?QueryZip=ocean).
18. RADNET, Information about source points of anthropogenic radioactivity (2003) (http://www.davistownmuseum.org/cbm/rad&e.html).
19. Ahmed Khan, A. (2001) Disposal of Radioactive Nuclear Waste Media Monitors Net (http://www.mediamonitors.net/ayazahmedkhan2.html).
20. Marine Radioactive Waste Dump Sites (2003) Antrophogenic radioactivity: Major Plume Source Points (http://www.davistownmuseum.org/cbm/rad&de.html).
21. Solholm, R. (2002) Nuclear Pollution in North Sea by oil industry, The Norway Post, 31 January.
22. IAEA (2001) Inventory of accidents and losses at sea involving radioactive material, IAEA Technical Document # 1242, Vienna.
23. Radioactive America (2003) Center for Defense Information, America Defense Monitor (http://www.cdi.org/adm/1341).
24. Nuclearfiles …(2003) http://www.nuclearfiles.org/hitikeline/nwa/short-list.htm.
25. Ikaheimonen, T.K. (2003) Determinantion of transuranic elements, their behaviour and sources in the aquatic environment, Radiation and Nuclear Safety Autority, Helsinki (http://www/stuk.fi/julkaisut/stuk-a/stuk-a194.pdf).
26. Sarkisov, A.A. (1999) Radioecological consequences of the Radiation Accident on Nuclear Submarine in Chazma Bay, The 4th International Conference on Environmental Radioactivity in the Arctic. Edinburgh, Scotland 20-23 September 1999, Extended Abstracts, 78.
27. Arkin, W., Hundler, J. (1989) Naval Nuclear Accidents: The Secret History, Greenpeace, 14, # 4, July/August 1989.
28. Tiwari, J., Gray, C.J. (2000) U.S. Nuclear Weapon Accidents (http://www.cdi.org/issues/nukeaccidents/accidents.htm).
29. Mackenzie, A.B. (2000) Environmental Radioactivity: Experience from the 20th century – trends and issues for the 21st century, The Science of the Total Environment, 249, 313 – 329.
30. Lisovsky, I.A., Blekher, A., Rubanov, S. et al. (1998) Description of the basic scripts of development of failures, resulting to excharge radionuclides in an environment, Part.1, INTAS Project 96-1802, Technical report B5 (direction 5), St.-Petersburg State Technical University, Saint-Petersburg.
31. Smith, J., Ellis, K., Polyak, L., Ivanov, G., Forman, S., Morgan, S. (2000) ^{239,240}Pu transport into the Arctic Ocean from underwater nuclear tests in Chernaya Bay, Novaya Zemlya, Continental Shelf Res,, 20, 255 – 279.
32. Matishov, G.G., Matishov, D.G., Pavlova, L.G., Rissanen, K., Szezpa, J. (1996) Radionuclides in Seas of the Western Arctic, Polar Geography, 20, # 1, 65-77.
33. Arctic…1997. (1997) Arctic Pollution Issue: A State of the Arctic Environment Report, Arctic Monitoring and Assessment Program XII, Oslo.
34. Ivanov, G.I. (1999) Assessment of Radioactive contamination in the Pechora Region, The 4th Intern. Conf. Environmental Radioactivity in the Arctic, Edinburgh, Scotland 20-23 September 1999, 259-262.
35. Badiaev, V.V. et al. (1990) Protection of Environment under exploitation of Nuclear Power Plants,

Energoatomizdat, Moscow (in Russian).
36. Povinec, P., Fowler, S., Baxter, M. (1996) Chernobyl and the marine environment: The radiological impact in context, IAEA Bulletin, 38, # 1 (http://www.iaea.org/worlatom/Periodicals/Bulletin/Bull381/chernoby.html).
37. McCarthy, W., Nicholls, T.M. (1990) Mass-spectrometric analysis of plutonium in soils near Sellafield, J. Environ. Radioactivity, 12, 1, 1-12.
38. Aston S.R., Stanners D.A. (1981) Plutonium transport to and deposition and immobility in Irish Sea intertidal sediments. Nature, 289, 581-582.
39. Oldfield, F., Richardson, N., Appleby, P.J. (1993) ^{241}Am and ^{137}Cs activity on fine grained saltmarsh sediments from parts of the N.E. Irish Sea Shoreline, J. Environ. Radioactivity, 19, 1, 1-24.
40. Sellafield (2001) Bellona working paper, 5. (http://www.bellona.no/energy/nuclear/sellafield/wp_5-2001/21895.html).
41. (www.greenpeace.org/~nuclear/ospar2000/html).
42. Nies, H., Harms, I.H., Bahe, C. et al (1998) Anthropogenic Radioactivity in the Nordic Seas and the Arctic Ocean – Results of a Joint Project, German Journal of Hydro graphy, 50, 4, 313-343
43. Masson, M., VanWeers, A.W., Groothuis, R.E.J., Dahlgaard, H., Ibbert, R.D., Leonard, K.S. (1995) Time series for sea water and seaweed of ^{99}Tc and ^{125}Sb originating from releases at La Hague, Journal of Marine Systems, vol. 6, 397 – 413.
44. Brown, J.E., Iospije, M., Kolstad, K.E., Lind B., Rudjord A.I., Strand, P. (1999) Temporal trends for ^{99}Tc in Norwegian Coastal Environments and Validation of a Marine Box Model, The 4th Intern. Conf. Environmental Radioactivity in the Arctic, Edinburgh, Scotland 20-23 September 1999, 104 - 107.
45. Raisbeck, G.M., Yiou, F., Christensen, G.C. (1999) ^{129}I/^{127}I in the Norwegian Coastal Current from 1980-1998, The 4th Int. Conf. On Environmental Radioactivity in the Arctic, Edinbburgh, Scotland 20-23 September 1999, 3—32.
46. Clover, Ch. (2003) Radioactive waste found in supermarket salmon (http://porta.telegraph.co.uk/news/main.jhtml?xml=/news/2003/06/23/nsalm23&s\).
47. Tanney, P. (2001) Northern Assembly calls for closure of Sellafield, The Irish Times, December 5.
48. Chugoku Newspaper (1992) Exposure Victims of Radiation Speak Out. Kodansha Int. Publ., 327 (http://www.amazon.com/exec/obidos/tg/detaill/-/4770016239?v=glance).
49. Staples, J. (2002) Nuclear plant admit illegal water-discharge. Scotsman.com 06.12.02 (http:www.Scotsman.com/).
50. Nielsen T., Kudrik I., Nikitin, A. (1996) The Russian Northern Fleet – Sourses to Radioactive Contamination, Report 2, April 1996, 168 (http://www.bellona.no/imaker?id=10090@sub=1).
51. Boehmer, N., Nikitin, A., Kudrik, T., Nielsen, Megovern, M., Zolotkov, A. (2001) The Arcemc Nuclear Challenge, Bellona Report, 3, Oslo (http://www.bellona.no.imaker?id=21133@sub+1).
52. ITAR/TASS (2001) Wednesday, March 14, 2001 8:04 AM EST, Murmansk (ITAR-TASS).
53. Matishov, D., Matishov, G., Namjatov, A. (1996) Radioactive contamination of sediments in the Kola Fiord near RTP "Atomflot", the base of the nuclear fleet, Proc.the NATO Advanced Reasearch Workshop "Rebucing Wastes from Decomissioned Nuclear Submarines in the Russain Northwest", 85 – 101.
54. Gorbunov, M. (2003) Mstitel' Okhotskogo morya (Avenger of the Okhotsk sea), Rossiiskaya Gazeta, July 30, 6 (in Russian).
55. Kudrik, I., Alimov, R., Digges, Ch. (2003) Two strontium powered lighthouses vandalised on the Kola Peninsula (http://www.nuclearNo.org).
56. Thief … (2003) Thief discards radioactive parts of Russian lighthouse in Gulf of Finland. No guards at lighthouses of Russian Navy Foreign, Helsinky Sanomaat (Foreign), April 16 (http://helsinki.hs.net).
57. http://www.nci.org/seatrans.htm
58. Bennet, G. (2003) South Africa: $1.3m Uranium Spill Improbable. (http://www.allAfrica.com:August25 http://www.moneyweb.co.za).
59. Sert, G. (1999) The Recovery Radioactive Sources after a Shipwreck. The Case of the Mont-Louis Cargo and the Implication of the M.S.C, Carla. (http://f40.iaea.org/search97cgi/s97-cgi?QueryZip=ocean).
60. Radiological assessment: Waste disposal in the Arctic Seas. Summary of results from an IAEA-supported study, IAEA Bulletin (1997) 39, # 1 (http://www.iaea.org/worlatom/Periodicals/Bulletin/Bull391/povoinec.html).
61. IAEA (1988) Assessing the impact of deep sea disposal of low level radioactive waste on living marine resources, Tech. Rep. Ser. 288, Vienna.

62. IAEA-TECDOC-725 (1993) Risk comparisons relevant to sea disposal of low level radioactive waste, IAEA Technical documents Series # 481, Vienna.
63. Yablokov, A. V. (2002) Myth on safety of the low radiation' doses, Center for Russian Environmental Policy, Moscow (in Russian).
64. Valles, C. (2002) Radioactive waste leak hit sanctuary at Farallons, ContraCosta Times (San Francisco), January 27

MARINE PROTECTED AREAS: A TOOL FOR COASTAL AREAS MANAGEMENT

C. F. BOUDOURESQUE
Centre of Oceanology of Marseilles, University of the Mediterranean, Luminy campus, 13288 Marseilles cedex 9, France

G. CADIOU
GIS Posidonie, Luminy campus, 13288 Marseilles cedex 9, France

L. LE DIRÉAC'H
GIS Posidonie, Luminy campus, 13288 Marseilles cedex 9, France

Abstract

Marine biodiversity is threatened by human impact. Though few marine species are regarded as being extinct due to Man, many species are critically endangered (e.g. the monk seal Monachus monachus), endangered (e.g. the Mediterranean giant limpet Patella ferruginea) or vulnerable, i.e. dwindling rapidly, although not threatened with extinction in the immediate future (e.g. the large mollusk Pinna nobilis). There are also threats to ecosystems (ecodiversity), such as, in the Mediterranean, the Lithophyllum byssoides rim and the seagrass Posidonia oceanica meadow. Marine Protected Areas (MPAs) were initially established to protect biodiversity via the removal of human exploitation and occupation. However, since the 1970s, the notion of MPA has moved on to a more general concept of nature conservation, then to a more dynamic one of nature management, within the framework of sustainable development. Today, the aims of MPAs are therefore six-fold: nature conservation, public education, reference areas for scientific research, tourism, export of fish eggs, larvae and adults to adjacent areas and finally management of the various uses of the sea (e.g. commercial fishing, recreational fishing, pleasure boating and tourism) in such a way that they do not conflict with each other or with conservation aims. Mediterranean MPAs, especially the Port-Cros National Park, illustrate the fact that they are rather characterized by the management of human activities than by a set of prohibitions and that there is no negative interaction between biodiversity conservation and artisanal fishing (i.e. small-scale commercial fishing), at least in the way it is done (i.e. with additional constraints to general regulations: mesh size, prohibition of trawling and long-lining, etc.). Consequently, MPAs are generally of benefit to the economy (e.g. commercial fishing and tourism industry), not only within MPAs but also in adjacent areas. They therefore constitute a powerful tool for integrated coastal management.

1. Introduction

The erosion of biodiversity (e.g. species diversity and ecosystem diversity) constitutes a major concern, both in the terrestrial and marine realm. The establishment of protected areas banning human activities was, from an historical point of view, the earliest response to their impact.

Today, especially with regard to the marine environment, the approach has totally changed. In the present paper, on the basis of examples mainly drawn from the experience of a Mediterranean Marine Protected Area (MPA), the Port-Cros National Park, we show that the efficiency of MPAs does not lie in the *a priori* prohibition of human activities but in their management, in such a way that they no longer conflict with each other or with nature conservation goals.

In the mind of the general public, MPAs are still often perceived as areas preserved from human presence. Here, we show that they actually constitute powerful economic tools, both for artisanal (i.e. small-scale) commercial fisheries and for the tourism industry.

2. The need for Marine Protected Areas

2.1. Marine biodiversity

Biological diversity (biodiversity) means the variety among living organisms from all sources including, *inter alia,* terrestrial, marine and other aquatic ecosystems and the complexes of which they are part. This includes: diversity within species, diversity between species, genera, families, phyla, etc., diversity between ecosystems, diversity between landscapes and functional diversity. Ecosystem diversity is often referred to as "**ecodiversity**". Within species diversity, one may distinguish **point diversity** (species number within a sample), α **diversity** (species number within a habitat or ecosystem in a given region), β **diversity** (the species turnover between adjacent habitats or sections of coastline), γ **diversity** (the number of species of a region, either defined on a political, geographical or biogeographical basis) and ϵ **diversity** (the number of species of a large geographical area, e.g. the Mediterranean basin) [62, 63]. There is no link between these levels of species diversity. For example, α diversity may be high and γ diversity low. Human impact may locally increase α diversity while diminishing γ diversity [81].

The overwhelming value of biodiversity, as an indication of environmental health and for the functioning of the biosphere, is now widely recognized, not only by academic scientists, but also by the mass media, decision makers and public opinion [82, 116].

Unfortunately, marine biodiversity has received only a very small fraction of the attention devoted to terrestrial environments. Not only do the species definitely recorded clearly represent only a small part of those that actually occur [46, 88, 107, 108], but the present status (how many? where? on the increase or on the decrease?) of most of them is poorly known [20, 23], with the exception of a few emblematic taxa (e.g. sea mammals, sea turtles, seagrasses, some fishes).

Rating the relative importance of human impact on biodiversity requires that the time needed for the impact to be reversed be taken into account (Alexandre Meinesz *in* [24]): one day to one month, one month to one year, one year to ten years (e.g. most of pollution events, including oil spills), ten years to one century (e.g. destruction of long-living species), one century to one millennium (e.g. destruction of the seagrass *Posidonia oceanica* meadow in the Mediterranean Sea) and finally more than one millennium, i.e. irreversible at human scale. Coastal development, species introduction and species extinction are so the greatest cause of concern, due to their irreversibility [25, 114].

2.2. Erosion of marine species diversity

The realization that marine species may become extinct is relatively recent. For example, in 1809, the French naturalist Jean-Baptiste de Lamarck wrote: *"Animals living in the water, especially the sea waters, are protected against the destruction of their species by man. Their multiplication is so rapid and their means of evading pursuit or traps are so great that there is no likelihood of his being able to destroy the entire species of any of these animals"* (translated from French). Along the same lines, in 1883, Thomas Huxley said (Address to the International Fisheries Exhibition in London): *"Any tendency to overfishing will meet with its natural check in the diminution of the supply (...), this check will always come into operation long before anything like permanent exhaustion has occurred"* [26, 61, 141]. Yet several marine species had already become extinct by that time (see below).

Species are classified as follows: extinct, extinct in the wild (only present in zoos or botanical gardens), threatened (either critically endangered, endangered, vulnerable or rare), of lower concern (i.e. whose populations seem to be in a normal state) and data deficient (i.e. whose present day status is unknown) [86, 87].

Modern day **extinctions** (neoextinctions) are for the most part due to human impact, as opposed to geological "natural" extinctions (paleoextinctions) [43]. A taxon is considered to be extinct when there is no reasonable doubt that the last individual has died, i.e. when individuals have not been located in the wild over a period of 50 years [86, 110].

Recent extinction rates in well documented groups (mammals and birds) are one hundred to one thousand times faster than the average background rates [10, 108]. Looking towards the immediate future, likely extinction rates of a factor of ten thousand above background can be expected. This represents a sixth great wave of extinction, fully comparable with the five major mass extinctions (the "Big Five") of the geological past: late Ordovician, late Devonian, late Permian, late Triassic and end-Cretaceous. However, this time it is different in that it results from the activities of a single other species, rather than from external environmental changes [16, 17, 108].

In contrast with terrestrial environments, very few marine species are regarded as being extinct. Examples are the Rhodobionta *Vanvoorstia bennetiana*, the eelgrass limpet *Lottia alveus*, the rocky shore limpet *Colisella edmitchelli*, the periwinkle *Littoraria flammea*, the horn snail *Cerithidea fuscata*, the Galápagos damselfish *Azurina eupalama*, the auk *Pinguinus impennis*, in the North Atlantic ocean (extinct in 1844), Steller's sea cow *Hydrodamalis gigas* in the eastern Pacific (extinct in 1768) and the Caribbean monk seal *Monachus tropicalis* [11, 40, 43, 65, 105, 115, 123, 141].

However, it is of interest to note that, if the definition of extinct species is applied (species that have not been located in the wild over a 50 year period), there may be hundreds of species of invertebrates or macrophytes that have not been recorded since the 19th century or the early 20th century. Are these species extinct, or is it simply an artefact due to the poor knowledge of many groups of marine organisms? We may have lost many more species than we suspect ("cryptic extinctions" [43]), and the expected extinction of the systematists will not make it easy to answer this question. As foretold by Carlton [43], *"the future historians of science may well find that a crisis that was upon us at the end of the 20th century was the extinction of the systematist, the extinction of the naturalist, the extinction of the biogeographer – those who would tell the tales of the potential demise of global marine diversity"*.

The Mediterranean monk seal *Monachus monachus* was formerly widespread around the whole Mediterranean Sea, the Black Sea and the Western Atlantic. It has become extinct in most of its range area. To date, according to the WWF, it is one of the ten species in the world that are most threatened with extinction (**critically endangered species**). It shares this status with *inter alia* the tiger *Panthera tigris*, the giant panda *Airulopoda melanoleuca*, the Javan rhinoceros *Rhinoceros sondaicus* and the Indus river dolphin *Platanista minor*. During the last 25 years, the total number of monk seals has dropped from 1 000 to about 300 individuals, and of these 150-200 (a rather optimistic census) are in the Mediterranean [3, 102, 103, 128, 131]. The reasons for the monk seal's decline are **(i)** the reduction of its natural habitat (beaches, caves) because of coastal development and tourism [119], **(ii)** overfishing of the fish stock on which it feeds, which leads to individuals being scattered and stealing fish from fishers' nets [31, 32], and **(iii)** its being destroyed by fishers [89]; this destruction is a consequence of the previous point. Despite hundreds of public awareness leaflets, legal protection in most of the Mediterranean countries and an impressive series of international Conventions aimed at its protection (e.g. Washington, Bern and Barcelona Conventions), the monk seal is still on the decline in the Mediterranean. This emphasizes the limited efficiency of legal protection of a species, when habitat and feeding resource are not preserved. The only site where a small population is on the increase is the Portuguese Island of Madeira, within a Marine Protected Area [42, 48, 124].

Endangered species are species which have disappeared from fairly extensive sectors and are threatened with extinction. However, in contrast with critically endangered species, strong protection measures are likely to save them. For example, a mollusk, the giant limpet *Patella ferruginea*, is on the brink of extinction. Formerly widespread throughout the western Mediterranean, where it is an endemic, it now only survives in sparse populations in Corsica, Sardinia, Tunisia, Algeria and southern Spain [30, 95, 96, 97, 126, 145]. Its decline has accelerated over the last 15 years. The reason for the disappearance of this large species (sometimes over 10 cm in diameter), which lives in the midlittoral zone (i.e. slightly above mean sea level), is its being gathered by humans either for consumption or for use as bait. In addition, individuals are male up to 4 cm in diameter and then become female; human gathering mainly hits larger individuals, i.e. female ones [57]. Finally, juvenile individuals often settle on adults, so that gathering adults may also remove juveniles [94]. In the Scandula marine Reserve (Corsica), the decline of *Patella ferruginea* is still going on (Table I). The reason might be that the prevailing current, running south to north, which comes from areas devoid of giant limpets, does not brings larvae into the reserve. As far as larvae produced by the individuals harbored by the reserve are concerned, they are swept along by the current to outside the reserve [98]. The fate of *P. ferruginea* illustrates the fact that protection of the habitat may prove to be inefficient, if the size of the protected area is small, which is the case of the Scandula marine Reserve.

Table I. Decline over time of the mean density of the giant limpet *Patella ferruginea* in two sites of the Scandula marine Reserve (Corsica). md = missing data. From [98].

Year	Mean number of individuals per 100 m of shoreline	
	Northern shore of Gargallu cape	Litizia cove
1983	md	25
1984	56	19
1987	md	21
1992	24	2

Vulnerable species are species which are still relatively common but whose populations are dwindling rapidly, although not threatened with extinction in the immediate future. Examples of species experiencing a steady and severe decline, at least in some parts of the Mediterranean, are the Stramenopiles (Fucophyceae) *Cystoseira amentacea, C. mediterranea, C spinosa* and *C. zosteroides* [9, 13, 18, 38], the Mollusks *Pinna nobilis,* especially in the north-western Mediterranean [146], *Luria lurida* and *Zonaria pyrum,* the seahorses *Hippocampus ramulosus* and *H. hippocampus* [38, 106], the dusky grouper *Epinephelus marginatus* in the north-western Mediterranean [44] and the sea-turtle *Caretta caretta* [64].

The noble pen shell *Pinna nobilis* is the largest Mediterranean mollusk: it can reach a size of up to 100 cm and its life span exceeds 20 years [146, 147, 152]. In the north-western Mediterranean, large adults are exceedingly rare, with mean densities of less than one individual per hectare [134]. The reasons for the decline of *P. nobilis* are: **(i)** Collection by divers for souvenirs. **(ii)** Gathering for human consumption (locally) and to feed fishes in fish farms (in Turkey). **(iii)** Breaking of shells by trawling. **(iv)** The decline of its main habitat, the meadows of the seagrass *Posidonia oceanica* (see below) [38, 146, 152].

The dusky grouper *Epinephelus marginatus* is unquestionably the most popular of littoral fish along the western Mediterranean coasts [70]. In the north-western Mediterranean, from Catalonia to Italy, it was common until the 1950s, but subsequently underwent a dramatic decline. In the 1980s, a very few individuals (only adults) survived in this area, probably migrants from southern areas [44]. Spear fishing and life traits were clearly the main reasons for this local near-extinction: **(i)** *Epinephelus marginatus* is rather easy to spear, due to its escape behavior to crevices, and is therefore very vulnerable to spear fishing. **(ii)** It is a long living species: up to 50 years. First sexual maturity is reached when individuals are 5 years old (40-50 cm long). **(iii)** It is a proterogynous hermaphrodite. Young individuals are female. Sex reversal occurs mainly when they are 14-17 years old (80-90 cm long): so older individuals are male. **(iv)** Finally, spawning is a very complex process which requires many individuals (reproductive aggregations) [38, 39, 44, 150, 151]. Since the late 1980s, some marine protected areas (MPAs; e.g. Medes Islands, Port-Cros, Scandula and Lavezzi Islands) have made possible the recovery of dense populations of *E. marginatus*, harboring both females and males, so that since the early 1990s, spawning events have been not uncommon within these MPAs. In addition, the spear fishing of this species has been prohibited in France since 1993. Since then, juveniles and subsequently adult dusky grouper have been observed outside the MPAs. Spear fishermen often say: *"The dusky grouper's recovery is not a consequence of the banning of spear fishing, but of the current warming of the Mediterranean Sea"*. The surface temperature of the Western Mediterranean did indeed rise (0.5-1.0°C) between the early 1960s and the late 1990s [19]. However, the dusky grouper was fairly common along the French coasts before the current warm climatic episode, which means that warming cannot explain its recovery, even if it may contribute to enhancing it.

2.3. Erosion of the marine ecodiversity

In the Mediterranean, the main threatened communities are the intertidal rims built by *Lithophyllum byssoides* (Rhodobionta), the *Neogoniolithon brassica-florida* (Rhodobionta) reefs, the vermetid (Gatropoda) platforms, the subtidal *Posidonia oceanica* seagrass meadow, several particular types of the *Posidonia oceanica* meadow (in particular barrier-reefs) and the coralligenous, a deep sciaphilous community built by encrusting calcareous Rhodobionta [28, 33, 34, 92, 112, 114].

The **Lithophyllum byssoides** rim is sensitive to pollution (especially hydrocarbons). The rims have died in French Catalonia, in the area of Marseilles (France) and in the Gulf of Palermo (Sicily): bio-erosion (perforating organisms) no longer being compensated for by bio-construction, the rims are progressively eroded and end up disappearing [93, 135]. Bearing in mind the slowness with which they are built up, this disappearance must be considered irreversible from the human point of view (even when the causes of its death are believed to have been removed).

Vermetid platforms are vulnerable to domestic pollution, low salinity rainwater and oil slicks. Sediment laden waters, as a consequence of coastal development (urbanization or construction of coastal roads), may kill vermetid formations by siltation [28, 92]. In addition, over-frequent walking over by tourists and amateur fishermen damage the vermetids [28].

The ***Posidonia oceanica*** **meadows** have dwindled considerably, in particular in the vicinity of the large urban centers. They are dwindling both at their lower limit (rising because of the water turbidity and the resulting deficit in light) and at intermediate depths. In Italy, Ligurian meadows have lost about 10-30% of their surface area [21]. In the Alicante region (Spain), 52% of the surface area has been lost [130]. In Marseilles (France), close to 90% of the meadows mapped by Marion in 1883 [104] have today disappeared. The causes are as follows [33, 34, 120, 121]: **(i)** Industrial and urban pollution (*P. oceanica* is very sensitive to this), in particular detergents and nutrients [6]. **(ii)** Turbidity, in reducing the limpidity of the water and the penetration of light to the deep. Phytoplanktonic blooms, whose intensity is accentuated by eutrophy, have the same impact. The result is a rising of the lower limit. **(iii)** Mooring of small boats [27]. **(iv)** Trawling. In the area of Alicante (Spain), it is responsible for almost half the surface area diminution in the meadow [130]. **(v)** Coastal development: ports, artificial beaches and reclamations over *P. oceanica* meadows [113]. **(vi)** Alteration of the sediment flow. A groyne perpendicular to the coastline results (in relation to coastal drift) in upstream hypersedimentation and a shortage in sediment (with baring of the rhizomes) downstream. The average maximum growth of orthotropic rhizomes being around

5-7 cm per year [29], the vegetative apexes are buried and die if the annual sediment input exceeds 5-7 cm. On the other hand, the bared rhizomes are vulnerable to water movement and to trawling. In both cases, the *P. oceanica* meadows can be destroyed.

Reduction in limpidity in waters (pollution, turbidity) and silting constitute the main threats to the **coralligenous community**. It is worth adding, locally, the over-frequent visits by scuba divers: erosion by contact of coralline Rhodobionta and Bryozoa (*Retepora* in particular), non-intentional breaking of gorgonians by beginners and deliberate tearing off of the red coral *Corallium rubrum* and the gorgonians *Eunicella* and *Paramuricea* [71, 76, 142].

Outside the Mediterranean, major threats concern tropical mangroves and coral reefs [58, 141, 143].

3. The early concept of protected areas

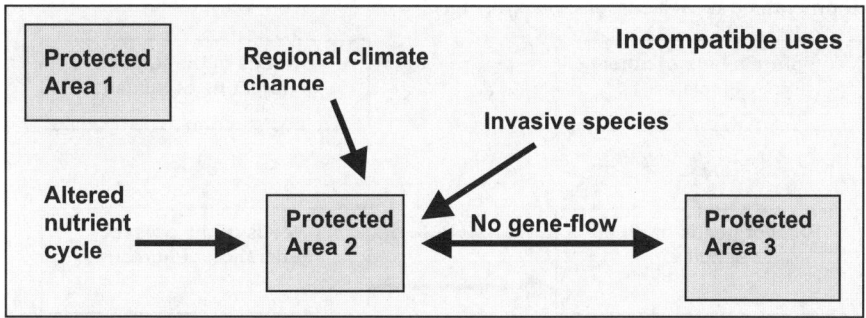

Fig. 1. Protected areas (e.g. 1 through 3) seen as "islands" of nature and tranquility surrounded by incompatible resource uses. Black arrows: negative impacts.

Until the late 1960s, the key concept behind protected areas was that they were areas not materially altered by human exploitation or occupation, and that steps should be taken by the competent authority to prevent or eliminate exploitation or occupation. So protected areas were seen as "islands" of nature and tranquility surrounded by incompatible resource uses [109, 127]. Yet such an "island" mentality is fatal in the long term because protected areas will not be able to conserve biodiversity if they are surrounded by degraded habitats that limit gene-flow, alter nutrient cycles, provide invasive species and cause regional climate change which may ultimately lead to the disappearance of these "island parks" (Fig. 1) [109].

Invasive species clearly illustrate this problem. The park boundary of the Port-Cros National Park (France) offered no protection from the immigration of *Caulerpa taxifolia* (Chlorobionta, Plantae), once it was present along the coasts of the French Riviera [35, 49, 137, 138].

4. The modern concept of protected areas

Since the 1970s, the notion of protected areas has moved on to a more general concept of nature conservation, then to a more dynamic one of **nature**

management. Protected areas therefore need to be part of a broader regional approach to land (and sea) management [2, 26, 109, 127]. Furthermore, it is recognized that conserving nature requires a flexible approach in which local people should not to be excluded *a priori*. This new perspective was first given full legitimacy in the World Conservation Strategy [85] and was developed into practical advice at the 3rd World National Parks Congress, held in Bali, Indonesia, in October 1982. The title of the congress proceedings ("National parks, conservation and development: the role of protected areas in sustaining society") gives a clear indication of the new direction being advocated [109]. This approach (sustainable development) was then popularized and formalized at the United Nations Conference on Environment and Development (UNCED) in Rio de Janeiro in June 1992: *"That range of activities and development which enables the needs of the present generation of humans and all other species to be met without jeopardizing the ability of the biosphere to support and supply the reasonably foreseeable future needs of humans and all other species"*. Sustainable development is thus a 4-corner concept (Fig. 2)

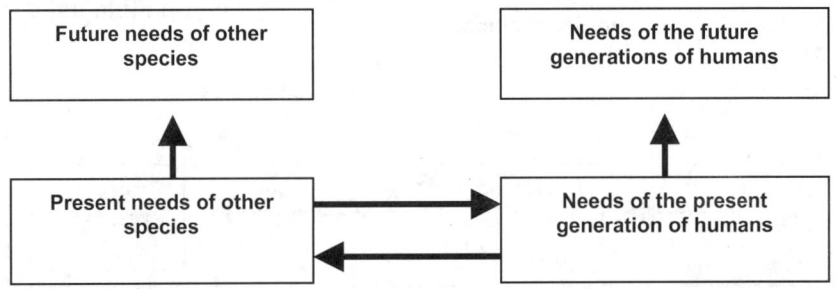

Fig. 2. The 4-corner concept of sustainable development. There is a symbiosis between meeting the needs of humans and those of other species, i.e. meeting the needs of other species helps supply those (economic, sociological and cultural) of humans.

Nowadays, the aims of the Marine Protected Areas (MPAs) are six-fold. **(i)** To set up conservatories for threatened species and habitats. **(ii)** To provide sites for public education on the environment (e.g. underwater nature trails, public information leaflets). **(iii)** To provide reference areas for scientific research. **(iv)** To provide attractive landscapes for tourism (bathing, pleasure craft, snorkeling, diving). **(v)** To establish no-take areas where fish density and sex-ratio make mating and spawning possible, and which subsequently export eggs, larvae and adults to surrounding unprotected areas and therefore enhance catches by fishermen. **(vi)** To manage the different uses of the sea (e.g. commercial fishing, recreational fishing, pleasure boating and tourism) in a rational way, so that they do not conflict with each other or with conservation aims [2, 26, 52, 56, 101, 140, 144].

Marine Protected Area are often perceived by the public at large as well as by the stakeholders and other users of coastal areas as a burdening **collection of prohibitions**. Possible constraints generated by a MPAs are as follows: **(i)** Prohibition of non-commercial collecting of fauna and flora. **(ii)** Prohibition of spear fishing. **(iii)** Prohibition of recreational angling. **(iv)** Prohibition of all forms, or only some forms (e.g. trawling), of commercial fishing. **(v)** Prohibition of scuba diving. **(vi)** Prohibition of pleasure craft mooring and anchoring. **(vii)** Prohibition of boating. **(viii)** Prohibition of bathing [24]. According to these constraints, **5 levels can be distinguished** (Table II). Usually, the area of a MPA is zoned in such a way that most MPAs include several levels (Fig. 3). It is worth noting that, even when there are no apparent differences in the regulations existing inside (level 1) and outside (level 0) an MPA, a major difference does exist: MPAs are usually the only sites where existing legislation is enforced (e.g. mesh size and prohibition of trawling close to the shore).

Table II. The levels of constraints outside (level 0) and within (levels 1 through 5) Marine Protected Areas. + = prohibited. - = non-prohibited. From [24].

Prohibition	5	4	3	2	1	0
Collection of fauna and flora	+	+	+	+	+	+/- a
Spear gun fishing	+	+	+	+	-	-
Recreational angling	+	+	+	-	-	-
Commercial fishing	+	+	- b	- b	-	-
Scuba diving	+	-	-	-	-	-
Pleasure craft mooring and anchoring	+	-	-	-	-	-
Boating	+/-	-	-	-	-	-
Bathing	+/-	-	-	-	-	-

(a) Depending upon local or national legislation. (b) Trawling sometimes prohibited.

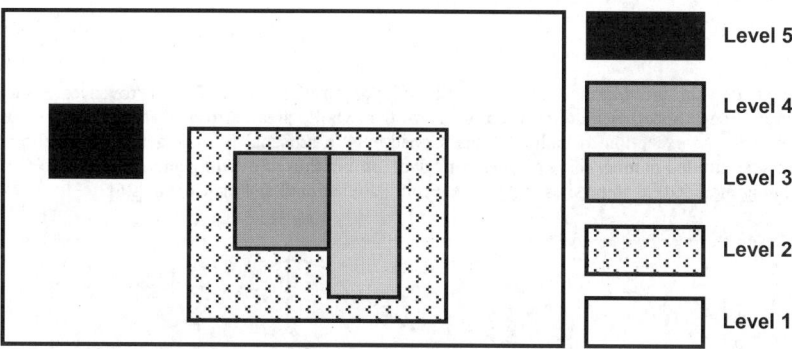

Fig 3. An MPA with zones at different levels of constraints (1 through 5).

In fact, with the exception of spear fishing, prohibitions usually concern only a small part of the MPA surface area. For example, if we consider the French Mediterranean Marine Protected Areas, commercial fishing is only banned in 7% of the total MPA surface area (Table III). This is the case in the Port-Cros National Park: artisanal fishing (i.e. small-scale commercial fishing) is possible in most of the area (1 380 ha), with the exception of a few hectares (Fig. 4, 5) [26].

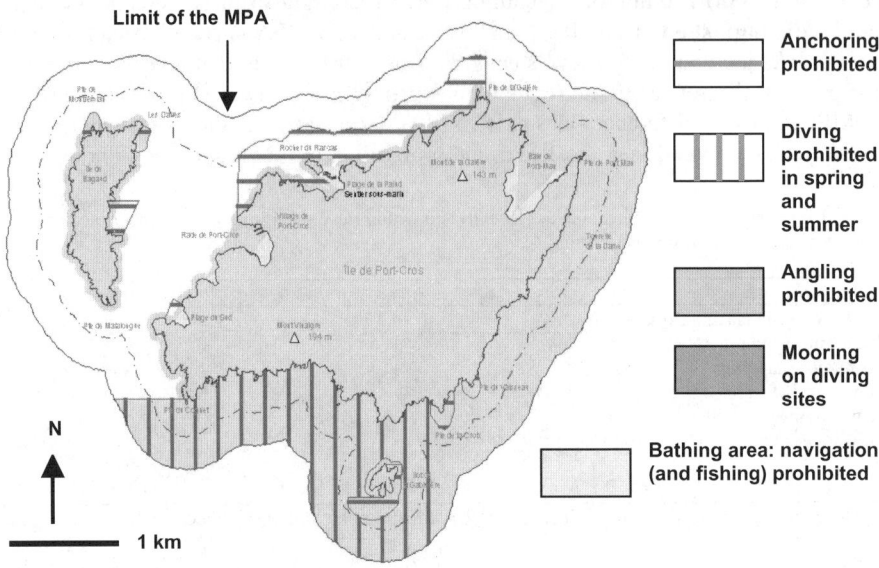

Fig. 4. Regulation in the Marine Protected Area of Port-Cros National Park (Provence, France, Mediterranean Sea). Spear fishing is prohibited in the whole area. Artisanal fishing authorized everywhere, with the exception of bathing areas. Constraints to artisanal fishing: general legislation + wider mesh size, limited number of gear types per fisher, prohibition of trawling and restriction of hook use (long-lining included) to some sites and places. From Parc national de Port-Cros *in* [26].

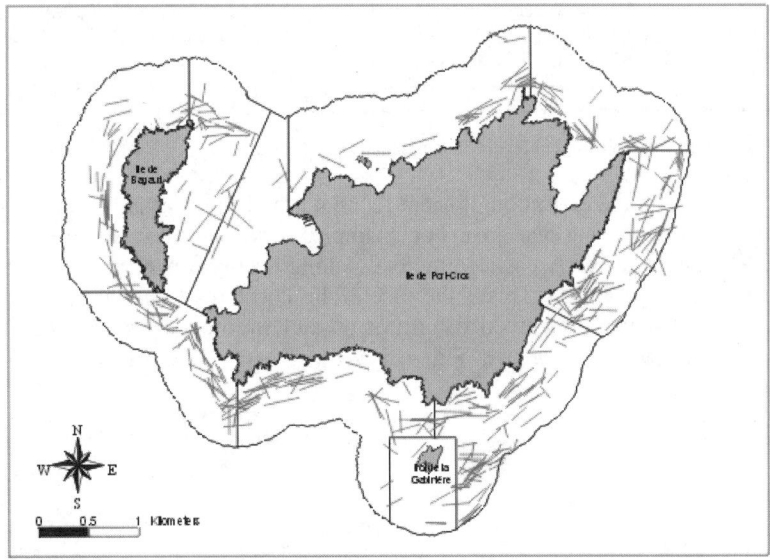

Fig. 5. Localization (red lines) of fishing nets in the Port-Cros National Park, between March and September 2001. Cumulated data from 63 daily surveys. From [41].

Table III. Main prohibited activities within French Mediterranean Marine Protected Areas, as a percentage of the total surface area (nearly 9 000 ha). From [79].

Prohibited activity	Percentage of the total surface area
Anchoring of pleasure boats	7%
Artisanal commercial fishing	7%
Scuba diving	17%
Recreational fishing (angling)	20%
Recreational fishing (spear fishing)	100%

5. Marine protected areas and coastal management

As far as fisheries are concerned, Marine Protected Areas can provide four basic benefits [91, 148]: **(i)** Protection of critical functions (such as spawning grounds, feeding grounds), **(ii)** protection of specific life stages (juvenile settlement, nursery grounds), **(iii)** provision of spillover of exploited species and **(iv)** provision of dispersion centers for supply of eggs and/or larvae (stock enhancement).

5.1. Interactions between artisanal fishing and nature conservation

The question which arises is twofold. Firstly, taking into account the constraints that characterize the practice of artisanal fishing within an MPA like Port-Cros National Park, does this activity threaten one of the major MPA's aims, nature conservation? Answering this question is of great concern since artisanal fishing is an activity that is profoundly rooted in Mediterranean customs and traditions. Secondly, do the constraints imposed on artisanal fishing hinder that activity, for example reduce the fishing effort, catch per unit effort (CPUE) and/or catch per surface area unit?

Benthic ecosystems in the Port-Cros National Park are healthy and habitat **diversity is well preserved**: e.g. meadows of the seagrass *Posidonia oceanica*, the coralligenous and sea-cave communities, and *Cystoseira* forests [5, 15, 73, 74, 83, 100]. Species diversity of macrophytes, invertebrates and fish is high [14, 55, 69, 80, 83, 118]. Of course, natural fluctuations may affect these populations, as a result of e.g. warm water episodes and diseases [7, 36, 59, 76, 122]. As far as emblematic species are concerned, the brown meager *Sciaena umbra* is not uncommon (1-4 individuals/ha) though less abundant than in other Mediterranean MPAs [75]. The population of the dusky grouper *Epinephelus marginatus* is in steady expansion[1] (Table IV) [72, 77, 78]. The mean density of the noble pen shell *Pinna nobilis* is 9-11 individuals/ha (adults), with much higher density, 100 individuals/ha, within its preferred habitat, the *Posidonia oceanica* meadow [111]. In non-protected areas of the north-western Mediterranean, mean adult density is less than one individual per hectare [134]. The success of Port-Cros as a hot-spot for scuba diving confirms the quality of its species, habitat and landscape (seascape) diversity, in particular fish density, which is particularly appealing for divers. All in all, on the basis of present day knowledge of Port-Cros biodiversity, it cannot be claimed that artisanal fishing, in the way it is done (see caption to Fig. 4), seriously hinders one of the aims of the MPA, biodiversity conservation.

---

[1] Clearly, the prohibition of spear fishing, together with restriction of long-lining and the banning of trawling play a major role in the expansion of *Epinephelus marginatus* population.

Table IV. Patterns of change over time of the population (number of individuals) of the dusky grouper (*Epinephelus marginatus*) of the Port-Cros National Park (ca 14 km², France, Mediterranean Sea), censused visually by snorkeling and scuba diving. md: missing data.

Year	Gabinière Island	Other sites	Total	References
1973	7	8-11	15-18	unpublished data *in* [72]
1983-1987	23-28	md	md	[139]
1988-1989	29-34	md	md	[66]
1993	34	52	86 (100[a])	[72, 77]
1996	84	76	160	[72, 77]
1999	156	143	299	[77]
2002	210	200	410	[78]

[a] Estimate.

To answer the question in reverse (i.e. do the constraints imposed on artisanal fishing within an MPA hinder that activity?) is more difficult. In the northwestern Mediterranean Sea, as well as in most coastal areas worldwide, quantitative data on artisanal fishing are scarce and quite difficult to compare, due e.g. to differences in the methods used, the sampling season, the type of gear taken into account, the target species and to the importance, usually unknown, of other catches (recreational fishing, trawling) and the target stock. In addition, the surface area of the regions studied is not a reliable datum, since it may include areas not suitable for artisanal fishing. In Port-Cros waters, the mean number of fishers was, in summer, 1.8/d in 2000 and 2.8/d in 2001 [41]. The fishing effort can be better estimated on the basis of the length of nets, or the number of 100 m net sections, per day and per ha (Table V). Comparison between Port-Cros MPA and the non-protected surrounding area shows that fishing effort is not lower, and may be higher, within the former than in the latter [41, Guerin, unpublished data]. As far as CPUE (catch per unit effort) is concerned, on the basis of available data, fish yield cannot be considered as lower at Port-Cros than in non-protected areas (Table V).

As a result, as far as the Port-Cros MPA is concerned, it seems that there is **no negative interaction** between biodiversity conservation and artisanal fishing within a MPA, at least in the way it is done there (see constraints in the caption to Fig. 4). This may be also the case in other Mediterranean MPAs (Table V).

5.2. Possible reasons for the absence of negative interaction

In the Mediterranean, which accounts for one third of total world tourism, recreational fishing catches, whether it be by spear fishing or angling, are far from negligible (2.8-8.4 t/km²/a; Table VI) as compared to those of the artisanal fishing industry (1.9-6.2 t/km²/a) [1, 18, 50, 51, 60], even if it must be emphasized that recreational fishing concerns a much smaller surface area than artisanal fishing. The overlap between the catches from spear fishing and angling is usually weak. In contrast, the overlap between recreational and artisanal fishing may be significant [51, but see 53].

At Port-Cros Island, spear fishing has been prohibited since the establishment of the National Park in 1963. Since 1999, angling has been prohibited from the coast to the offshore limit of the park marine area (East and South) and from the coast to 50 m offshore (North and West) (Fig. 4). If one considers catches from recreational fishing in other coastal areas, this prohibition can be seen to significantly relieve the fishing pressure on the Port-Cros fish stock.

Table V. Data on artisanal fishing in some localities of the French Mediterranean coast. Fishing effort: number of 100 m net sections per ha and per year (or day). CPUE: catch per unit effort, i.e. kg per 100 m of fishing net and per day (= per outing). md: missing data. upd: unpublished data.

Locality	MPA	Surface area	Fishing effort	CPUE	Total catch	References
Côte Bleue (Provence)	No [a]	ca 13 km²	0.07/ha/d (summer) [b]	0.7-2.5 kg/ 100 m/d (summer)	md	[53]
Côte Bleue (Provence)	No [a]	-	md	0.7 kg/100 m/d (summer)	md	[90]
Riou Archipelago (Provence)	No	21 km²	13.3/ha/a [b]	md	md	[18]
Port-Cros (French Riviera)	Yes	14 km²	5.4/ha/a [c]	1.2 kg/100 m/d [de] 0.8-1.2 kg/100 m/d [e] 1.4-1.5 kg/100 m/d [d]	1.9-3.2 kg/ha/a [ef]	[41, 67, Guerin, upd]
Galeria-Ghjiru-lata (Corsica)	Yes No [a]	md md	md md	1.1 kg/100m/d (April) 1.0 kg/100 m/d (July)	md md	[99]
NW Corsica	No	md	md	0.9 kg/100 m/d	md	[136]
Lavezzi Islands (Corsica)	Yes	37 km²	7.9/ha/a	0.8-0.9 kg/ 100 m/d	6.2 kg/ha/a	[50]
Bonifaziu straits (Corsica)	Yes	800 km²	md	1.4-2.9 kg/100 m/d	md	[54]

[a] An MPA is present in the vicinity ("Parc Marin de la Côte Bleue"). [b] Calculated from the author's data. [c] Based upon [67]. [d] Year 2001. [e] Only target species. *Conger conger* and *Muraena helena*, for example, are not taken into account. [f] Under-evaluation: value based upon the fishing log books of 6 out of the 9 fishers who were observed fishing in Port-Cros waters.

Table VI. Catches from recreational fishing. The studied surface area or shore length are mentioned (in brackets). md: missing data.

Locality	Spear fishing (t/km²/a)	Angling from the shore (t/km/a)	Angling from a boat (t/km²/a)	Total recreational fishing (t/km²/a)	References
Rayol-Canadel (French Riviera)	1.3 (2.4 km²)	< 0.1 [a] (7.5 km)	0.2 (7.5 km²)	2.8 [b]	[45]
Port-Cros MPA (French Riviera)	Prohibited	0.2 [cd] (26 km)	0.4 [cde] (14 km²)	8.4 [b]	[47]
Riou Archipelago (Provence)	1.3 (8.5 km²)	0.1 (26 km)	1.3 (21 km²)	6.3 [b]	[51]
Riou Archipelago (Provence)	md	md	1.1 [c] (21 km²)	md	[22]
Cerbère-Banyuls MPA (French Catalonia)	Prohibited	0.1 [c] (5 km)	0.5-1.1 [c] (5.8 km²)	4.5-5.1 [b]	[4]

[a] Exact value: 0.033 t/km/a. [b] At places where the different types of recreational fishing coexist, and assuming that angling from the shore concerns a 25 m wide littoral belt. [c] Calculated from the author's data. [d] Later on (1999), angling from the shore was banned and angling from a boat restricted to some areas (see Fig. 4). [e] A lower value is mentioned by [51], due to miscalculation.

Furthermore, it must be emphasized that an apex predator, the monk seal *Monachus monachus*, was formerly present in the area of Port-Cros National Park [102, 103]. In the absence of this fish-eating seal, a no-take area for artisanal fishermen would not represent a natural environment since the seal's preys could proliferate and cause a shift in the natural equilibrium. Fishermen may therefore contribute to mitigating the ecological impact of the monk seal's local extinction.

Finally, the actual implementation of the general legislation within an MPA, and the additional fishing constraints specific to the MPA (wider mesh size, banning of trawling, etc.), put an end to overfishing increase the fish stock and subsequently catches [e.g.149].

5.3. Marine Protected Areas enhance fishery yield in adjacent areas

A large number of littoral fish species undergo a sex change over time. Some of these are male when young, then become female. In contrast, others are initially female and subsequently become male. A consequence of these features is that, in overfished populations (Fig. 6, left), one of the sexes can either become scarce or even absent. In addition, large individuals may be lacking; the number of eggs laid by fish steadily increases with size and age, so that a young female may lay up to 200-fold less eggs than an old one. Within MPAs (Fig. 6, right), the simultaneous presence of females and males makes sexual reproduction possible and old females are present. In addition, in some species, nuptial pairing not only involves a mating pair but also requires the participation of a large number of individuals, both adults and juveniles, a process only possible when fish density is high and all age classes are present (i.e. diversified demographic structure). As a consequence, MPAs export huge amounts of eggs and larvae to neighboring areas (Fig. 6, left and center). Furthermore, due to overpopulation, individuals continuously leave the MPA to occupy unprotected adjacent areas (spillover), where they can be caught by fishermen. In this way, MPAs can substantially contribute to maintaining profitable yields in the regional fishing industry [8, 12, 37, 68, 91, 117, 125, 148]. In the Mediterranean, it has been estimated (or hypothesized) that a large number of small MPAs, 200 to 1 000 ha, spaced 10 to 20 km apart, has a more positive effect than a few very large MPAs [24].

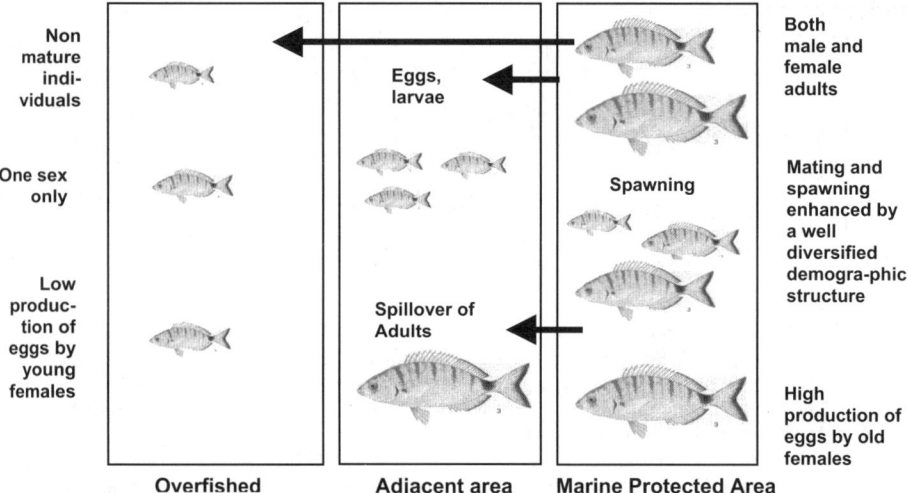

Fig. 6. Reproduction of fish, as a function of the population demographic structure in an overfished area, a Marine Protected Area (MPA) and an adjacent area. The MPA exports adults (spillover), eggs and larvae. From [37], redrawn.

Is this a more or less theoretical view, or does this work? Many concrete examples suggest that this definitely works. In Spain, three years after the establishment of the Tabarca no-take MPA (Alicante, Mediterranean), catches of high selling price fish species (e.g. *Sparus aurata*) **increased twofold** (Fig. 7) [129]. On St Lucia (Caribbean), the Soufrière Marine Management Area was set up in 1995. It encompassed a network of no-take areas (35% of coral fishing grounds). Six years later, in 2001, catch per unit effort

(CPUE) increased by 80% for small traps and by 36% for large traps; mean catch per trip increased respectively by 90% and 46%, though fishing effort remained stable [140].

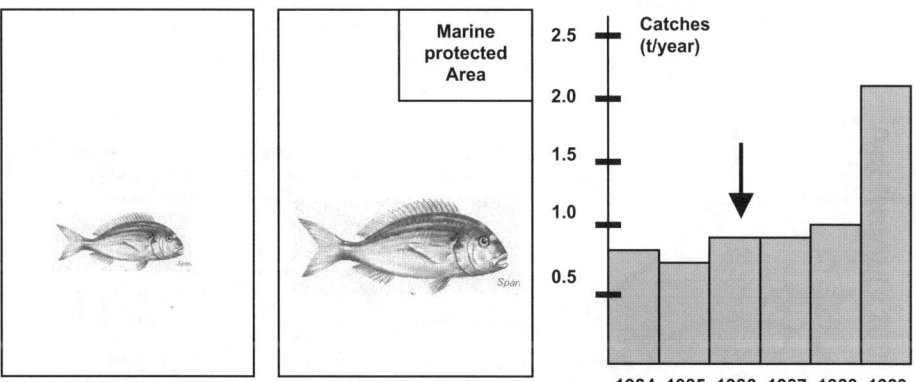

Fig. 7. Left: Catches of artisanal (commercial) fishing in an overfished area near Alicante (Spain). Center: catches in the remaining fished area after the establishment of a no-take MPA. Right: change over time of catches of *Sparus aurata*. Arrow: the year of establishment of the MPA. From [129], redrawn.

The effectiveness of **larvae exportation** by a MPA was evidenced by Francour and Le Diréac'h (*in* [79]) downstream of the Scandula MPA, Corsica (Table VII). The analysis of harvesting models shows that no-take MPAs are part of an optimum harvest designed to maximize yield [e.g. 68, 117].

Table VII. Evidence of fish larvae (*Diplodus annularis*) exportation from the Scandula MPA (Corsica, Mediterranean Sea). From Francour and Le Diréac'h (*in* [79]).

Current	Locality	Protection level	Mean juvenile density/10m²
North	Galeria Gulf	Unprotected	1.12
	Elbu Bay	MPA (no recreational fishing)	0.47
	Gargalu	No-take MPA	0.34
South	Portu Gulf	Unprotected	0.22

5.4. Marine protected Areas: a tool for integrated coastal management

Within an MPA, the zoning of the human activities, i.e. the separation of conflicting activities in specialized zones (e.g. bathing, scuba diving, pleasure boat mooring, recreational fishing) makes it possible to manage user conflicts and results in an optimization of human activities for the benefit of both stakeholders and nature conservation (Fig. 8). In this way, an MPA constitutes a scale model of what should be a regional integrated coastal management policy, including MPAs and unprotected areas.

Obviously, MPAs, together with regional integrated management of user conflicts, result in economic benefits, both for fishermen and the tourism industry, in such a way that there should no longer be a need to try to set off environmental values against economic values [8, 24, 129, 132, 133, 148]. For example, it has been estimated that the tiny (20 km² of land and sea) Port-Cros National Park, French Riviera, produces, directly and indirectly, a mean annual turnover of 300 M€ per year [26, 84]. The total gross revenue generated by the Bonaire Marine

Park (Caribbean Sea) was estimated at 23 M€ per year in 1991. This Park also generated substantial employment with up to 750 local workers and 240 foreign workers in park associated activities [91]. In Australia, the Great Barrier Reef attracts about 1.8 million tourists valued at over $A 1 000 million per year, compared to estimates of $A 360 million for the annual worth of Great Barrier Reef fisheries [91].

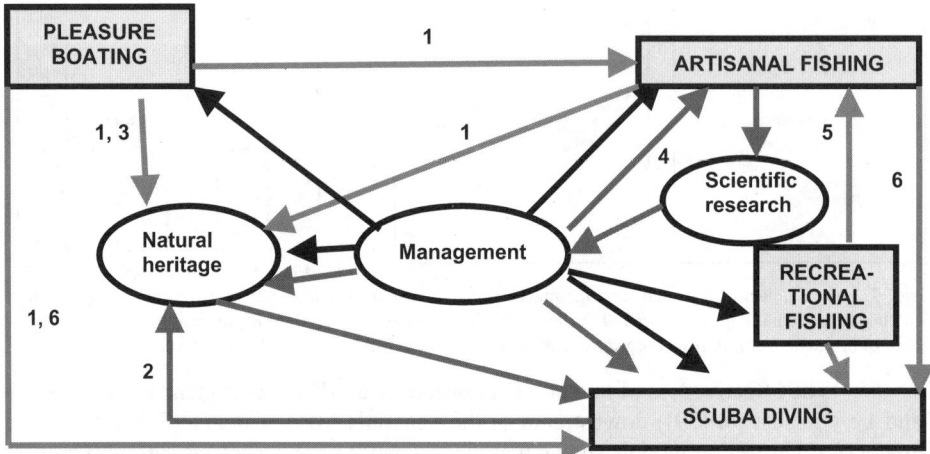

Fig. 8. Integrated management of user (yellow boxes) conflicts and nature conservation within a Marine Protected Area, the Port-Cros National Park. Red arrows: negative interactions. Blue arrows: positive interactions. Black arrows: management. 1: Dissemination of introduced species (e.g. *Caulerpa taxifolia*); 2: Localization of introduced species stands; 3: Impact of mooring (e.g. on the seagrass *Posidonia oceanica* and the coralligenous community); 4: increase of fish stock; 5: competition for fish stock harvesting; 6: effect on Scuba divers safety. From [26].

6. Conclusions

The concept of sustainable development means that there is a symbiosis between supplying the needs of humans and nature conservation. Accordingly, man, especially local people, should not be excluded *a priori* from Marine Protected Areas (MPAs).

Although present day scientific data related to MPAs deal more with nature conservation than with the economy, there is growing evidence that MPAs constitute a powerful tool not only for natural heritage conservation but also for economic development (tourism industry, artisanal fisheries) and regional integrated management of user conflicts, at least in temperate and warm seas.

The *à la mode* new concept of the Ecosystem Approach of Fisheries (EAF) may be considered as an offshore generalization of the MPA's experience and success story worldwide.

Many more MPAs should therefore be set up, for the benefit both of nature conservation and of economic development.

7. Acknowledgements

The authors are indebted to Nicolas Gérardin and Philippe Robert (Port-Cros National Park) for providing regulation data concerning the Port-Cros National Park, and to Michael Paul for improving the English text.

8. References

1. Aboussouan, A. and Boutin, C. (1993) La pêche professionnelle dans les eaux du Parc national de Port-Cros. Parc national de Port-Cros et GIS Posidonie publ., Marseilles, 16 pp + Appendices 1-3.
2. Agardy, T.S. (1997) Marine protected areas and ocean conservation. Academic Press publ., San Diego, 244 pp.
3. Anselin, A., Van Der Elst, M. des N., Beudels, R.C. and Devillers P. (1990) Analyse descriptive et projet pilote préparatoire à une stratégie pour la conservation du phoque moine en Méditerranée (*Monachus monachus*). Commission des Communautés européennes, Environnement et Qualité de vie, Rapport EUR 13448 FR, 62 pp.
4. Athias-Binche, F. (1996) Impact d'une réserve marine sur les zones périphériques. Domaine public maritime et littoro-bathyal. Programme NATMAR (1992-1996), rapport final, Banyuls-sur-Mer, Fr., 27 pp.
5. Augier, H. and Boudouresque, C.F. (1976) Végétation marine de l'île de Port-Cros (Parc national). XIII. Documents pour la carte des peuplements benthiques. *Trav. sci. Parc nation. Port-Cros*, Fr. 2, 9-22.
6. Augier, H., Monnier-Besombes, G. and Sigoillot, G. (1984) Influence des détergents sur *Posidonia oceanica* (L.) Delile, *in* C.F Boudouresque., A. Jeudy de Grissac and J. Olivier (eds.), *First International Workshop on Posidonia oceanica beds*, GIS Posidonie publ., Marseilles, pp. 407-418.
7. Azzolina, J.F., Boudouresque, C.F. and Nédélec, H. (1983) Seasonal and year-to-year changes of the edible sea-urchin *Paracentrotus lividus* populations in the bay of Port-Cros (Var, France). *Rapp. P.V. Réun. Commiss. internation. Explor. sci. Médit* 28 (3), 265-266.
8. Badalamenti, F., Ramos, A.A., Voultsiadou, E., Sanchez-Lizaso, L.J., Danna, G., Pipitone, C., Mas-Fernández, J.A.R., Whitmarsh, D. and Riggio, S. (2000) Cultural and socio-economic impacts of Mediterranean marine protected areas. *Environmental Conservation* 27, 110-125.
9. Ballesteros, E., Sala, E., Garrabou, J. and Zabala, M. (1998) Community structure and frond size distribution of a deep water stand of *Cystoseira spinosa* (Phaeophyta) in the Northwestren Mediterranean. *Eur. J. Phycol.* 33, 121-128.
10. Balmford, A. (1996) Extinction filters and current resilience: the significance of past selection pressures for conservation biology. *Trends Ecol. Evol.* 11 (5), 193-196.
11. Barber, B.J. (1997) Impacts of bivalve introductions on marine ecosystems: a review. *Bulletin nation. Res. Inst. Aquac.*, suppl. 3, 141-153.
12. Bell, J.D. (1992) The use of marine protected areas to enhance fisheries, *in* Economic impact of the Mediterranean coastal protected areas, Ajaccio, 26-28 Septembre 1991, *Medpan News*, Fr. 3, 33-39.
13. Bellan-Santini, D. (1966) Influence des eaux polluées sur la faune et la flore marines benthiques dans la région marseillaise. *Techn. Sci. municipales*, Fr. 61 (7), 285-292.
14. Belsher, T., Augier, H., Boudouresque, C.F. and Coppejans, E. (1976) Inventaire des algues marines benthiques de la rade et des îles d'Hyères (Méditerranée, France). *Trav. sci. Parc nation. Port-Cros*, Fr. 2, 39-89.
15. Belsher, T. and Houlgatte, E. (2001) Carte de l'herbier à *Posidonia oceanica* et des principaux faciès sédimentaires des fonds sous-marins du Parc national de Port-Cros (France). IFREMER Publ., Brest, 1 map.
16. Benton, M..J. (1994) Palaeontological data and identifying mass extinctions. *Trends Ecol. Evol.* 9 (5), 181-185.
17. Benton, M..J. (1995) Diversification and extinction in the history of life. *Science* 268, 52-58.
18. Bernard, G., Bonhomme, P. and Daniel B. (1998) Archipel de Riou: étude socio-économique sur la pêche, la plaisance, la plongée et la chasse sous-marine. Périodes estivale et hivernale. Ville de Marseille and GIS Posidonie, GIS Posidonie publ., Marseilles, 154 pp + Appendices.
19. Béthoux, J.P., Gentili, B. and Tailliez, D. (1998) Warming and freshwater budget change in the Mediterranean since the 1940s, their possible relation to the greenhouse effect. *Geophysical Res. Letters* 25 (7), 1023-1026.

20. Bianchi, C.N. and Morri, C. (2000) Marine biodiversity of the Mediterranean Sea: situation, problems and prospects for future research; *Mar. Poll. Bull.* 40 (5), 367-376.
21. Bianchi, C.N. and Peirano, A. (1995) Atlante delle Fanerogame marine della Liguria. *Posidonia oceanica* e *Cymodocea nodosa.* Centro Ricerche Ambiente Marino, ENEA publ., La Spezzia, 146 pp.
22. Bonhomme, P., Bernard, G., Daniel, B. and Boudouresque, C.F. (1999) Archipel de Riou: édude socio-économique sur la plaisance, la pêche amateur, la plongée et la chasse sous-marine. Période de printemps. Synthèse sur un cycle annuel juillet 97-juin 98. Ville de Marseille and GIS Posidonie, GIS Posidonie publ., Marseilles, 83 pp.
23. Boudouresque, C.F. (1995) The marine biodiversity in the Mediterranean: status of species, populations and communities. United Nations Environment Programme, Mediterranean Action Plan, Regional Activity Centre for Specially Protected Areas, Expert Meeting on endangered species in the Mediterranean, Montpellier, France, 22-25 November 1995, UNEP(OCA)MED WG 100/inf.3, 46 pp.
24. Boudouresque, C.F. (1996) Impact de l'homme et conservation du milieu marin en Méditerranée. GIS Posidonie publ., Marseilles (ISBN 2-905540-21-4), 243 pp.
25. Boudouresque, C.F. (2002a) The spread of a non native species, *Caulerpa taxifolia.* Impact on the Mediterranean biodiversity and possible economic consequences, in F. Di Castri and V. Balaji (eds.), *Tourism, Biodiversity and Information,* Backhuys publ., Leiden, pp. 75-87.
26. Boudouresque, C.F. (2002b) Concilier protection et usages du milieu marin: l'expérience du Parc national de Port-Cros. *La Jaune et la Rouge,* Fr. 575, 31-35.
27. Boudouresque, C.F., Arrighi, F., Finelli, F. and Lefèvre, J.R. (1995a) Arrachage des faisceaux de *Posidonia oceanica* par les ancres: un protocole d'étude. *Rapp. P.V. Réun. Commiss. internation. Explor. sci. Médit.* 34, 21.
28. Boudouresque, C.F., Ballesteros, E., Ben Maiz, N., Boisset, F., Bouladier, E., Cinelli, F., Cirik, S., Cormaci, M., Jeudy de Grissac, A., Laborel, J., Lanfranco, E., Lundberg, B., Mayhoub, H., Meinesz, A., Panayotidis, P., Semroud, R., Sinnassamy, J.M., Span, A. and Vuignier, G. (1990) Livre rouge "Gérard Vuignier" des végétaux, peuplements et paysages marins menacés de Méditerranée. Programme des Nations Unies pour l'Environnement publ., 250 pp.
29. Boudouresque, C.F., Jeudy de Grissac, A. and Meinesz, A. (1984) Relations entre la sédimentation et l'allongement des rhizomes orthotropes de *Posidonia oceanica* dans la baie d'Elbu (Corse), in C.F. Boudouresque, A. Jeudy de Grissac and J. Olivier (eds.), *First International Workshop on Posidonia oceanica beds,* GIS Posidonie publ., Marseilles, pp. 185-191.
30. Boudouresque, C.F. and Laborel-Deguen, F. (1986) *Patella ferruginea, in* C.F. Boudouresque, J.G. Harmelin and A. Jeudy de Grissac (eds.), *Le Benthos marin de l'île de Zembra (Parc National, Tunisie).* UNEP-IUCN-RAC/SPA, GIS Posidonie publ., Marseilles, pp. 105-110.
31. Boudouresque, C.F. and Lefèvre, J.R. (1988) Nouvelles données sur le statut du phoque moine *Monachus monachus* dans la région d'Oran (Algérie). GIS Posidonie publ., Marseilles, 30 pp.
32. Boudouresque, C.F. and Lefèvre, J.R. (1992) Ressources alimentaires, phoque moine (*Monachus monachus*) et stratégie de protection, *in Environmental Encounters,* Antalya, Turkey, 1-4 May 1991, Council of Europe publ., 13, pp. 73-78.
33. Boudouresque, C.F. and Meinesz, A. (1982) Découverte de l'herbier de Posidonie. *Cah. Parc nation. Port-Cros,* Fr. 4, 3 + 79 pp.
34. Boudouresque, C.F., Meinesz, A., Ledoyer, M. and Vitiello, P. (1994) Les herbiers à phanérogames marines, *in* D. Bellan-Santini, J.C. Lacaze and C. Poizat (eds.), *Les biocénoses marines et littorales de Méditerranée, synthèse, menaces et perspectives.* Muséum National d'Histoire Naturelle publ., Paris, pp. 98-118.
35. Boudouresque, C.F., Meinesz, A., Ribera, M.A. and Ballesteros, E. (1995b) Spread of the green alga *Caulerpa taxifolia* (Caulerpales, Chlorophyta) in the Mediterranean: possible consequences of a major ecological event. *Scientia marina* 59 (suppl. 1), 21-29.
36. Boudouresque, C.F., Nédélec, H. and Shepherd, S.A. (1981) The decline of a population of the seaurchin *Paracentrotus lividus* in the bay of Port-Cros (Var). *Rapp. P.V. Réun. Commiss. internation. Explor. sci. Médit.* 27 (2), 223-224.
37. Boudouresque, C.F. and Ribera, M.A. (1995) Les espèces et les espaces protégés marins en Méditerranée. Situation actuelle, problèmes et priorités, *in Les zones protégées en Méditerranée: espaces, espèces et instruments d'application des conventions et protocoles de la Méditerranée,* Tunis, nov. 1993, CERP-CEM-ISPROM publ., Tun., pp. 93-142.

38. Boudouresque, C.F., Van Klaveren, M.C. and Van Klaveren, P. (1996) Proposal for a list of threatened or endangered marine and brackish species (plants, invertebrates, fish, turtles and mammals) for inclusion in appendices I, II and III of the Bern Convention. *Council of Europe, Document S/TPVS96/TPVS48E, 96A*, 138 pp.
39. Bruslé, J. (1985) Exposé synoptique des données biologiques sur les mérous *Epinephelus aeneus* (Geoffroy Saint Hilaire, 1809) et *Epinephelus guaza* (Linnaeus, 1758) de l'Océan Atlantique et de la Méditerranée. Synopsis sur les pêches, FAO publ., Rome, 129, 64 pp.
40. Busch, B.C. (1985) The war against the seals, a history of the north American seal fishery. McGill-Queen's University Press, Kingston and Montreal, 374 pp.
41. Cadiou, G., Le Direach, L., Bernard, G. and Boudouresque, C.F. (2002) Suivi de l'effort de pêche professionnelle dans les eaux du Parc national de Port-Cros. Année 2001. Parc national de Port-Cros and GIS Posidonie, GIS Posidonie publ., Marseilles, 42 pp. + Appendices.
42. Caltagirone, A. (1995) The Mediterranean monk seal. RAC/SPA publ., Tunis, 71 pp + 7 pl.
43. Carlton, J.T. (1993) Neoextinctions of marine invertebrates. *Amer. Zool.* 33, 499-509.
44. Chauvet, C. (1991) Statut d'*Epinephelus guaza* (Linnaeus, 1758) et éléments de dynamique des populations méditerranéenne et atlantique, *in* C.F. Boudouresque, M. Avon and V. Gravez (eds.), *Les espèces marines à protéger en Méditerranée*, GIS Posidonie publ., Marseilles, pp. 255-275.
45. Chavoin, O. and Boudouresque, C.F. (1997) Données préliminaires sur la fréquentation plaisancière, la pêche amateur, la plongée et le tourisme balnéaire au Rayol-Canadel-sur-Mer et à Cavalaire (Var, France). GIS Posidonie publ., Marseilles, 98 pp + 7 Appendices.
46. Clark, J.A. and May, R.M. (2002) Taxonomic bias in conservation research. *Science* 297, 191-192.
47. Combelles, S. (1991) Pêche amateur dans les eaux du Parc national de Port-Cros; Rapport d'enquête. Document préalable au rapport final. Contrat d'Etude Parc national de Port-Cros N° 8604983400 PC, Hyères, 16 pp.
48. Costa Neves, H. (1992) The monk seal (*Monachus monachus*). Conservation and monitoring on the Desertas Islands, Madeira, *in Environmental Encounters,* Antalya, Turkey, 1-4 May, 1991, Council of Europe publ., 19, pp. 21-24.
49. Cottalorda, J.M., Robert, P., Charbonnel, E., Dimeet, J., Menager, V., Tillman, M., Vaugelas, J. de and Volto, E. (1996) Eradication de la colonie de *Caulerpa taxifolia* découverte en 1994 dans les eaux du parc National de Port-Cros (Var, France), *in* M.A. Ribera, E. Ballesteros, C.F. Boudouresque, A. Gómez A. and V. Gravez (eds.), *Second international workshop on Caulerpa taxifolia*, Univ. Barcelona publ., pp. 149-155.
50. Culioli, J.M. (1995) La pêche professionnelle dans la réserve naturelle des îles Lavezzi (Corse): efforts de productions. Mémoire DES, University of Montpellier II, Fr., 148 pp.
51. Daniel, B., Bonhomme, P., Guillaume, B. and Boudouresque, C.F. (1998) La pêche amateur dans l'archipel de Riou (Marseille, Méditerranée occidentale). Analyse des pratiques. Essai de quantification de l'effort de pêche et des captures. Ville de Marseille and GIS Posidonie publ., Marseilles, 63 pp + Appendices.
52. Dayton, P.K., Sala, E., Tegner, M.J. and Thrush, S. (2000) Marine reserves: parks, baselines, and fishery enhancement. *Bull. mar. Sci.* 66 (3), 617-634.
53. Delaunay, A. (2003) La pêche professionnelle "aux petits métiers" et la pêche amateur autour de la réserve marine de Carry-le-Rouet (Bouches-du-Rhône, France). Mémoire DESS Economie et Environnement, University of Aix-Marseille 2, Fr., 72 pp.
54. Ehlinger, L. (2001) Rendement des filets trémails dans la Réserve naturelle des Bouches de Bonifacio (Corse). Evolution et optimisation. Mémoire Diplôme Etudes sup. approf., University of Corsica, 50 pp. + Appendices.
55. Francour, P. and Chauvet, C. (1993) Présence de *Epinephelus alexandrinus* (Valenciennes, 1828) dans la zone maritime du Parc national de Port-Cros. *Sci. Rep. Port-Cros nation. Park,* Fr. 15, 279-283.
56. Francour, P., Harmelin, J.G., Pollard, D. and Sartoretto, S. (2001) A review of marine protected areas in the northwestern Mediterranean region: siting, usage, zonation and management. *Aquatic Conserv.: mar. freshw. Ecosyst.* 11, 155-188.
57. Frenkiel, L. (1975) Contribution à l'étude des cycles de reproduction des Patellidae en Algérie. *Pubbl. Staz. zool. Napoli* 39 (suppl.), 153-189.
58. Gardner, T.A., Côté, I.M., Gill, J.A., Grant, A. and Watkinson, A.R. (2003) Long-term region-wide declines in Caribbean corals. *Science* 301, 958-960.

59. Garrabou, J., Perez, T., Sartoretto, S. and Harmelin, J.G. (2001) Mass mortality event in red coral *Corallium rubrum* populations in the Provence region (France, NW Mediterranean). *Mar. Ecol. Progr. Ser.* 217, 263-272.
60. Geronimi, I. (1988) Introduction à la pêche artisanale de la région Calvi-Galeria. Présentation de l'activité et étude de la production halieutique. Etude des paramètres de croissance de trois espèces d'intérêt économique. Mémoire de Maîtrise, University of Corsica, 30 pp + 8 pp.
61. Gould, S.J. (1991) On the lost of a limpet. *Natural History* 100, 22-27.
62. Gray, J.S. (2000) The measurement of marine species diversity, with an application to the benthic fauna of the Norwegian continental shelf. *J. exp. mar. Biol. Ecol.* 250, 23-49.
63. Gray, J.S. (2001) Marine diversity: the paradigms in patterns of species richness examined. *Scientia marina* 65 (suppl. 2), 41-56.
64. Groombridge, B. (1990) Les tortues marines en Méditerranée: distribution, populations, protection. Conseil de l'Europe publ., Strasbourg, 116 pp.
65. Groombridge, B. (1993) 1994 IUCN red list of threatened animals. IUCN publ., Gland, Switzerland, lvi + 286 pp.
66. Groupe d'Étude du Mérou (1996) Le mérou brun en Méditerranée. GEM publ., Hyères, Fr., 28 pp.
67. Guerin, B. (2003) Approche descriptive de l'activité de pêche aux "petits métiers". Le cas des îles d'Hyères (Var, France). Mémoire Diplôme Agronomie Approfondie, Spécialité halieutique, Univ. Rennes, Fr., 50 pp.
68. Hall, S.D. (1998) Closed areas for fisheries management. The case consolidates. *Trends Evol. Ecol.* 13 (8), 297-298.
69. Harmelin, J.G. (1973) Bryozoaires de l'herbier de Posidonies de l'île de Port-Cros. *Rapp. P.V. Réun. Commiss. internation. Explor. sci. Médit.* 21 (9), 675-677.
70. Harmelin, J.G. (1993) Invitation sous l'écume. *Cah. Parc nation. Port-Cros,* Fr. 10, 1-83.
71. Harmelin, J.G. (1995) Gorgones. Les plus beaux ornements de Méditerranée sont-ils menacés? *Océanorama,* Fr, 24, 3-9.
72. Harmelin, J.G. (1999) Visual assessment of indicator fish species in Mediterranean Marine Protected Areas. *Naturalista sicil.* 23 (suppl.), 83-104.
73. Harmelin, J.G. (2003) Biodiversité des habitats cryptiques marins du parc national de Port-Cros (Méditerranée, France). Assemblages de bryozoaires d'une grotte sous-marines et des faces inférieures de pierres. *Sci. Rep. Port-Cros nation. Park,* Fr. 19, 101-115.
74. Harmelin, J.G., Boury-Esnault, N., Fichez, R., Vacelet, J. and Zibrowius, H. (2003a) Peuplement de la grotte sous-marine de l'île de Bagaud (Parc national de Port-Cros, France, Méditerranée). *Sci. Rep. Port-Cros nation. Park,* Fr. 19, 117-134.
75. Harmelin, J.G. and Marinopoulos, J. (1993) Recensement de la population de corbs (*Sciaena umbra* Linnaeus, 1758: Pisces) du Parc national de Port-Cros (Méditerranée, France) par inventaires visuels. *Sci. Rep. Port-Cros nation. Park,* Fr. 15, 265-276.
76. Harmelin, J.G. and Marinopoulos, J. (1994) Population structure and partial mortality of the gorgonian *Paramuricea clavata* (Risso) in the North-Western Mediterranean (France, Port-Cros Island). *Marine Life* 4 (1), 5-13.
77. Harmelin, J.G. and Robert, P. (2001) Evolution récente de la population du mérou brun (*Epinephelus marginatus*) dans le Parc national de Port-Cros (France, Méditerranée). *Sci. Rep. Port-Cros nation. Park,* Fr. 18, 149-161.
78. Harmelin, J.G., Robert, P. and Cantou, M. (2003b) Recensement de la population de mérou brun (*Epinephelus marginatus*) du Parc national de Port-Cros: premiers résultats de la campagne "Philippe Tailliez" du 7 au 11 octobre 2002. Parc national de Port-Cros and Groupe d'Étude du Mérou, Fr., 11 pp.
79. Harmelin, J.G., Sartoretto, S., Francour, P., Boudouresque, C.F., Bellan-Santini, D. and Vacelet, J. (1998) Création d'une aire marine protégée dans l'archipel de Riou: proposition de plans de gestion. Direction de l'Environnement et des Déchets, Ville de Marseille and Centre d'Océanologie de Marseille, 198 pp.
80. Harmelin-Vivien, M. (1982) Ichtyofaune des herbiers de Posidonies du Parc national de Port-Cros. I. Composition et variations spatio-temporelles. *Trav. sci. Parc nation. Port-Cros,* Fr. 8, 69-92.
81. Harmelin-Vivien, M., Francour, P. and Harmelin, J.G. (1999) Impact of *Caulerpa taxifolia* on Mediterranean fish assemblages: a six year study, *in Proceedings of the workshop on invasive Caulerpa in the Mediterranean.* Heraklion, Crete, Greece, 18-20 March 1998. UNEP publ., Athens, Greece, pp. 127-138.

82. Henry, M., Stevens, H. and Carson, W.P. (2001) Phenological complementarity, species diversity and ecosystem function. *Oikos* 92, 291-296.
83. Hereu, B., Zabala, E. and Ballesteros, E. (2003) On the occurrence of a population of *Cystoseira zosteroides* Turner and *Cystoseira funkii* Schiffner *ex* Gerloff *et* Nizamuddin (Cystoseiraceae, Fucophyceae) in Port-Cros National Park (Northwestern Mediterranean, France). *Sci. Rep. Port-Cros nation. Park*, Fr. 19, 93-99.
84. IRAP, 1999. Etude des retombées du Parc national sur l'activité économique et sur l'emploi. Parc National de Port-Cros and IRAP publ., Annecy, Fr., 76 pp.
85. IUCN, 1980. The World Conservation Strategy: living resource conservation for sustainable development. IUCN/UNEP/WWF, IUCN publ., Gland, Switerland.
86. IUCN, 1994. Catégories de l'IUCN pour les Listes Rouges. IUCN Species Survival Commission, IUCN publ., Gland, Switerland, 21 pp.
87. IUCN, 2000. IUCN guidelines for the prevention of biodiversity loss caused by alien invasive species. SSC Invasive Species Specialist Group. Univ. Auckland publ., New-Zealand, 15 pp.
88. Jackson, J.B.C. and Johnson, K.G. (2001) Measuring past biodiversity. *Science* 293, 2401-2404.
89. Jacobs, J. and Panou, A. (1988) Conservation of the Mediterranean monk seal, *Monachus monachus*, in Kefalonia, Ithaca and Lefkada Isl., Ionian Sea, Greece. Institut royal des Sci. nat. Belgique, Projet ACE 6611/28, 221 pp.
90. Jouvenel, J.Y. and Bachet, F. (2002) Programme de suivi des peuplements ichtyologiques de la réserve marine Richard Fouque du Cap Couronne. Rapport final, bilan 1995 à 2001. Aquafish Technology and Parc marin de la Côte Bleue, Fr., 25 pp.
91. Kenchington, R., Ward, T. and Hegerl, E. (2003) The benefit of marine protected areas. Commonwealth of Australia publ., 20 pp.
92. Laborel, J. (1987) Marine biogenic constructions in the Mediterranean. *Sci. Rep. Port-Cros nation. Park*, Fr. 13, 97-126.
93. Laborel, J., Boudouresque, C.F. and Laborel-Deguen, F. (1994) Les bioconcrétionnements littoraux de Méditerranée, *in* D. Bellan-Santini, J.C. Lacaze and C. Poizat (eds.), *Les biocénoses marines et littorales de Méditerranée, synthèse, menaces et perspectives,* Muséum National d'Histoire Naturelle publ., Paris, pp. 88-97.
94. Laborel-Deguen, F. (1985) Biologie et répartition de *Patella ferruginea. Trav. sci. Parc nat. rég. Rés. nat. Corse,* Fr. 2, 41-48.
95. Laborel-Deguen, F. and Laborel, J. (1990) Nouvelles données sur la patelle géante *Patella ferruginea* Gmelin en Méditerranée. I. Statut, répartition et étude des populations. *Haliotis* 10, 41-54.
96. Laborel-Deguen, F. and Laborel, J. (1991a) Statut de *Patella ferruginea* Gmelin en Méditerranée, *in* C.F. Boudouresque, M. Avon and V. Gravez (eds.), *Les espèces marines à protéger en Méditerranée,* GIS Posidonie publ., Marseilles, pp. 91-103.
97. Laborel-Deguen, F. and Laborel, J. (1991b) Nouvelles observations sur la population de *Patella ferruginea* Gmelin de Corse, *in* C.F. Boudouresque, M. Avon and V. Gravez (eds.), *Les espèces marines à protéger en Méditerranée,* GIS Posidonie publ., Marseilles, pp. 105-110.
98. Laborel-Deguen, F., Laborel, J. and Morhange, C. (1993) Appauvrissement des populations de la patelle géante *Patella ferruginea* Gmel. (Mollusca, Gasteropoda, Prosobranchiata) des côtes de la Réserve marine de Scandola (Corse du Sud) et du Cap Corse (Haute Corse). *Trav. sci. Parc nat. rég. Rés. nat. Corse* 41, 25-32.
99. Le Diréach, L., Cadiou, G. and Boudouresque, C.F. (2002) Mise en place d'un suivi de l'effort de pêche professionnelle dans la réserve naturelle de Scandola (Corse). Données 2000-2001. Parc naturel régional de Corse and GIS Posidonie, GIS Posidonie publ., Marseilles, 63 pp.
100. Loquès, F., Bellone, E., Meinesz, A. and Villette, M. (1995) Cartographie sous-marine du Parc national de Port-Cros (Var, France). II. La zone protégée de la baie de La Palud. *Sci. Rep. Port-Cros nation. Park,* Fr. 16, 129-133 + 1 map.
101. Malakoff, D. (2001) Reserves found to aid fisheries. *Science* 294, 1807-1809.
102. Marchessaux, D. (1989a) Recherches sur la biologie, l'écologie et le statut du phoque moine *Monachus monachus.* GIS Posidonie publ., Marseilles, 280 pp.
103. Marchessaux, D. (1989b) Distribution et statut des populations du phoque moine *Monachus monachus* (Hermann, 1779). *Mammalia* 53 (4), 621-642.
104. Marion, A.F. (1883) Esquisse d'une topographie zoologique du golfe de Marseille. *Ann. Mus. Hist. nat. Marseille* 1, 6-108.

105. Marion, R. and Sylvestre, J.P. (1993) Guide des Otaries, Phoques et Siréniens. Delachaux and Niestlé publ., Switzerland, 159 pp.
106. Maurin, H. and Keith, P. (1994) Le livre rouge. Inventaire de la flore menacée de France. Nathan publ., Paris,: 175 pp.
107. May, R. (1997) L'inventaire des espèces vivantes. *L'évolution.* Dossier Hors-série *Pour la Science,* Fr., 40-47.
108. May, R.M. (1999) What we do and do not know about the diversity of life on Earth *in* A. FARINA (ed.), *Perspectives in Ecology,* Backhuys publ., Leiden, pp. 33-40.
109. McNeely, J.A. (1994) Protected areas for the twenty-first century: working to provide benefits for Society. *Unasylva 176,* 45, 3-7.
110. McNeely, J.A., Miller, K.R., Reid, W.V., Mittermeier, R.A. and Werner, T.B. (1990) Conserving the world's biological diversity. IUCN publ., Switzerland.
111. Medioni, E. and Vicente, N. (2003) Cinétique des populations de *Pinna nobilis* dans les zones interdites au mouillage et des zones autorisées. Parc national de Port-Cros and University of Aix-Marseille III, Fr., 15 pp. + 4 pl.
112. Meinesz, A., Astier, J.M., Bodoy, A., Cristiani, G. and Lefèvre, J.R. (1982) Impact de l'aménagement du domaine maritime sur l'étage infralittoral des Bouches du Rhône (France, Méditerranée occidentale). *Vie Milieu* 32 (2), 115-124.
113. Meinesz, A. and Lefèvre, J.R. (1978) Destruction de l'étage infralittoral des Alpes-Maritimes (France) et de Monaco par les restructurations de rivage. *Bull. Ecol.* 9 (3), 259-276.
114. Meinesz, A., Lefèvre, J.R. and Astier, J.M. (1991) Impact of coastal development on the infralittoral zone along the southern Mediterranean shore of continental France. *Mar. Poll. Bull.* 23, 343-347.
115. Millar, A.J.K. (2001) The world's first recorded extinction of a seaweed, *in 17th international Seaweed Symp.,* Cape-Town, South Africa. Abstracts, 94.
116. Naeem, S. and Li, S. (1997) Biodiversity enhances ecosystem reliability. *Nature* 390, 507-509.
117. Neubert, M.G. (2003) Marine reserves and optimal harvesting. *Ecology Letters* 6, 843-849.
118. Noël, P. (2003) Les crustacés du Parc national de Port-Cros et de la région des îles d'Hyères (Méditerranée), France. Etat actuel des connaissances. *Sci. Rep. Port-Cros nation. Park,* Fr. 19, 135-306.
119. Oztürk, B. (1992) Aeniz foku *Monachus monachus.* Anahtar KITAPLAR publ., Istanbul, 215 pp.
120. Paillard, M., Gravez, V., Clabaut, P., Walker, P., Blanc, J.J., Boudouresque, C.F., Belsher, T., Urscheler, F., Poydenot, F., Sinnassamy, J.M., Augris, C., Peyronnet, J.P., Kessler, M., Augustin, J.M., Le Drezen, E., Prudhomme, C., Raillard, J.M., Pergent, G., Hoareau, A. and Charbonnel, E. (1993) Cartographie de l'herbier de Posidonie et des fonds marins environnants de Toulon à Hyères (Var, France). Reconnaissance par sonar latéral et photographie aérienne. Notice de présentation. IFREMER and GIS Posidonie publ., Marseilles, 36 pp. + 3 maps.
121. Pérès, J.M. (1984) La régression des herbiers à *Posidonia oceanica, in* C.F. Boudouresque, A. Jeudy de Grissac and J. Olivier (eds.), *First international Workshop on Posidonia oceanica beds,* GIS Posidonie publ., Marseilles, pp. 445-454.
122. Perez, T., Garrabou, J., Sartoretto, S., Harmelin, J.G., Francour, P. and Vacelet, J. (2000) Mortalité massive d'invertébrés marins: un événement sans précédent en Méditerranée nord-occidentale. *C.R. Acad. Sci., Life Sci.* 323, 853-865.
123. Phillips, J.A. (1998) Marine conservation initiatives in Australia: their relevance to the conservation of macroalgae. *Botanica marina* 41, 95-103.
124. Pires, R. and Costa Neves, H. (2000) Monk seal sightings on open beaches in the Desertas Islands, Madeira Archipelago. *Monachus Guardian,* Canada 3 (1), 70-71.
125. Planes, S., Galzin, R., García-Rubies, A., Goni, R., Harmelin, J.G., Le Diréach, L., Lenfant, P.and Quetglas, A. (2000) Effects of marine protected areas on recruitment processes with special reference to Mediterranean littoral ecosystems. *Environmental Conservation* 27 (2), 126-143.
126. Porcheddu, A. and Milella, I. (1991) Aperçu sur l'écologie et sur la distribution de *Patella ferruginea* (L.) Gmelin 1791 en mers italiennes, *in* C.F. Boudouresque, M. Avon and V. Gravez (eds.), *Les espèces marines à protéger en Méditerranée,* Gis Posidonie publ., Marseilles, pp. 119-128.
127. Raffin, J.P. (2001) De la "mise à part" au "vivre avec": approche d'une histoire des concepts de protection de la nature. *La Jaune et la Rouge,* Fr. 566, 45-47.
128. Ramade, F. (1990) Conservation des écosystèmes méditerranéens: enjeux et perspectives. Fascicules du Plan Bleu, PNUE-CAR/PB, Diff. Economica, Paris, xvi + 144 pp.

129. Ramos, A.A. (1992) Impact biologique et économique de la réserve marine de Tabarca (Alicante, Sud-Est de l'Espagne). *Economic impact of the Meditarranean coastal protected areas, Mepan News,* Fr. 3, 59-66.
130. Ramos-Esplà, A.A., Aranda, A., Gras, D. and Guillèn, J.E. (1994) Impactos sobre las praderas de *Posidonia oceanica* (L.) Delile en el SE español: necesidad de establecer herramientas de ordenamiento y gestión del litoral, *in Actes du colloque scientifique "Pour qui la Méditeranée au 21° siècle? Villes des rivages et environnement littoral en Méditeranée",* Okéanos, Montpellier, pp. 64-69.
131. Reijnders, P.J.H. (1997) Seal specialist group. *Species* 29, 49-50.
132. Ribera Siguan, M.A. (1992a) Réserve des îles Medes et fréquentation touristique régionale. *Economic impact of the Mediterranean coastal protected areas,* Ajaccio, 26-28 September 1991, *Medpan News,* Fr. 3, 51-57.
133. Ribera Siguan, M.A. (1992b) La réserve marine des îles Medes: bilan d'un succès imprévu, *in Atti del 2° Convegno internazionale "Parchi marini del Mediterraneo, Problemi e Prospettive",* San Teodoro, 17-19 May 1991, Ital., pp. 152-161.
134. Richardson, C.A., Kennedy, H., Duarte, C.M., Kennedy, D.P. and Proud, S.V. (1999) Age and growth of the fan mussel *Pinna nobilis* from south-east Spanish Mediterranean seagrass (*Posidonia oceanica*) meadows. *Marine Biology* 133, 205-212.
135. Riggio, S., Calvo, S., Fradà-Orestano, C., Chemello, R. and Arculeo, M. (1994) La dégradation du milieu dans le golfe de Palerme (Sicile Nord-Ouest) et les perspectives d'assainissement, *in* Actes du colloque scientifique *"Pour qui la Méditerranée au 21° siècle? Villes des rivages et environnement littoral en Méditerranée",* Okéanos, Montpellier, pp. 82-89.
136. Riutort, J.J. (1989) Première estimation des captures et de l'effort de pêche déployé par les "petits métiers" sur le littoral Nord-Ouest de la Corse. Etude de la biologie des principales espèces cibles. Station de recherche sous-marine et océanographique Stareso publ., Calvi, Corsica, 133 + 18 pp.
137. Robert, P. (1996) Recherche de l'algue *Caulerpa taxifolia* dans les eaux du Parc National de Port-Cros, *in* M.A. Ribera, E. Ballesteros, C.F. Boudouresque, A. Gómez and V. Gravez (eds.), *Second international workshop on Caulerpa taxifolia,* Univ. Barcelona publ., Spain, pp. 99-100.
138. Robert, P. and Gravez, V. (1998) Contrôle de l'algue *Caulerpa taxifolia* dans le Parc National de Port-Cros (Var, France), *in* C.F. Boudouresque, V. Gravez, A. Meinesz and F. Palluy (eds.), *Third international workshop on Caulerpa taxifolia,* GIS Posidonie publ., Marseilles, pp. 79-87.
139. Robert, P., Perrocheau, D., Gerardin, N. and Vix, J.M. (1987) Comptage des mérous de l'îlot de La Gabinière, Parc national de Port-Cros, été 1983. *Trav. sci. Parc nation. Port-Cros,* Fr. 13, 129-131.
140. Roberts, C.M., Bohnsack, J.A., Gell, F., Hawkins, J.P. and Goodridge, R. (2001) Effects of marine reserve on adjacent fisheries. *Science* 294, 1920-1923.
141. Roberts, C.M. and Hawkins, J.P. (1999) Extinction risk in the sea. *Trends Ecol. Evol* 14 (6), 241-246.
142. Sala, E., Garrabou, J. and Zabala, M. (1996) Effects of diver frequentation on Mediterranean sublittoral populations of the bryozoan *Pentapora fascialis. Mar. Biol.* 126, 451-459.
143. Salvat, B., Haapkilä, J. and Schrimm, M. (2002) Coral reef protected areas in international instruments. World Heritage Convention, World network of Biosphere Reserves, Ramsar Convention, CRIOBE-EPHE, Moorea, French Polynesia, xi + 196 pp.
144. Sumaila, U.R., Guénette, S., Alder, J.and Chuenpagdee, R. (2000) Addressing ecosystem effects of fishing using marine protected areas. *ICES J. mar. Sci.* 57, 752-760.
145. Templado, J. (1995) La conservación del medio marino en España. *Fronteras Ciencia Tecnol,* Spain 9, 22-26.
146. Vicente, N. and Moreteau, J.C. (1991) Statut de *Pinna nobilis* L. en Méditerranée (Mollusque Eulamellibranche), *in* C.F. Boudouresque, M. Avon and V. Gravez (eds.), *Les espèces marines à protéger en Méditerranée.* GIS Posidonie publ., Marseilles, pp. 159-168.
147. Vicente, N., Moreteau, J.C. and Escoubet, P. (1980) Etude de l'évolution d'une population de *Pinna nobilis* L. (Mollusque eulamellibranche) au large de l'anse de la Palud (Parc national sous-marin de Port-Cros, France). *Trav. sci. Parc nation. Port-Cros,* Fr. 6, 39-67.
148. Ward, T. and Hegerl, E. (2003) Marine Protected Areas in ecosystem-based management of fisheries. Natural Heritage Trust, Commonwealth of Australia publ., 66 pp.
149. Welcomme, R.L. (1999) A review of a model for quantitative evaluation of exploitation levels in multispecific fisheries. *Fisheries Managem. Ecol.* 6, 1-19.

150. Zabala, M., García-Rubies, A., Louisy, P. and Sala, E. (1997a) Spawning behaviour of the Mediterranean dusky grouper *Epinephelus marginatus* (Lowe, 1834) (Pisces, Serranidae) in the Medes Islands Marine Reserve (NW Mediterranean, Spain). *Scientia marina* 61 (1), 65-77.
151. Zabala, M., Louisy, P., García-Rubies, A. and Gracia, V. (1997b) Socio-behavioural context of reproduction in the Mediterranean dusky grouper *Epinephelus marginatus* (Lowe, 1834) (Pisces, Serranidae) in the Medes Islands marine reserve (NW Mediterranean, Spain). *Scientia marina* 61 (1), 79-89.
152. Zavodnik, D., Hrs-Brenko, M. and Legac, M. (1991) Synopsis on the fan shell *Pinna nobilis* L. in the eastern Adriatic Sea, *in*: C.F. Boudouresque, M. Avon and V. Gravez (eds.), *Colloque international "Les espèces marines à protéger en Méditerranée"*. GIS Posidonie publ., Marseilles, pp. 169-178

DAMAGE CONTROL IN THE COASTAL ZONE: IMPROVING WATER QUALITY BY HARVESTING AQUACULTURE-DERIVED NUTRIENTS

Dror L. Angel
Woods Hole Oceanographic Institute
Woods Hole, Massachusetts, USA
dror@mit.edu

Timor Katz and Noa Eden
Interuniversity Institute of Eilat
P.O.B 469, Eilat 88103 Israel

Ehud Spanier
The Leon Recanati Centre for Maritime Studies
University of Haifa, 31905 Israel

Kenny D. Black
Scottish Association for Marine Science
Oban PA37 1QA, UK

Abstract

There is a clear relationship between nutrient enrichment and the eventual deterioration of coastal water quality (eutrophication). Eutrophication occurs when the rate of nutrient supply exceeds its transformation or removal rate and excess nutrients stimulate excess biological production. This imbalance may be corrected to varying degrees by enhancing certain natural biological/ecological attributes (ecosystem services), such as the removal of particles by means of filter feeding animals. This topic is addressed by means of a case study - aquaculture in coastal waters. We begin by examining the impacts related to the release of nutrient-rich effluents from a commercial fish farm in the oligotrophic Gulf of Aqaba. This is followed by description of 3 approaches that have been tested as means to capture and remove aquaculture effluents, as an example of how we may enhance the sustainability or reduce the environmental impacts of commercial activities. These approaches include placement of detritus feeding grey-mullets in benthic enclosures on the organically enriched seafloor below commercial fish cages, mooring artificial reefs as benthic biofilters next to fish farms and deployment of pelagic biofilters in the water column, adjacent to fish cages.

Introduction

The most heavily populated areas worldwide are the coastal regions and the environmental implications of such human density are manifold (Small and Nicholls, 2003). Nutrient enrichment, eutrophication and such phenomena as water column hypoxia, benthic anoxia and harmful algal blooms are among the biggest problems confronting managers in estuarine and coastal waters (Bricker et al. 1999, NRC 2000). Among the various activities contributing nutrients to the coastal nutrient pool, fish farming has been implicated as an increasingly important factor (McGinn 1998). Although the scale of marine aquaculture in most European waters is small, relative to other nutrient-producing activities, it is a rapidly growing industry and must be considered and monitored, alongside other nutrient-emitters in the coastal zone.

Aquaculture may serve as a good model to quantitatively examine the in-situ, large-scale environmental impacts of marine nutrient enrichment. Most fish farms accurately document their feed inputs and for many of the commercial feeds the nutrient release rates have been determined. Thus, aquaculture is very conducive to the study of point-source nutrient inputs and implications. In the following we will use a case study - aquaculture in the Gulf of Aqaba - to explore some ways in which effluents released from cage aquaculture affect their surroundings. Moreover, we will describe several approaches whereby these effluents may be captured and harvested, thereby reducing: a) their flux into the marine environment and b) their environmental impacts.

Study area and site

The Gulf of Aqaba is a semi-enclosed sea extending northeastward from the Red Sea and surrounded by deserts. Four countries share this unique body of water: Saudi Arabia, Jordan, Israel and Egypt. The Gulf is very narrow, reaching only 26 km wide at its widest point, and very deep, with an average depth of 800 m, and reaching 1,800 m deep in some areas. The waters of the Gulf of Aqaba are warm (annual range is 20 - 26 C) and highly saline (41 psu), due to the high evaporation rates (1 cm/d) (Reiss & Hottinger 1984). The Gulf boasts some of the most diverse, complex and attractive coral reefs worldwide, consisting of more than 300 sub-species of coral, over 1,000 species of fish and many more invertebrate taxa. These are also the world's northern-most coral reef ecosystems, due most likely to the surrounding desert areas. While the reefs are biologically mature and stable, they are also vulnerable to a number of environmental hazards. The foreseeable dangers to the Gulf of Aqaba reefs include hydrochemical (oil) spills, organic enrichment (mainly from sewage), inorganic nutrients (mostly phosphates), tourist activities and coastal development (Atkinson et al. 2001).

Two commercial net pen fish farms, *Ardag* and *DagSuf*, operate at the northern end of the Gulf of Eilat. In our studies, we have focused the bulk of our work on the *Ardag* farm which is closer to shore. The *Ardag* fish farm is situated 300-500 m from shore and close to the Israeli-Jordanian border. The mean annual current velocity is less than 10 cm s^{-1}, though in winter flow rate may occasionally be as rapid as 35 cm s^{-1} (Brenner et al. 1988, 1991). The dominant component of the flow at the fish farm location is east to west, however there are frequent erratic changes in both flow direction and velocity. The water temperature ranges from 21°C in winter to 26°C in summer (Reiss and Hottinger 1984). The *Ardag* farm consists of three parallel pontoons (pontoon length, 150-200m), each with 10 pairs of cylindrical floating net cages (mean dimensions: 13m diameter, 10m deep), moored perpendicular to the predominant current direction (Katz et al. 2002, Eden et al. 2003). The major fish species reared in the cages is gilthead seabream, *Sparus aurata*, stocked at between 20 to 25 kg·m^{-3}. The natural, unenriched sediments near the farm consist of fine sand that support a wide variety of invertebrates and *Halophila stipulacea* (seagrass) beds (Fishelson, 1971). The organically-enriched sediments, below the fish cages, are often covered by microbial mats that consist mainly of benthic sulfur bacteria (*Beggiatoa* spp.) and cyanobacteria (Angel et al. 1995).

Aquaculture impacts in the Gulf of Aqaba

The influence of the fish farms on the marine environment is largely related to the effluents released from the cages to the surrounding waters. Between 1996 and 1999, nutritional studies showed that 1.79 ton of feed were required to produce 1 ton of gilthead sea bream (Lupatsch and Kissil 1998), however the actual feed conversion ratio was between 2.3 - 2.6. The nutrient budget calculated from these studies indicated that 68% of the nitrogen (N), 29% of the phosphorus (P) and 56% of the carbon provided in the feed was released to the environment as dissolved compounds, while 9% of the N, 42% of the P and 15% of the C in the feed descended to the seafloor as particulate matter.

In an attempt to get a general notion regarding the dispersal of dissolved compounds from the fish farms to the surrounding waters, a hydrodynamic model (Princeton Ocean Model) was employed with a passive tracer, using data from the Gulf of Aqaba. This model yielded worst-case scenarios since it focused on dissolved nitrogen which is not an inert tracer at all (Angel et al. 1998). The model predicted very limited local impact during the winter months due to high background nitrogen levels in this season (deep vertical mixing) yet up to a fourfold increase in local N levels during summer, when the water column is stratified and N levels are barely measureable. The model also showed that the impact decreases with distance from the fish farms.

The flux of particulate organic carbon (POC) to the sediment, as measured under the Ardag farm, varied between 4.5-12.7 g C M^{-2} d^{-1} (Angel et al. 1995, 1998). The input of particulate organic matter (POM) to the sediment as measured by loss-on-ignition (LOI) varies seasonally with higher input in summer than in winter months. Unpublished data indicate that despite increased fish production, there was a decrease in summer LOI levels between 1990-1999, possibly resulting from improvements in feed quality and feeding strategies. The studies conducted below the Ardag fish farm indicated that farm-impacted sediments are restricted to the immediate vicinity of the farm (Angel et al. 1995, Angel et al. 1998, Eden et al. 2003). Decomposition rates of the organic matter in the enriched sediments under the Ardag fish cages based on oxygen demand were

2.7±0.5, 5.8±3.1 and 5.4±1.3 gC m^{-2} d^{-1} for July, August and September 1998, respectively (Katz 2000), very similar to the mean rate (5.3 g C m^{-2} d^{-1}) previously determined by Angel et al. (1995) based on ammonia flux data. Although sedimentation rates were higher than decomposition rates, Angel et al. (1995) did not find accumulation of organic matter in the sediments under the fish farm and they suggested that occasional strong bottom currents and bioturbation may be responsible for lateral transport and additional OM decomposition.

Due to the organic enrichment the sediments under the fish cages contain high concentrations of dissolved nitrogen, phosphorus and hydrogen sulfide and low concentrations of dissolved oxygen (DO). Concentrations of dissolved nitrogen, phosphorus and hydrogen sulfide in the sediment porewater decreased while the concentrations of DO at the sediment surface increased with increasing distance from the farm (Angel et al. 1995, Eden et al. 2003). In summer (May-October) the porewater nutrient levels in the sediment below the farms were higher than the winter concentrations (November-March). Undisturbed sediments did not contain measurable H_2S in the top 3 cm. Ammonia, nitrate, nitrite and phosphate concentrations in the water column near the fish farms were generally under detection limit (Angel et al. 1998). The ammonia concentrations recorded on the Jordanian side of the northern shore of the Gulf of Eilat (nearshore) were consistently higher than the concentrations measured at an offshore reference station during 1994-1995 (Badran and Foster 1998). Ammonia concentrations at this station reached a peak of 0.12-0.25 µM between March and June 1995; whereas nitrite and nitrate peaked at 0.25-0.38 µM and 1.2µM, respectively. Angel et al. (1998) measured 0-40 nM (0 - 0.04µM) phosphate in the vicinity of the fish farms, which was not significantly higher than that recorded at an offshore reference station. The phosphate concentrations recorded at a nearby sampling station on the Jordanian side of the north shore were somewhat higher (100 nM) in spring and lower again in summer (30 nM) (Badran and Foster 1998).

On several occasions, chlorophyll *a* concentrations were measured near the fish farms and these were 2-4 times higher around the fish cages (0.35-0.55 µg/l) as compared with an open water reference station (0.15-0.35 µg/l) (Angel et al. 1998). In subsequent diurnal studies carried out in October 2002 and August 2003 there were not significant differences between chl a concentrations measured at the fish cages and at a reference site (Angel and Katz, unpublished). Badran and Foster (1998) observed high chl *a* concentrations (>0.60 µg l^{-1}) towards the end of winter-early spring and low concentrations during summer at the northern tip of the Gulf of Aqaba, on the Jordanian side of the border. Similarly high levels of chlorophyll were not detected at any of the other stations sampled, including the port of Aqaba, the industrial area, an offshore station and a distant coral reef. Long term monitoring in the middle of the Gulf of Eilat (this station is located several km south of the north shore where the fish farms are situated) revealed a gradual increase in chl *a* concentrations, from 0.1 µg l^{-1} in 1995 to 0.4 µg l^{-1} in 2000 (Genin and Zakai 2000). It was argued that this increase correlated well with the increase in commercial fish production from @1000 ton $year^{-1}$ in 1995 to about 2000 ton year^{-1} in 2000, but the expansion of the towns of Eilat and Aqaba, and all related activities, during this period were ignored.

The enriched sediments under Ardag fish farm are usually covered by *Beggiatoa* (bacterial) mats while the sediments up to 100m away from the farm are covered by pigmented mats comprised of a complex microbial community, including *Beggiatoa* spp., filamentous cyanobacteria, euglenoids, pennate diatoms and ciliates (Angel et al. 1992, 1995). The mats undergo seasonal changes wherein the filamentous cyanobacteria dominate in summer and fall and *Beggiatoa* spp. and *Euglena* spp. dominate during winter and spring. Occasionally during winter strong bottom currents erode the mats, strongly reducing the organic content of the surface sediments, however the mat communities generally reestablish and recover at the begining of spring.

Live benthic foraminifera were found under the Ardag fish farm in the early 1990's and peak abundances were recorded near, but not below the farm (Angel et al. 2000). The abundance of live foraminifera showed negative correlation with sediment porewater ammonia values but no correlation was found with any of the other geochemical variables. Strauss (2000) found that meiofaunal assemblages under the Ardag farm were dominated by Nematoda, although Ciliophora and Turbellaria were also occasionally abundant. In the unenriched sediments, Nematoda, Harpacticoid Copepoda and Gastrotricha were the major taxa. The over-all meiofauna density, taxonomic richness and taxonomic diversity increased with increasing distance from the fish cages. Increased abundances of the opportunistic mud snail *Nassarius sinusigerus* were observed in the vicinity of the Ardag fish farm since 1995 (Eden et al. 2003). A peak in abundance was found some distance from the farm and its position was apparently related to the geochemical conditions in the sediment. The Eilat net-cage fish farms attract many pelagic and reef fishes by virtue of the shelter and nutritional opportunities they provide. Angel et al. (1995) observed large groups of goatfish (*Parupeneus forsskalii*) and rabbitfish (*Siganus rivulatus*) feeding from the sediment in the highly impacted sediments below the fish cages.

The fish reared in the cages occasionally experience outbreaks of parasitic diseases and there is concern that these may spread to the wild fish living just outside the net cages (Diamant et al. 1999). Net cage mariculture may also expose wild fish populations to new diseases and may amplify existing diseases. The pathogenic *Mycobacterium* was not detected among *S. rivulatus* during studies conducted in the 1970's and 1980's. However, since 1990 when it was first detected in farmed sea bass, *Dicentrarchus labrax* (Colorni 1992) the prevalence of the disease increased dramatically in both farmed fish and in wild rabbitfish (Diamant et al. 2000). The pathogen was identified as *Mycobacterium marinum* (Knibb et al. 1993) and may be a local strain (Ucko and Kvitt 2000).

Although there have been numerous "releases" over the past decade (mainly the result of fish cages breaching during storms, and cage nets being torn by predators and human poachers) of farm-reared gilthead seabream to the Gulf, it is not clear whether these fish have had any effect on the icthyofauna or other marine communities in the Gulf of

Aqaba. *Sparus aurata* is a native of the Mediterranean Sea and is probably not one of the Lessepsian migrants that entered the Gulf of Aqaba via the Suez Canal (i.e. it is not endemic to the Gulf of Aqaba).

Reducing aquaculture effects on the environment

As indicated above, most of the nitrogen and phosphorus in the food that is offered to fish reared in net pens is not retained by the fish, but is released to the surrounding waters. The bulk of the nitrogen effluent is released as dissolved forms - mainly ammonia and urea - that are rapidly assimilated by the surrounding biota. The remainder is released from the fish cages as dissolved and particulate organic N that may be taken up by suspension and filter feeders or fall to the seafloor, where it may enter the benthic food web. Phosphorus effluents follow similar pathways and are partitioned to water column and benthic food webs. In the following, 3 scenarios will be described wherein fish farm effluents may be captured and harvested in order to reduce the enrichment of the marine environment in these nutrients.

1. MULLET-ENHANCED BIOREMEDIATION OF AQUACULTURE-ENRICHED SEDIMENTS

As described above, one of the more prominent effects of intensive fish farming is an organically-enriched seafloor under the cages. Several approaches have been proposed to deal with the accumulated organic material (reviewed in Beveridge 1996), including: a) physical removal by suctioning methods, b) mechanical "harrowing" of the sediments to oxygenate these and c) site rotation. In the early 1990's, Porter et al. (1996) carried out preliminary observations on the use of the grey mullet, *Mugil cephalus* to reduce the benthic organic load under fish cages and observed that the idea had merit. This approach was subsequently re-examined in the context of a larger experimental study. Katz et al. (2002) argued that the grey mullet has commercial value and by deploying this fish in benthic enclosures on the seafloor, below commercial fish cages, this detritivore fish may take up sediment organic matter (harvesting the fish constitutes removal of fish farm effluents) while serving as a supplementary "crop" (i.e. this may act as an incentive for farmers to deploy them below fish cages).

The aim of this study was to determine whether, and to what extent, mullet activity (feeding, digging and swimming) could reduce the organic load, diminish sediment hydrogen sulfide levels and increase the dissolved oxygen level in the seabed, thereby reducing the impact of fish farms on the benthos. In 1998, four 1 m^3 benthic enclosures (open on the bottom so mullets could feed from the sediments) were stocked with grey mullets, while 3 identical enclosures without mullets, and three 1 m^2 patches of bare sediment, served as references. Sediment cores were taken from the treatments initially and after 70 days to determine whether there were changes in dissolved oxygen, total dissolved sulfides, water content and organic matter, and macrofauna abundance, as a result of mullet presence/activity. Chemical analyses indicated that there was significant improvement in the sediments (lower sulfide and higher oxygen levels) within the mullet enclosures in comparison to the references. The benthic chemical changes were accompanied by a surprising increase in the population of mud snails (*Nassarius sinusigerus*) exclusively within the mullet enclosures; a likely product of improved sediment conditions, or exclusion of snail predators (or both). Although the mullets were not fed in the course of this 2-month *in situ* experiment, they grew significantly and clearly ingested and assimilated particulate organic matter from the enriched seafloor (Lupatsch et al. 2003).

While foraging for food, the mullets stirred up the sediment and ultimately caused the removal of the upper 5cm layer of the organically-enriched sediments. The sediment removal from the square meter plot was equivalent to 2.6 kg organic carbon and corresponded to a mean carbon removal rate of 20.6 g m^{-2} d^{-1}. Although this trial has not been upscaled, the results suggest that deployment of active detritivores, such as grey mullets, in benthic enclosures under net cage fish farms may be a viable means, with an important economic spin, to reduce some of the detrimental impacts of intensive mariculture.

2. BENTHIC BIOFILTERS BELOW FISH FARMS

The seabed under intensive net-cage fish farms is often barren and devoid of macro-invertebrates (Pearson and Black 2001) in comparison to the nearby unenriched sediments that support undisturbed endemic invertebrate communities. Whereas this is generally true of the seafloor under the Ardag farm, structures and surfaces that are suspended above the seafloor support diverse communities of invertebrates and fish (cite). These observations served as the basis for an experiment (Angel et al. 2002) to test the feasibility of creating artificial reefs below fish farms as a means to trap some of the aquaculture effluents and to boost the locally-depressed biological diversity.

Two triangular-shaped structures (volume 8.2 m^3) constructed from porous durable polyethylene were deployed in March 1999 at 20 m depth; one below the Ardag farm and the other 500 m west of this farm in order to monitor the colonization of these reefs by the local fauna and to determine whether the reef community can remove fish farm effluents from the water. Both reefs became rapidly colonized by a wide variety of organisms. Despite the differences between the fish farm and the reference sites, the communities that developed on the reefs were quite similar. During the first year fish abundances ranged 518–1185 individuals per reef and the number of species ranged between 25–42 species per reef. In addition, the reef communities included benthic algae, bryozoa, tunicates, bivalves, polychaetes, sponges, anemones, crustaceans, sea urchins, gastropods, crinoids and corals.

The biofiltration potential (ability to trap planktonic particles) of the reefs was measured by sampling water upstream and downstream of the structures and recording differences in chlorophyll a concentration in these. Chlorophyll a concentrations were reduced on the downstream side by as much as 15–35% as compared to ambient concentrations and the reefs removed particles with highest efficiency at intermediate current speeds (3–10 cm s^{-1}). In conclusion, the reef structures provided a platform for settlement and recruitment of a diverse community of benthic and pelagic organisms that would not have developed on the seabed and thereby boosted local biodiversity. The reefs also enabled the recruitment of stony corals; a locally-endangered taxon. In addition, the organisms associated with the benthic structures consumed planktonic organisms (and probably detrital particles released from the fish farm too; see below) and as such, served as a barrier reducing the broadcast of organic compounds associated with fish farm effluents.

Whereas the benthic biofilter concept (Angel et al. 2002) is promising as a means to limit fish farm impacts, it cannot yield net removal of aquaculture effluent from the sea because the biomass associated with the artificial structures is not harvested. Once the associated biomass reaches a steady state, the benthic biofilters serve as a conduit for carbon, nutrients and energy emitted from the fish cages. This conclusion prompted the design of a follow-up study to examine the performance of harvestable biofilters deployed in the water column around fish farms.

3. PELAGIC BIOFILTERS AROUND AQUACULTURE NET CAGES

Although the seafloor under intensive aquaculture cages is often organically enriched, nutrient budgets indicate that the bulk of the nutrients released from fish farming are not in the form of large particles but rather as dissolved compound or very fine particles with a wider range of dispersion (e.g. Lupatsch and Kissil 1998). In order to capture the aquaculture nutrients before they are dispersed, it is possible to establish biofilters in the water column, around the fish cages. It was this rationale, as well as the need to harvest the biofilter biomass (see above) that lead to the formation of the interdisciplinary project *BIOFAQs*; a study to examine the use of removable pelagic structures to harvest aquaculture effluents (Black 2003).

In order to examine this concept along a pan-European transect, 4 study sites were selected, including Oban, Scotland; Piran, Slovenia; Crete, Greece; Eilat, Israel. At each of these sites 8 sets of pelagic biofilter arrays were deployed; 4 arrays adjacent to a fish farm and another 4 arrays at a nearby reference site (Spanier et al. 2003). Biofilter arrays consisted of a series of net-mesh cylinders attached to a horizontal and a vertical framework and moored in an upright position in the water column by means of buoys and floats. The composition and biomass of the communities that developed on the biofilters were recorded during the first year as well as their biofiltration activity. In the following, we will only present data from the Eilat study site.

There was a steady temporal increase in biofilter biomass at the fish farm site, following deployment, up to an upper limit, whereafter excess biomass sloughed off. From a practical point of view, this finding serves as an indication of the appropriate time (< 9 months) to harvest the biofilters in order to optimize removal of farm effluents. It appears that it does not pay to harvest the biofilters prematurely since the developed communities are more efficient at effluent removal than the "young" ones. In contrast to the fish farm site, biofilters at the reference site were practically devoid of biomass during the first 6 months; probably the result of intense grazing activity (Spanier et al. 2003). Because we did not quantify grazing rates, it is impossible to accurately assess biomass accumulation rates. Moreover, it is not clear whether there were different grazing pressures on the biofouling communities at the fish farm site as compared to the reference site, but in the absence of any information, we must assume that the grazing rates were similar. It is also noteworthy that natural grazing on the biofilter communities leads to loss of harvestable material from these structures to the surrounding environment.

Biofouling communities on the cylindrical biofilters were very diverse. The community succession began with macroalgae which eventually became overgrown by successive layers of sessile invertebrates, including bryozoans, polychaetes, sponges, tunicates, hydrozoans (Spanier et al. 2003). Motile invertebrates, including sea urchins, gastropods and crustaceans generally appeared intermittently at later stages, and many of the biofilters had schools of fish associated with them. Although many of the biofilters were covered by communities that were predominantly heterotrophic in composition, there were biofilters that supported considerable macroalgal and coralline algal biomass, but there was a clear decrease in the relative abundance of algae with depth.

One of the central questions in this study was whether the biofouling communities associated with the biofilters consumed fish farm effluents. Short-term grazing experiments indicated that there was rapid uptake of phytoplankton by the biofilter communities at both the fish farm and the reference sites. However, the biofouling growth rate was much higher at the fish farm site. Because there was no difference in phytoplankton abundance between the fish farm and reference sites, this suggests that the biofouling communities on the biofilters at the fish farm site thrive on non-living farm-derived detritus. This deductive evidence was supported by another set of unrelated data. A stable isotope study was conducted to trace the flow of nitrogen from the farmed fish to the surrounding environment (Lojen et al. 2003, Lojen et al. 2004). This study indicated that as much as 60 % of the food assimilated by the biofilter communities, at the fish farm site originated in the fish farm effluents.

This study has clearly shown that biofilters can trap fish farm effluents, but the question then is whether the amount of nutrients retrieved is significant relative to the fluxes released. Division of the fish farm nutrient release rates by the calculated biofilter uptake/removal rates yields numbers of biofilters that are unwieldy and therefore impractical, however, this calculation assumes 100% nutrient removal. If we attempt to remove a more modest proportion of effluent nutrients and couple that effort with cultivation of economically-lucrative biofiltering organisms (e.g. mussels or seaweeds) on artificial substrates around fish farms, this may be a more attainable objective with both environmental value and economic incentive.

Summary and conclusions

Net cage aquaculture has the potential to impact its surroundings, yet if best-practice measures are exercised (proper site selection, prudent husbandry, efficient management, etc.) these impacts may be substantially reduced. In addition to preventive measures, it is possible to capture the effluents discharged from fish cages by employing a variety of innovative actions to reduce organic loading in the fish farm environs, while providing aquaculturists with an economic incentive. Such activities include the use of detritivorous fish or invertebrates to harvest particulate organic matter from enriched seabeds; deployment of artificial reefs under fish farms to provide recruitment surface area and habitat to organisms that can help process the farm effluents; establishment of harvestable "benthic communities" in the water column around fish cages to trap and assimilate suspended and dissolved farm effluents. It is possible, therefore, that the provision of solid substrates (and possibly also seeding these with desireable biofiltering organisms) to enhance biological filtering activity around finfish farms, accompanied by harvesting of this biomass, will make coastal aquaculture a more acceptable and sustainable industry.

Literature cited

Angel, D.L., Krost, P., Zuber, D. and A. Neori. 1992. Microbial mats mediate the benthic turnover of organic matter in polluted sediments in the Gulf of Aqaba. L. Saar and G. Kissil (eds), Proceedings of the U.S-Israel Workshop on Mariculture and the Environment, pp. 66-73. Eilat, Israel - June 8 - 10, 1992

Angel, D.L., Krost, P. and H. Gordin. 1995. Benthic implications of net cage aquaculture in the oligotrophic Gulf of Aqaba. *European Aquaculture Society* 25: 129-173.

Angel, D. L., Post, A., Brenner, S., Eden, N., Katz, T., Cicelsky, A., and Lupatsch, I. 1998. Environmental impact assessment of the Ardag net cage fish farm on the northern Gulf of Aqaba. (In Hebrew). 95 pp. A report prepared for the Ardag Fish Company.

Angel, D.L., Verghese, S., Lee, J.J., Saleh, A.M., Zuber, D., Lindell, D. and A. Symons. 2000. Impact of a net cage fish farm on the distribution of benthic foraminifera in the northern Gulf of Eilat (Aqaba, Red Sea). *Journal of Foraminiferal Research* 30: 54-65.

Angel, D.L., Eden, N., Breitstein, S., Yurman, A., Katz, T., and E. Spanier. 2002. In situ biofiltration: a means to limit the dispersal of effluents from marine finfish cage aquaculture. *Hydrobiologia* 469: 1-10.

Angel, D.L., Eden, N., Katz, T., and E. Spanier. 2003. Mariculture-environment interactions and biofiltration in oligotrophic Red Sea waters. *Annales Ser Hist Nat* 13: 33-36.

Atkinson, M.J., Birk, Y. and H. Rosenthal. 2001. Evaluating pollution in the Gulf of Eilat. Report for the Israeli Ministries of Infrastructure, Environment and Agriculture, Jerusalem

Badran, M.I. and P. Foster. 1998. Environmental quality of the Jordania costal waters of the Gulf of Aqaba. *Aquatic Ecosystem Health and Management* 1: 75-89.

Beveridge, M.C.M. 1996. Cage aquaculture. 2nd Edition. Fishing News Books, Oxford

Black, K.D. 2003. The BIOFAQs project: BIOFiltration and AQuaculture: an evaluation of substrate deployment performance within mariculture developments. *Annales Ser Hist Nat* 13: 1-2.

Bongiorni, L., Shafir, S., Angel, D.L. and B. Rinkevich. 2003. Survival, growth and reproduction of hermatypic corals subjected to *in situ* fish farm nutrient enrichment. *Marine Ecology Progress Series* 253: 137–144.

Brenner, S., Rosentroub, Z., and Y. Bishop. 1988. Current measurements in the Gulf of Elat. Israel Oceanographic & Limnological Research (IOLR) Report H3/88

Brenner, S., Rosentroub, Z., and Y. Bishop. 1991. Current measurments in the Gulf of Elat 1990/91. Israel Oceanographic & Limnological Research (IOLR) Report H12/91.

Bricker, S.B., Clement, C.G., Pirhalla, D.E., Orlando, S.P. and D.G.G. Farrow. 1999. National estuarine eutrophication assessment: effects of nutrient enrichment in the nation's estuaries. Special Projects Office and the National Centers for Coastal Ocean Science, National Ocean Service, National Oceanic and Atmospheric Administration, Silver Spring, Maryland.

Colorni, A. 1992. A systematic mycobacteriosis in the European sea bass *Dicentrarchus labrax* cultured in Eilat (Red Sea). *Bamidgeh-Israel Journal of Aquaculture* 44: 75-81.

Diamant, A., Banet, A., Paperna, I., von Westernhagen, H., Broeg, K., Kruener, G., Koerting, W. and S. Zander. 1999. The use of fish metabolic, pathological and parasitological indices in pollution monitoring II. The Red Sea and Mediterranean, *Helgolander Marine Research* 53: 195-208.

Diamant, A., Banet, A., Ucko, M., Colorni, A., Knibb, W. and H. Kvitt. 2000. Mycobacteriosis in wild rabbitfish *Siganus rivulatus* associated with cage farming in the Gulf of Eilat, Red Sea. *Diseases of Aquatic Organisms* 39: 211-219.

Eden, N., Katz, T. and D.L. Angel. 2003. The impact of net cage fish farms on *Nassarius* (Niotha) *sinusigerus* distribution in the Gulf of Eilat (Aqaba), Red Sea. *Marine Ecology Progress Series* 263:139-147.

Fishelson, L. 1971. Ecology and distribution of the benthic fauna in the shallow waters of the Red Sea. *Marine Biology* 10: 113-133.

Genin, A. and D. Zakai. 2000. The coral reef and water quality in the northern Gulf of Eilat: updated data, Spring 2000. Internal Report of the Interuniversity Institute of Eilat, Israel (in Hebrew).

Katz T. 2000. Reduction of organic load in sediments under a marine fish farm by striped grey mullets. MSc thesis, Hebrew University, Jerusalem

Katz, T., Herut, B., Genin, A. and D.L. Angel. 2002. Grey mullets ameliorate organically-enriched sediments below a fish farm in the oligotrophic Gulf of Aqaba (Red Sea). *Marine Ecology Progress Series* 234: 205-214.

Knibb, W., Colorni, A., Ankaoua, M., Lindell, D., Diamant, A. and H. Gordin. 1993. Detection and identification of a pathogenic marine mycobacterium from the European sea bass *Dicentrarchus labrax* using polymerase chain reaction and direct sequencing of 16s rDNA sequences. *Molecular Marine Biology and Biotechnology* 2: 225-232.

Lojen, S., Angel, D.L., Katz, T., Tsapakis, M., Kovac, N. and A. Malej. 2003. ^{15}N enrichment in fouling communities influenced by organic waste deriving from fish farms. *Annales Ser Hist Nat* 13: 9-12.

Lojen, S., Spanier, E., Tsemel, A., Katz, T., Eden, N. and D.L. Angel. 2004. δ^{15}N as a natural tracer of particulate nitrogen effluents released from marine aquaculture. *Marine Ecology Progress Series* (in review).

Lupatsch, I. and Kissil, G.Wm. 1998. Predicting aquaculture waste from gilthead seabream (*Sparus aurata*) culture using a nutritional approach. *Aquatic Living Resources*, 11: 265-268.

Lupatsch, I., Katz, T. and D.L. Angel. 2003. Removal of fish farm effluents by grey mullets: a nutritional approach. *Aquatic Living Resources* 34: 1367-1377.

McGinn, A.P. 1998. Blue revolution: the promises and pitfalls of fish farming. *World Watch* 11: 1-10.

National Research Council (NRC). 2000. Clean coastal waters: understanding and reducing the effects of nutrient pollution. National Academy Press, Washington, D.C.

Pearson, T.H. and K.D. Black. 2001. The environmental impacts of marine fish cage culture. In: K.D. Black (ed), Environmental Impacts of Aquaculture. Sheffield Academic Press.

Porter, C.B., Krost, P., Gordin, H. and D.L. Angel. 1996. Preliminary assessment of grey mullet (*Mugil cephalus*) as a forager of organically enriched sediments below marine fish farms. *Israel Journal of Aquaculture-Bamidgeh* 48:47-55.

Reiss, Z. and L. Hottinger. 1984. The Gulf of Aqaba: ecological micropaleontology. Springer-Verlag, New-York

Small, C. and R.J. Nicholls. 2003. A global analysis of human settlement in coastal zones. *Journal of Coastal Research* 19:584-599.

Spanier, E., Tsemel, A., Lubinevski, H., Roitemberg, A., Yurman, A., Breitstein, S., Angel, D., Eden, N. and T. Katz. 2003. Can open water bio-filters be used for the reduction of the environmental impact of finfish net cage aquaculture in the coastal waters of Israel? *Annales Ser Hist Nat* 13: 25-28.

Strauss, M. 2000. Benthic meiofauna along an organic enrichment gradient in the sediments below the Ardag fish farm. MSc thesis, Tel Aviv University, Ramat Aviv

Ucko, M. and H. Kvitt. 2000. Israel Oceanographic & Limnological Research Annual Report E13/2000, Genetics Department, pp. 127-130.

A MODULAR STRATEGY FOR RECOVERY AND MANAGEMENT OF BIOMASS YIELDS IN LARGE MARINE ECOSYSTEMS

Kenneth SHERMAN
USDOC/ NOAA, NMFS Narragansett Laboratory, 28 Tarzwell Drive, Narragansett, RI 02882 USA

1. Abstract

During the decade since the 1992 United Nations Conference on Environment and Development, considerable movement has been made by international organizations engaged in ocean affairs to move nations toward adopting ecosystem-based assessment and management strategies. The 191 nations, including 82 heads of state, participating in the 2002 World Summit on Sustainable Development in Johannesburg agreed to a plan of implementation (POI) that encourages nations to apply the ecosystem approach to marine resource assessment and management practices by 2010, and maintain or restore fish stocks to levels that can produce maximum sustainable yield levels by 2015. To achieve these targets will require an improved understanding and assessment of the effects of physical, biological and human forcing causing changes in biomass yields of large marine ecosystems (LMEs). An international financial mechanism, the Global Environment Facility (GEF), is assisting developing countries in meeting the Summit targets by supporting LME assessment and management projects. Of the 29 LMEs for which published case study information is available for analyses of principal forces driving changes in biomass yields, fishing effort was the primary forcing mechanism in 14 LMEs; climate forcing was the principal factor in 13 LMEs, eutrophication in one case and the data were inconclusive in another. Fishing effort was a secondary driver of change in biomass yields in the LMEs driven by climate forcing. Mitigating actions for reducing fishing effort to promote recovery of lost biomass yield is proving successful in one case study. Actions for improving forecasts of oceanographic conditions affecting fish stocks are underway in four GEF supported LME projects (e.g. Humboldt Current, Canary Current, Guinea Current, Benguela Current); measures to assess and manage excessive fishing effort are planned for 8 LME projects, eutrophication reduction and control in another, and 6 LMEs with relatively stable decadal biomass yields appear suitable for mandating precautionary total allowable catch (TAC) levels. The GEF-LME projects include countries that contributed to 45% of global marine biomass yields in 1999.

2. Ecosystem Based Fishery Assessments

Countries around the globe are concerned about the degraded condition of their coastal ecosystems from excessive fishing effort, degraded habitats, eutrophication, pollution aerosol contamination, and emerging diseases. To redress these issues, and move countries toward more sustainable use of ocean resources, 82 heads of state and senior government representatives from 191 countries attending the World Summit on Sustainable Development held in Johannesburg in September 2002, agreed to a Plan of Implementation (POI). One of the more important actions in the Summit POI is the commitment to move ahead to meet two important fisheries related targets: (1) to introduce ecosystem-based assessment and management practices by 2010; and (2) restore depleted fish stocks to maximum sustainable yield levels by 2015. Although advances were made toward similar objectives in the 10 years since the 1992 UN Conference on Sustainable Development in Rio (UNCED), the general results, were limited. Among the most positive outcomes was the establishment of the Global Environmental Facility (GEF) as a financial mechanism to assist developing nations address global environmental issues affecting resource sustainability.

Following a three-year pilot phase, the Global Environment Facility was formally launched to forge cooperation and finance actions in the context of sustainable development that address critical threats to the global environment including: unsustainable fishing practices, biodiversity loss, climate change, degradation of international waters, ozone depletion, and persistent organic pollutants. Activities concerning land degradation, primarily desertification and deforestation as they relate to these threats, are also addressed. GEF projects are implemented by the United Nations Development Program (UNDP), UN Environment Program (UNEP), UN Industrial Development Program (UNIDO) and the World Bank and expanded opportunities exist for other executing agencies.

During its first decade, GEF allocated $US 3.2 billion in grant financing, supplemented by more than $US 8 billion in additional financing, for 800 projects in 156 developing countries and those in economic transition. All six thematic areas of GEF, including the land degradation cross-cutting theme, have implications for coastal and marine ecosystems. Priorities have been established by the GEF Council in its Operational Strategy adopted in 1995. Of the six areas, the biodiversity and international waters focal areas have been utilized by developing States more than other areas to address marine resources and environmental issues. The biodiversity programs were developed with guidance from the Conferences of the Parties (COPs) of the Convention on Biological Diversity (CBD) while the international waters focal area was designed to be consistent with Chapters 17 and 18 of UNCED Agenda 21.

In the 2nd report of the UN Consultative Process on Ocean Affairs [1] the GEF's role in assisting developing countries to address concerns of the marine environment was highlighted. Reference was made to the GEF's use of "Large Marine Ecosystems" for building capacity of States to utilize sound science in improving management of the coastal and marine environment. Large Marine Ecosystems (LMEs) are relatively large regions of ocean space equal to or greater than 200,000 km^2 characterized by unique bathymetry, hydrography, and trophically dependent populations [2][3].

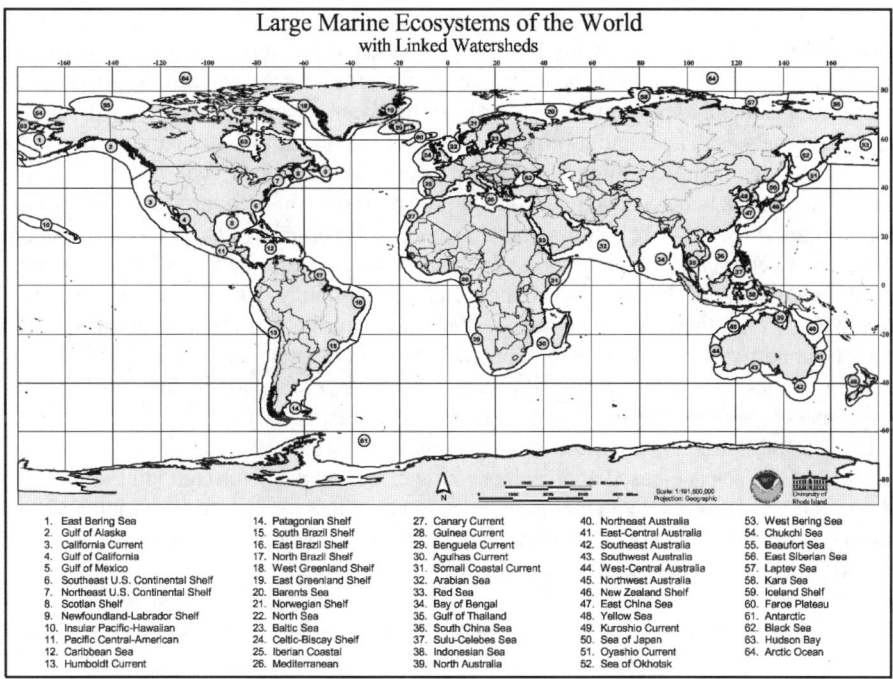

Figure 1. Large marine ecosystems of the world and linked watersheds

Since 1993, the National Oceanic and Atmospheric Administration (NOAA-Fisheries) has been cooperating with the GEF, the International Union for the Conservation of Nature (the World Conservation Union), the Intergovernmental Oceanographic Commission of UNESCO (IOC) and several UN agencies (UNIDO, UNDP, UNEP, FAO) in an action program to assist developing countries in planning and implementing an ecosystem-based strategy that is focused on Large Marine Ecosystems (LMEs) as the principal assessment and management unit for coastal ocean resources. NOAA contributes scientific and technical assistance and expertise to aid developing countries in reaching the targets of UNCED and the targets of the Johannesburg Summit. With the evolving movement toward ecosystem-based management, stimulated by UNCED and reinforced and targeted at the 2002 Summit, there is a shift evolving from relatively short-term single species assessment and management toward large-scale, multispecies and multi-sectoral long term management of the goods and services of the world's LMEs. The geographic area of the LME, its coastal area, and contributing basins constitute the place-based area for assisting countries to understand linkages among root causes of degradation and integrating needed changes in sectoral economic activities. The 64 LMEs, located around the margins of the ocean basins, serve as global units to initiate capacity building and to bring science to pragmatic use in improving the management of coastal and marine ecosystems (Figure 1).

The GEF Operational Strategy recommends that nations sharing an LME begin to address coastal and marine issues by jointly undertaking strategic processes for analyzing factual, scientific information on transboundary concerns, their root causes, and setting priorities for action on transboundary concerns. This process has been referred to as a Transboundary Diagnostic Analysis (TDA) and it provides a useful mechanism to foster participation at all levels. Countries then determine the national and regional policy, legal, and institutional reforms and investments needed to address the priorities in a country-driven Strategic Action Program (SAP). This allows sound science to become the basis for policy-making and fosters a geographic location upon which an ecosystem-based approach to assessment and management can be developed, and more importantly, can be used to engage stakeholders within the geographic area so that they contribute to the dialogue and in the end they support the ecosystem-based approach that can be pragmatically implemented by the communities and governments involved. Without such participative processes to engage specific stakeholders in a place-based setting, marine science has often remained confined to the marine science community or has not been embraced in policy-making. Furthermore, the science-based approach encourages transparency through joint monitoring and assessment processes, including joint assessment cruises for countries sharing an LME with a level of transparency that builds trust among nations over time and can overcome the barrier of false information being reported.

Developing countries and those in economic transition have requested and received GEF support for LME projects through the International Waters focal area of the GEF. The approved GEF-LME projects include developing nations and those in economic transition as well as other OECD countries since the living resources, the pollution loading and the critical habitats have transboundary implications across rich and poor nations alike. A total of $500 million in costs from the North and South is currently being invested in the global Network of LME projects as of July 2003 in 10 projects in 72 countries with $225 million in GEF grant finance. An additional 7 LME projects are under preparation in 54 different nations. Currently, 126 different countries are engaged in the LME global Network of projects (Table 1). With OECD countries involved that share the LMEs with the GEF recipient nations, expectations are that reforms will take place in both the North and the South in order to operationalize this ecosystem-based approach to managing human activities in the different economic sectors that contribute to place-specific degradation of the LME and adjacent waters.

4. LME Modules

A five-module approach to the assessment and management of LMEs has been proven to be useful in ecosystem-based projects in the United States and elsewhere. The modules are customized to fit the situation within the context of the transboundary diagnostic analysis (TDA) process and the strategic action plan (SAP) development process for the groups of nations sharing the particular LME based on available information and capacity. These processes are critical to integrate science into management in a practical way and to establish governance regimes appropriate for the particular situation. The five modules consist of 3 that are science-based activities focused on: productivity, fish/fisheries, pollution/ecosystem health; the other two,

Table 1. Countries Participating in the global Network of GEF/Large Marine Ecosystem Projects where stewardship ministries have agreed to initiate ecosystem-based LME assessment and management practices.

Approved GEF Projects	
LME	Countries
Gulf of Guinea (6)............................	Benin, Cameroon, Côte d'Ivoire, Ghana, Nigeria, Togo[a]
Yellow Sea (2)..................................	China, Korea
Patagonia Shelf/Maritime Front (2).............. Argentina, Uruguay	
Baltic (9)..	Denmark, Estonia, Finland, Germany, Latvia, Lithuania, Poland, Russia, Sweden
Benguela Current (3)............................	Angola,[b] Namibia, South Africa[b]
South China Sea (7)..............................	Cambodia, China, Indonesia, Malaysia, Philippines, Thailand, Vietnam
Black Sea (6).......................................	Bulgaria, Georgia, Romania, Russian Federation, Turkey,[b] Ukraine
Mediterranean (19)..............................	Albania, Algeria, Bosnia-Herzegovina, Croatia, Egypt,[b] France, Greece, Israel, Italy, Lebanon, Libya, Morocco,[b] Slovenia, Spain, Syria, Tunisia, Turkey, Yugoslavia, Portugal
Red Sea (7)..	Djibouti, Egypt, Jordan, Saudi Arabia, Somalia, Sudan, Yemen
Western Pacific Warm Water Pool-SIDS (13)...	Cook Islands, Micronesia, Fuji, Kiribati, Marshall Islands, Nauru, Niue, Papua New Guinea, Samoa, Solomon Islands, Tonga, Tuvalu, Vanuatu
	Total number of countries: 72[c]
GEF Projects in the Preparation Stage	
Canary Current (7)...............................	Cape Verde, Gambia, Guinea,[b] Guinea-Bissau,[b] Mauritania, Morocco, Senegal
Bay of Bengal (8)................................	Bangladesh, India, Indonesia, Malaysia, Maldives, Myanmar, Sri Lanka, Thailand
Humboldt Current (2)...........................	Chile, Peru
Guinea Current (16).............................	Angola, Benin, Cameroon, Congo, Democratic Republic of the Congo, Côte d'Ivoire, Gabon, Ghana, Equatorial Guinea, Guinea, Guinea-Bissau, Liberia, Nigeria, Sao Tome and Principe, Sierra Leone, Togo
Gulf of Mexico (3)..............................	Cuba,[b] Mexico,[b] United States
Agulhus/Somali Currents (8).................	Comoros, Kenya, Madagascar, Mauritius, Mozambique, Seychelles, South Africa, Tanzania
Caribbean LME (23)...........................	Antigua and Barbuda, The Bahamas, Barbados, Belize, Columbia, Costa Rica, Cuba, Grenada, Dominica, Dominican Republic, Guatemala, Haiti, Honduras, Jamaica, Mexico, Nicaragua, Panama, Puerto Rico, Saint Kitts and Nevis, Saint Lucia, Saint Vincent and the Grenadines, Trinidad and Tobago, Venezuela
	Total number of countries: 54[c]

[a]The six countries participating in the Gulf of Guinea project also appear in a GEF/LME project in the preparatory phase
[b]Countries that are participating in more than one GEF/LME project
[c]Adjusted for multiple listings

socio-economics and governance, are focused on socioeconomic benefits to be derived from a more sustainable resource base and implementing governance mechanisms for providing stakeholders and stewardship interests with legal and administrative support for ecosystem-based management practices. The first four modules support the TDA process while the governance module is associated with periodic updating of the Strategic Action Program or SAP. Adaptive management regimes are encouraged through periodic assessment processes (TDA updates) and updating of SAPs as gaps are filled.

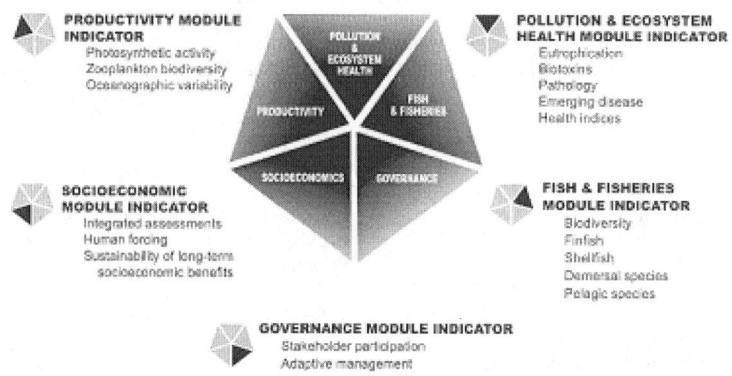

Figure 2. The five-module approach to large marine ecosystem assessments and management

Productivity Module
Productivity can be related to the carrying capacity of an ecosystem for supporting fish resources [4]. Recently, scientists have reported that the maximum global level of primary productivity for supporting the average annual world catch of fisheries has been reached, and further large-scale "unmanaged" increases in fisheries yields from marine ecosystems are likely to be at trophic levels below fish in the marine food web [5]. Measuring ecosystem productivity also can serve as a useful indication of the growing problem of coastal eutrophication. In several LMEs, excessive nutrient loadings of coastal waters have been related to algal blooms implicated in mass mortalities of living resources, emergence of pathogens (e.g., cholera, vibrios, red tides, paralytic shellfish toxins), and explosive growth of non-indigenous species [6].

The ecosystem parameters measured in the productivity module are zooplankton biodiversity and information on species composition, zooplankton biomass, water column structure, photosynthetically active radiation (PAR), transparency, chlorophyll-a, NO_2, NO_3, and primary production. Plankton of LMEs have been measured by deploying Continuous Plankton Recorder (CPR) systems monthly across ecosystems from commercial vessels of opportunity over decadal time scales. Advanced plankton recorders can be fitted with sensors for temperature, salinity, chlorophyll, nitrate/nitrite, petroleum, hydrocarbons, light, bioluminescence, and primary productivity, providing the means for in situ monitoring and the calibration of satellite-derived oceanographic conditions relating to changes in phytoplankton, zooplankton, primary productivity, species composition and dominance, and long-term changes in the physical and nutrient characteristics of the LME and in the biofeedback of plankton to the stress of environmental change[7][8].

Fish and fisheries module
Changes in biodiversity among the dominant species within fish communities of LMEs have resulted from: excessive exploitation, naturally occurring environmental shifts in climate regime, or coastal pollution. Changes in the biodiversity of a fish community can generate cascading effects up the food web to apex predators and down the food web to plankton components of the ecosystem. The Fish and Fisheries module includes fisheries-independent bottom-trawl surveys and acoustic surveys for pelagic species to obtain time-series information on changes in fish biodiversity and abundance levels. Standardized sampling procedures, when deployed from small calibrated trawlers, can provide important information on diverse changes in fish species [9]. Fish catch provides biological samples for stock assessments, stomach analyses, age, growth, fecundity, and size comparisons; data for clarifying and quantifying multispecies trophic relationships; and the collection of samples for monitoring coastal pollution. Samples of trawl-caught fish can be used to monitor pathological conditions that may be associated with coastal pollution and can be used as platforms for obtaining water, sediment, and benthic samples for monitoring harmful algal blooms, diseases, anoxia, and changes in benthic communities.

Pollution and ecosystem health module
In several LMEs, pollution and eutrophication have been important driving forces of changes in biomass yields. Assessing the changing status of pollution and health of the entire LME is scientifically challenging. Ecosystem "health" is a concept of wide interest for which a single precise scientific definition is problematical. The health paradigm is based on multiple-state comparisons of ecosystem resilience and stability and is an evolving concept that has been the subject of a number of meetings[10]. To be healthy and sustainable, an ecosystem must maintain its metabolic activity level and its internal structure and organization, and must resist external stress over time and space scales relevant to the ecosystem [11]). The ecosystem sampling strategies are focused on parameters related to overexploitation, species protected by legislative authority (marine mammals), and other key biological and physical components at the lower end of the food web (plankton, nutrients, hydrography) as noted by Sherman (1994) [12].
Fish, benthic invertebrates, and other biological indicator species are used in the Pollution and Ecosystem Health module to measure pollution effects on the ecosystem, including the bivalve monitoring strategy of "Mussel-Watch;" the pathobiological examination of fish; and the estuarine and nearshore monitoring of contaminants and contaminant effects in the water column, substrate, and in selected groups of organisms. The routes of bioaccumulation and trophic transfer of contaminants are assessed, and critical life history stages and selected food web organisms are examined for parameters that indicate exposure to, and effects of, contaminants. Effects of

impaired reproductive capacity, organ disease, and impaired growth from contaminants are measured. Assessments are made of contaminant impacts at the individual species and population levels. Implementation of protocols to assess the frequency and effect of harmful algal blooms, emergent diseases and multiple marine ecological disturbances [13] are included in the pollution module.

Socioeconomic module
This module is characterized by its emphasis on practical applications of its scientific findings in managing an LME and on the explicit integration of economic analysis with science-based assessments to assure that prospective management measures are cost-effective. Economists and policy analysts work closely with ecologists and other scientists to identify and evaluate management options that are both scientifically credible and economically practical with regard to the use of ecosystem goods and services.

Designed to respond adaptively to enhanced scientific information, socioeconomic considerations must be closely integrated with science. This component of the LME approach to marine resources management has recently been described as the human dimensions of LMEs. A framework has been developed by the Department of Natural Resource Economics at the University of Rhode Island for monitoring and assessment of the human dimensions of an LME and the socioeconomic considerations important to the implementation of an adaptive management approach for an LME [14]. One of the more critical considerations, a methodology for considering economic valuations of LME goods and services has been developed around the use of interaction matrices for describing the relationships between ecological state and the economic consequences of change and is included in the framework.

Governance module
The Governance module is evolving based on demonstrations now underway among ecosystems to be managed from a more holistic perspective than generally practiced in the past. In projects supported by GEF- for the Yellow Sea ecosystem, the Guinea Current LME, and the Benguela LME - agreements have been reached among the environmental ministers of the countries bordering these LMEs to enter into joint resource assessment and management activities as part of building institutions. Among other LMEs, the Great Barrier Reef ecosystem is being managed from an ecosystem-based perspective; the Antarctic marine ecosystem is also being managed from an ecosystem perspective under the Commission for the Conservation of Antarctic Marine Living Resources (CCAMLR). Governance profiles of LMEs are being explored to determine their utility in promoting long-term sustainability of ecosystem resources[15].

5. Driving Forces of Biomass Yields in LMEs

To assist stewardship agencies in the implementation of ecosystem-based assessment and management practices, the transboundary diagnostic analyses (TDAs) are focused on the root causes of trends in LME biomass yields. In addition, information on principal driving forces of biomass yields imbedded in 29 invited LME case studies conducted by a group of marine resource experts has been analyzed. A list of the principal investigators constituting an "expert-systems analysis" appearing in 12 peer-reviewed and published LME volumes is given in Tables 2a and 2b together with information on the annual biomass yields of the 29 LMEs in millions of metric tons. The biomass yields are based on the mid-point value in 1995 of a decadal trend in LME yields compiled by FAO for the period 1990 through 1999 [16]. Biomass yield data for 4 LMEs not included in the FAO report were taken from published LME case studies. Based on the "expert systems analyses," principal and secondary driving forces were assigned to each LME using four categories (climate, fisheries, eutrophication, inconclusive)(Table 2a).

Of the 29 LME case studies, 13 were assigned to climate forcing as the principal drivers of change in biomass yield, 14 are listed as fisheries driven, one is principally driven by eutrophication, and the results of the analysis as presented in another case study is listed as inconclusive. In all but one case, where climate forcing was the principal driver of changing biomass yield, fisheries were secondary drivers in 13 LMEs. In the case of the Mediterranean LME, the secondary driver is eutrophication according to Caddy [17].

The contribution to the annual global biomass yields of the 29 LMEs amounts to 54.4 mmt or 64 percent based on the average annual 5-year yield from 1995 to 1999 of 85mmt [16]. From a global management perspective, it would appear that nearly half of this LME yield (27.0 mmt) will require significant focus on improvements in forecasting of the climate signal, whereas an estimated 24.8 mmt. of the biomass yield will need to be subjected to a principal management focus on catch-control combined with a secondary effort on forecasting the effects of climate forcing in an effort to recover depleted fish stocks and achieve maximum sustainable yield levels (Table 3).

The influence of climate forcing in biomass yields based on the expert analysis of Lluch-Belda et al. (2003)[18] is illustrated in Figure 3 for the California Current LME. Climate forcing for the Humboldt Current LME is analyzed by Wolff et al. (2003)[20], and in the analyses for the Iceland Shelf LME by Astthorsson and Vilhjálmsson (2002)[21]. In contrast, the argument for urgent reduction in fishing effort is supported by the data presented for the US Northeast Shelf LME by the NEFSC (1999)[22] and in the analyses for the Gulf of Thailand presented by Pauly and Chuenpagdee (2003)[23].

The observation that excessive fishing effort can alter the structure of the ecosystem, resulting in a shift from relatively high-priced, large, long-living demersal species, down the food chain toward lesser-valued, smaller, short-lived pelagic species [22] is supported by the LME data on species biomass yields. Evidence from the East China Sea, Yellow Sea, and Gulf of Thailand, suggests that these three LMEs are approaching a critical state of change, wherein recovery to a previous ratio of demersal to pelagic species may become problematic. In all three cases, the fisheries are now being directed toward fish protein being provided by catches of smaller species of low value [24][23][25]

Table 2a. Primary and Secondary Driving Forces of LME Biomass Yields

PRIMARY AND SECONDARY DRIVING FORCES OF LME BIOMASS YIELDS (Annual biomass yield levels based on 1990-1999 mid-decadal data (1995) from FAO 2003)					
LME	Primary	Secondary	Level mmt	Expert Assessments	Vol. #
Humboldt Current	climate	fishing	16.0	Alheit and Bernal Wolff et al.	5 12
South China Sea	fishing	climate	10.0	Pauly and Christensen	5
East China Sea	fishing	climate	3.8	Chen and Shen	8
North Sea	fishing	climate	3.5	McGlade	11
Eastern Bering Sea	-	-	2.1	Schumacher et al.	12
Bay of Bengal	fishing	climate	2.0	Dwividi Hazizi	5 7
Okhotsk Sea (based on mid-decadal 1972 data on fishing yields from 1962 to 1982)	climate	fishing	2.0	Kusnetsov et al.	5
Canary Current	climate	fishing	1.8	Roy and Cury Bas	12 5
Norwegian Shelf	climate	fishing	1.5	Ellertsen et al. Blindheim and Skjoldal	3 5
Iceland Shelf	climate	fishing	1.3	Astthorsson and Vilhjálmsson	11
Benguela Current	climate	fishing	1.2	Crawford et al. Shannon and O'Toole	2 12
Gulf of Thailand	fishing	climate	1.1	Pauly and Chuenpagdee	12
Mediterranean	fishing	eutrophication	1.1	Caddy	5
Sea of Japan (based on mid-decadal data 1985 from Sea of Japan 1980-1990)	climate	fishing	1.0	Terazaki	
Gulf of Mexico	fishing	climate	0.9	Richards and McGowan Brown et al. Shipp	2 4 9
Guinea Current	climate	fishing	0.9	Binet and Marchal Koranteng and McGlade	5 11
Baltic Sea	fishing	eutrophication	0.8	Kullenberg Jansson	1 12
California Current	climate	fishing	0.7	MacCall Lluch-Belda et al.	1 12
U.S. Northeast Shelf	fishing	climate	0.7	Sissenwine Murawski Sherman et al.	1 6 11
Scotian Shelf	fishing	climate	0.7	Zwanenburg et al. Zwanenburg	11 12
Black Sea	eutrophication	fishing	0.5	Caddy Daskalov	5 12
Barents Sea	climate	fishing	0.5	Skjoldal and Rey Borisov Blindheim and Skjoldal Matishov et al.	2 4 5 12
Caribbean Sea	fishing	climate	0.4	Richards and Bohnsack	3
Iberian Coastal	climate	fishing	0.3	Wyatt and Perez-Gandaras	2
Newfoundland-Labrador	fishing	climate	0.2	Rice et al.	11
Yellow Sea (based on mid-decadal data for demersal species for the Yellow Sea, 1952 to 1992)	fishing	climate	0.2	Tang Tang	2 12
Great Barrier Reef	fishing	climate	0.1	Brodie	12
West Greenland Shelf	climate	fishing	0.1	Hovgard and Buch Pederson and Rice	3 11
Faroe Plateau	climate	fishing	0.1	Gaard et al.	11

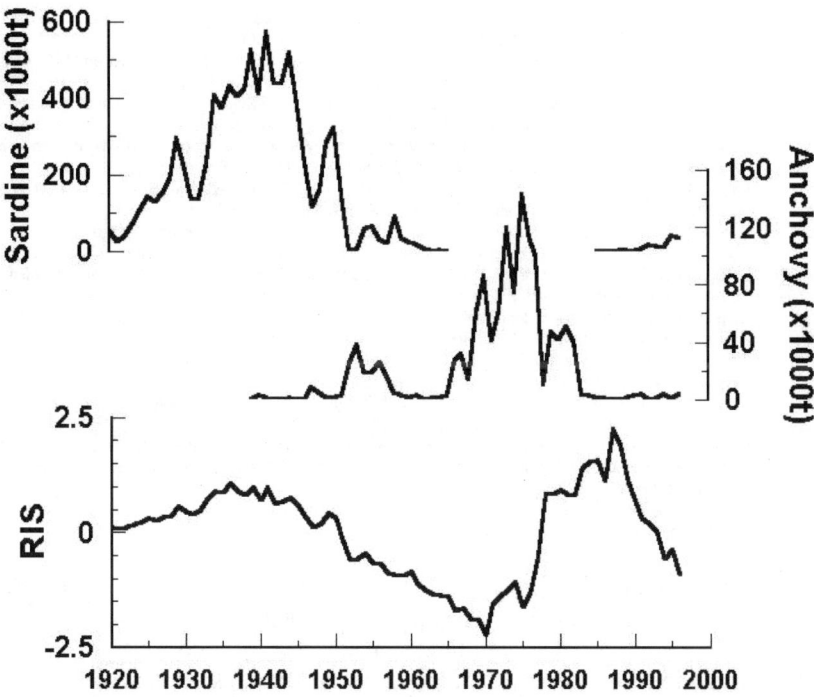

Figure 3.   Historic sardine (upper) and anchovy (middle) catches (1920-2000) from the California Current System, and the Regime Indicator Series (RIS: lower). Catch data were obtained from Schwarzlose et al. 1999. RIS is a composite series reflecting synchronous variability of sardine and anchovy populations of the Japan, California, Benguela and Humboldt currents. Modified from Lluch-Cota D.B. et al. 1997.

The species change in biomass yields of the Yellow Sea as shown in Figure 4 represents an extreme case wherein the annual demersal species biomass yield was reduced from 200,000 mt in 1955 to less than 25,000 mt. through 1980. The fisheries then targeted anchovy and, between 1990 and 1995, landings of anchovy reached an historic high of 500,000 mt.

6. Recovering Fisheries Biomass

The GEF/LME projects presently funded or in the pipeline for funding in Africa, Asia, Latin America and eastern Europe, represent a growing Network of marine scientists, marine managers and ministerial leaders who are engaged in pursuing the ecosystem and fishery recovery goals. The significant annual global biomass yields of marine fisheries from ecosystems in the GEF-LME Network of 44.8% provides a firm basis for moving toward the Summit goal for introducing an ecosystem-based assessment and management approach to global fisheries by 2010, and fishing MSY levels by 2015 (Table 4).

Table 2b. LME Published Volumes

Vol.1	Variability and Management of Large Marine Ecosystems. Edited by K. Sherman and L. M. Alexander. AAAS Selected Symposium 99. Westview Press, Inc., Boulder, CO, 1986. 319 p.
Vol.2	Biomass Yields and Geography of Large Marine Ecosystems. Edited by K. Sherman and L.M. Alexander. AAAS Selected Symposium 111. Westview Press, Inc., Boulder, CO, 1989. 493p.
Vol.3	Large Marine Ecosystems: Patterns, Processes, and Yields. Edited by K. Sherman, L.M. Alexander, and B.D. Gold. AAAS Symposium. AAAS, Washington, DC, 1990. 242 p.
Vol.4	Food Chains, Yields, Models, and Management of Large Marine Ecosystems. Edited by K. Sherman, L.M. Alexander, and B.D. Gold. AAAS Symposium. Westview Press, Inc., Boulder, CO, 1991. 320 p.
Vol.5	Large Marine Ecosystems: Stress, Mitigation, and Sustainability. Edited by K. Sherman, L.M. Alexander, and B.D. Gold. AAAS Press, Washington, DC, 1992. 376 p.
Vol.6	The Northeast Shelf Ecosystem: Assessment, Sustainability, and Management. Edited by K. Sherman, N.A. Jaworski, and T. J. Smayda. Blackwell Science, Inc., Cambridge, MA, 1996. 564 p.
Vol.7	Large Marine Ecosystems of the Indian Ocean: Assessment, Sustainability, and Management. Edited by K. Sherman, E.N. Okemwa, and M.J. Ntiba. Blackwell Science, Inc., Malden, MA, 1998. 394 p.
Vol.8	Large Marine Ecosystems of the Pacific Rim: Assessment, Sustainability, and Management. Edited by K.Sherman and Q. Tang. Blackwell Science, Inc., Malden, MA. 1999, 455 p.
Vol.9	The Gulf of Mexico Large Marine Ecosystem: Assessment, Sustainability, and Management. Edited by H. Kumpf, K. Stiedinger, and K. Sherman. Blackwell Science, Inc., Malden, MA, 1999. 736 p.
Vol.10	Large Marine Ecosystems of the North Atlantic: Changing States and Sustainabiity. Edited by H.R. Skjoldal and K. Sherman. Elsevier, Amsterdam and New York. 2002. 449 p.
Vol.11	Gulf of Guinea Large Marine Ecosystem: Environmental Forcing and Sustainable Development of Marine Resources. Edited by J. McGlade, P. Cury, K. Koranteng, N.J. Hardman-Mountford. Elsevier Science, Amsterdam and New York. 2002.
Vol.12	Large Marine Ecosystems of the World: Trends in Exploitation, Protection, and Research. Edited by G. Hempel and K. Sherman 2003.

Even now there is immediate applicability to reaching Summit fishery goals. The FAO Code of Conduct for Responsible Fishery practice of 2002 www.fao.org/FI/agreem/codecond/ficonde.asp argues for moving forward with a "precautionary approach" to fisheries sustainability given a situation wherein available information can be used to recommend a more conservative approach to fish and fisheries total allowable catch levels (TAC) than has been the general practice over the past several decades. Based on the decadal profile of LME biomass yields from 1990 to 1999 [16], it appears that the yields of total biomass and the biomass of 11 species groups of 6 LMEs have been relatively stable or have

Table 3. LMEs and their principal driving forces based on 29 case studies

Climate forcing		Fisheries Forcing		Other:	
LME	YieldMMT	LME	YieldMMT	LME	Yield MMT
Humboldt Current	16.0	South China Sea	10.0	Eastern Bering Sea[1]	2.1
Canary Current	1.8	North Sea	3.5	Black Sea[2]	0.52
Norwegian Sea	1.52	East China Sea	2.8		
Okhotsdk Sea	1.5	Bay of Bengal	2.2		
Iceland Shelf	1.3	Gulf of Thailand	1.12		
Benguela Current	1.21	Mediterranean	1.1		
Sea of Japan	1.0	Gulf of Mexico	0.91		
Guinea Current	0.9	Baltic	0.8		
California Current	0.7	U.S. Northeast Shelf	0.71		

[1] Inconclusive
[2] Eutrophication

Barents Sea	0.5	Scotian Shelf	0.7		
Iberian Coastal	0.35	Caribbean Sea	0.4		
West Greenland Shelf	0.1	Newfoundland-Labrador Shelf	0.23		
Faroe Plateau	<0.1	Yellow Sea	0.2		
		Great Barrier Reef	0.12		
	26.98 **31.74%**		24.77 **29.14%**		2.62 **3.08%**
TOTAL: 54.37 mmt Or 64% of world marine biomass					

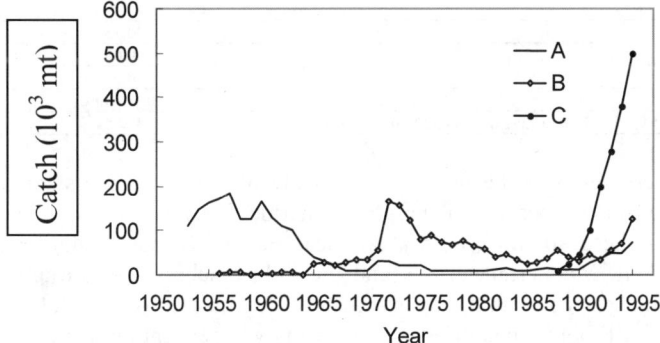

Figure 4. Annual catch of dominant species: (A) small yellow croaker and hairtail, (B) Pacific herring and Japanese mackerel, and (C) anchovy and half-fin anchovy indicating the long term shift in dominant biomass yield from demersal species prior to 1970 to smaller, less desirable pelagic species from the 1970s to 1995 (from Tang 2003)[25].

Table 4. Fisheries Biomass Yields of LMEs with GEF-LME Projects

Fisheries Biomass Yields of LMEs Where Stewardship Ministries are Implementing or Planning GEF-LME Projects [3,4]	
LME	Reported 1999 Annual Biomass Yield
South China Sea	13.9
Humboldt Current	12.0
Bay of Bengal	2.3
Patagonian Shelf	1.7
Canary Current	1.6
Benguela Current	1.1
Guinea Current	1.0
Mediterranean Sea	1.0
Gulf of Mexico	1.0
Baltic Sea	0.9
Yellow Sea[5]	0.6
Black Sea	0.5
Caribbean Sea	0.35
Red Sea	0.08
Agulhas/Somali Currents	0.07
TOTAL:	38.10 mmt
Percentage of Global Marine Yield	44.8%

shown marginal increases over the decade. The yield of marine biomass for these 6 LMEs was 8.1 mmt, or 9.5 percent of the global marine fisheries yield in 1999. The countries bordering these six LMEs—Arabian Sea, Bay of Bengal, Indonesian Sea, Northeast Brazil Shelf, Mediterranean Sea and the Sulu-Celebes Sea—are among the world's most populous, representing approximately one-quarter of the total human population. These LME border countries are increasingly dependent on marine fisheries for food security and for national and international trade. Given the risks of "fishing-down-the-food-chain," it would appear opportune for the stewardship agencies responsible for the fisheries of the bordering countries to consider options for mandating precautionary total allowable catch levels during a period of relative biomass stability.

Evidence for species recovery following a significant reduction in fishing effort through mandated actions is encouraging. Following management actions to reduce fishing the robust condition of the U.S. Northeast Shelf ecosystem with regard to the average annual level of primary productivity ($350 gCm^2 \cong yr$), stable annual average levels of zooplankton ($33 \ cc/100m^3$), and a relatively stable oceanographic regime [26], contributed to: (1) a relatively rapid recovery of depleted herring and mackerel stocks[, with the cessation of foreign fisheries in the mid-1970s[9]; and 2) initiation of the recovery of depleted yellowtail flounder and haddock stocks following a mandated 1994 reduction in fishing effort (Figures 5a and 5b)[26].

---

[3] Data from FAO Fisheries Technical Paper 435 (Garibaldi and Limongelli 2003 [16])
[4] No biomass yield data available for the Western Pacific Warm Water Pool
[5] Biomass Yield Data for 1995 from Tang 2003[25]

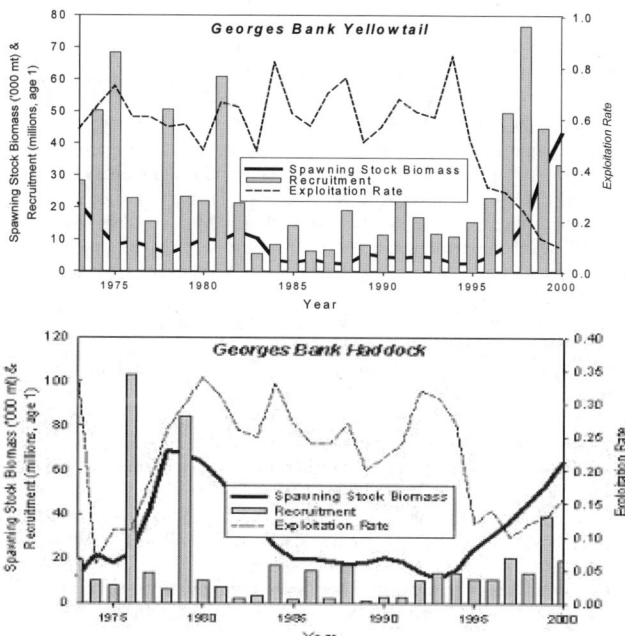

Figure 5. Trends in spawning stock biomass (ssb) and recruitment in relation to reductions in exploitation rate (fishing effort) for two commercially important species inhabiting the Georges Bank sub-area of the Northeast Shelf ecosystem, yellowtail flounder (a) and haddock (b).

Three LMEs remain at high risk for fisheries biomass recovery expressed as a pre-1960s ratio of demersal to pelagic species—Gulf of Thailand, East China Sea and Yellow Sea. However, mitigation actions have been initiated by the People's Republic of China toward recovery by mandating 60 to 90 day closures to fishing in the Yellow Sea and East China Sea during summer months[25]. The country-driven planning and implementation documents supporting the ecosystem approach to LME assessment and management practices can be found at http://www.iwlearn.org.

7. References

[1] UNEP (United Nations Environment Program). 2001. Economic reforms, trade liberalization, and the environment: a synthesis of UNEP country projects. Division of Technology, Industry, and Economics, Geneva, Switzerland. 21p.
[2] Sherman, K. 1991. Sustainability of resources in large marine ecosystems. In: Sherman, K., L.M. Alexander, Barry D. Gold, eds. Food Chains, Yields, Models, and Management of Large Marine Ecosystems. Westview Press. 1-34. 320p.
[3] Sherman, K. 1993. Emerging theoretical basis for monitoring changing states (health) of large marine ecosystems. U.S. Dept. Com. NOAA Tech. Mem. NMFS-F/NEC-100.
[4] Pauly, D., Christensen V. 1995. Primary production required to sustain global fisheries. Nature 374:255-7.
[5] Beddington, J.R. 1995. The primary requirements. Nature 374:213-14.
[6] Epstein, P.R. 1993. Algal blooms and public health. World Resource Review 5(2):190-206.
[7] Berman, M.S. and K. Sherman. 2001. A towed body sampler for monitoring marine ecosystems. Sea Technology 42(9):48-52.
[8] Aiken, J., R. Pollard, R. Williams, G. Griffiths, I. Bellan. 1999. Measurements of the upper ocean structure using towed profiling systems. In: Sherman, K. and Q. Tang, eds. Large Marine Ecosystems

of the Pacific Rim: Assessment, Sustainability, and Management. Malden: Blackwell Science, Inc. 346-362.
[9] Sherman K, J. Green, A. Solow, S.A. Murawski, J. Kane, J. Jossi, and W. Smith. 1996. Zooplnkton prey field variability during collapse and recovery of pelagic fish in the northeast Shelf ecosystem. In: K. Sherman, N.A. Jaworski, T.J. Smayda, eds. The Northeast Shelf Ecosystem: Assessment, Sustainability, and Management. Blackwell Science. 217-235..564p.
[10] NOAA (National Oceanic and Atmospheric Administration). 1993. Emerging theoretical basis for monitoring the changing states (Health) of large marine ecosystems. Summary report of two workshops: 23 April 1992, National Marine Fisheries Service, Narragansett, Rhode Island, and 11-12 July 1992,Cornell University, Ithaca, New York. NOAA Technical Memorandum MNFS-F/NEC-100.
[11] Costanza, R. 1992. Toward an operational definition of ecosystem health. In: Costanza R, B.G. Norton, B.D. Haskell, eds. Ecosystem Health: New Goals for Environmental Management. Island Press, Washington DC. 239-256.
[12] Sherman, K. 1994. Sustainability, biomass yields, and health of coastal ecosystems: an ecological perspective. Marine Ecology Progress Series 112:277-301.
[13] Sherman, B. 2000. Marine ecosystem health as an expression of morbidity, mortality, and disease events. Marine Pollution Bulletin 41(1-6):232-54.
[14] Sutinen, J. editor. 2000. A framework for monitoring and assessing socioeconomics and governance of large marine ecosystems. NOAA Technical Memorandum. NMFS-NE-158. 32 p.
[15] Juda, L. and T. Hennessey. 2001.Governance profiles and the management of the uses of large marine ecosystems. Ocean Development and International Law. 32:41-67.
[16] Garibaldi, L. and L. Limongelli. 2003. Trends in oceanic captures and clustering of large marine ecosystems: Two studies based on the FAO capture database. FAO Fisheries Technical paper 435. Food and Agriculture Organization of the United Nations. Rome.71p.
[17] Caddy, J.F. 1993. Contrast between recent fishery trends and evidence for nutrient enrichment in two large marine ecosystems: the Mediterranean and the Black Seas. In: Sherman, K., L.M. Alexander and B.D. Gold, eds. Large Marine Ecosystems: Stress, Mitigation and Sustainability. AAAS Press. Washington, D.C. 376p.
[18] Lluch-Belda, D., D. B. Lluch-Cota and S.E. Lluch-Cota. 2003. Figure 9, p.212 from Chapter 9, Interannual variability impacts on the California Current large marine ecosystem. In: Hempel, G. and K. Sherman, eds. Large Marine Ecosystems of the World: Trends in exploitation, protection and research.. Elsevier Science. Netherlands, London, New York, Tokyo. 423p.
[19] Schwarzlose, R.A., J. Alheit, T. Baumgartner, R. Cloete, R.J.M. Crawford, W.J. Fletcher, Y. Green-Ruiz, E. Hagen, T. Kawasaki, D. Lluch-Belda, S.E. Lluch-Cota, A.D. MacCall, Y. Matsuura, M.O. Nevárez-Martínez, R.H. Parrish, C. Roy, R. Serra, K.V. Shust, N.M. Ward and J.Z. Zuzunaga. 1999. Worldwide large-scale fluctuations of sardine and anchovy populations. S. Afr. J. Mar. Sci. 21:289-347.
[20] Wolff, M., C. Wosnitza-Mendo and J. Mendo. 2003. Figure 6, p.287 from Chapter 12, The Humboldt Current LME. In: Hempel, G. and K. Sherman, eds. Large Marine Ecosystems of the World: Trends in exploitation, protection and research. Elsevier Science. Netherlands, London, New York, Tokyo. 423p.
[21] Astthorsson, O.S. and H. Vilhjálmsson. 2002. Figures 19 and 20, p.240 from Iceland Shelf large marine ecosystem. In: Sherman, K. and H.R. Skjoldal, eds. Large Marine Ecosystems of the North Atlantic: Changing States and Sustainability. Elsevier Science. Netherlands, London, New York, Tokyo. 449p.
[22] NEFSC. 1999. G. Atlantic Herring. Report of the 27th Northeast Regional Stock Assessment Workshop (27th SAW). Stock Assessment Review Committee (SARC) Consensus Summary of Assessments. Woods Hole Laboratory Reference Document No. 98-15.
[23] Pauly, D. and R. Chuenpagdee. 2003. Development of fisheries in the Gulf of Thailand large marine ecosystem: Analysis of an unplanned experiment. In: Hempel, G. and K. Sherman, eds. Large Marine Ecosystems of the World: Trends in exploitation, protection and research. 337-354.
[24] Chen, Ya-Qu and Xin-Qiang Shen. 1999. Changes in the biomass of the East China Sea ecosystem. In: Sherman, K. and Q. Tang, eds. Large Marine Ecosystems of the Pacific Rim: Assessment, Sustainability and Management. Blackwell Science. Malden, Massachusetts. 221-239.
[25] Tang, Q. 2003. Figure 10, p.137 from Chapter 6, The Yellow Sea and mitigation action. In: Hempel, G. and K. Sherman, eds. Large Marine Ecosystems of the World: Trends in exploitation, protection and research.. Elsevier Science. Netherlands, London, New York, Tokyo. 423p.
[26] Sherman, K., J. Kane, S. Murawski, W. Overholtz and A. Solow. 2002. In: Sherman, K. and H.R. Skjoldal, eds. Large Marine Ecosystems of the North Atlantic: Changing States and Sustainability. 195-215.

EXPRESS METHODS OF NONDESTRUCTIVE CONTROL FOR PHYSIOLOGICAL STATE OF ALGAE

T.V.PARSHIKOVA
Kiev National University named Taras Shevchenko,
Vladimirskaya st., 64, 01017, Kiev, UKRAINE

Introduction

In recent years algae have been widely used in mariculture as primary producers of organic material and oxygen, and utilizers of CO_2 in water ecosystems. In connection with this special interest the search is on for express methods to control its growth and productivity.

The aims of our investigations were the selection and approval of experimental methods, which would make it possible to receive the necessary information on the living cells of algae in normal conditions as well as under the effect of stress-factors. Surfactants were used as stress-factors, which can be found in significant volume in littoral water as a result of the rejection of sewage or of superficial runoffs, heavy metals (Cr^{6+} for example) and natural compounds with surface activity (such as alginic, miristinic acids). The methods used are laser-doppler spectroscopy, extremely high frequency dielectrometry (EHF-dielectrometry), differential fluorimetry and photofading of pigments [1].

The experiments were carried out over 17 species of industrially valuable algae cultures :*Chlorophyta*, *Cyanophyta* (*Cyanobacteria*), *Rhodophyta* and naturally occurring algae from the Dnieper Basin and the Sea of Azov (Ukraine). The experiments investigated movable, unmovable and attached to substratum algae.

1. Laser – Doppler Spectroscopy (LDS)

Figure 1 [2] shows a block diagram of informational-measuring systems on the basis of laser Doppler spectrometer. LDS makes it possible to estimate the speed of cell movement (μm/s) and its moving energy potential (relative units) as well as the reactions under the effect of stress-factors. 300-500 μl of algae suspension with cell concentration from 0.5 to 200 billion cells per 1 ml were placed in a cuvette. Measurements for 1 sample lasted up to 3 minutes. Measurement error does not exceed 1-3% for all parameters.

This highly sensitive research method was also used successfully for unphotosynthesizing but movable cells, such as protozoa and spermatozoa [3].

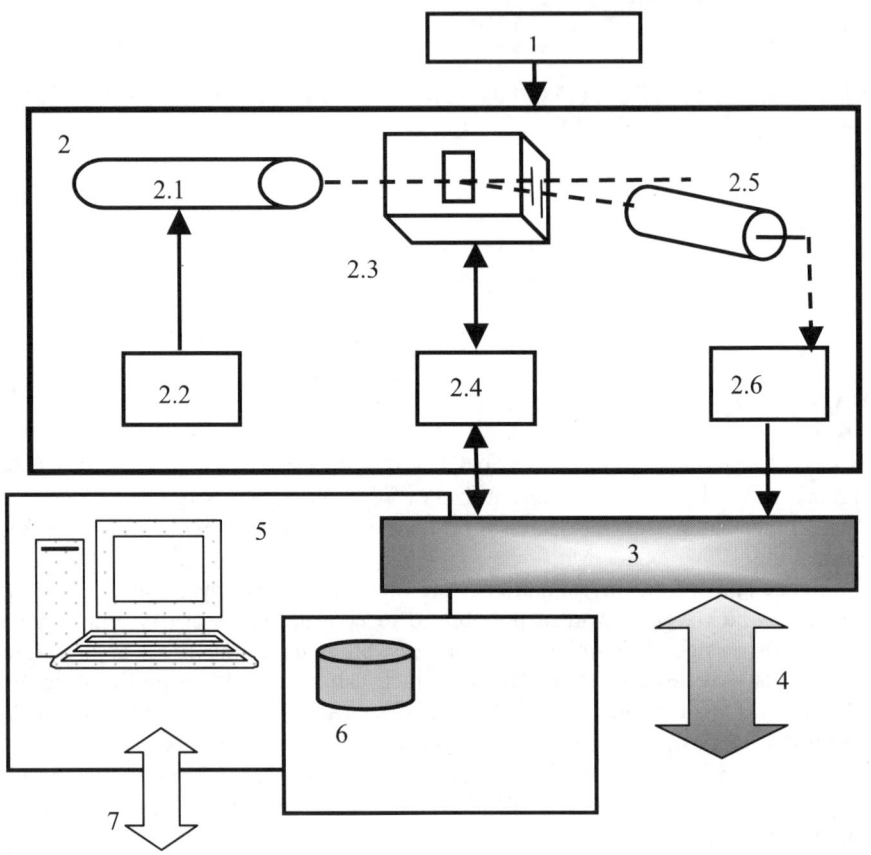

Figure 1: Laser-doppler spectrometer.
1. Electronic module of a network or autonomous power supply.
2. Laser-optical measuring module.
   2.1. Laser. 2.2. Laser power supply unit. 2.3. Temperature-controlled zone of a measurement. 2.4. Electronic control of thermal stability. 2.5. Photodetector. 2.6. Electronic amplifier.
3. Multichannel analog-digital signal processor.
4. Universal interface with auxiliary and other specialized measuring systems of monitoring.
5. Computer.
6. Specialized software.
7. Channels of exchange and connection of a complex with standard networks.

When a cationic surfactant (CS) – catamine (alkyldimethylbenzilammonium chloride, SIGMA, USA) — is introduced into the cultural medium, the mobile algae cells (*Chlamydomonas reinhardtii* Dang.) try to vacate the zone of contact with the surfactant. The results of measuring energy potential and motion speed add considerable support to this fact. From the first minutes of contact with the surfactant, the cells mobilize their energy potential sharply and increase their motion speed as shown. As can be seen from Figure 2, the algae cells' motion speed and energy potential increase up to 30-40% in 1-3 hours after CS addition. However, it is a short-term effect. Sequent decreasing was then supervised for levels of these indices [4].

Introducing such systems in the technological processes of companies whose production is harmful to the environment will allow to inspect and prevent toxic environmental pollutions, use efficient cleaning structures, improve production and increase its ecological purity.

2. Extremely High Frequency Dielectrometry (EHF-dielectrometry)

Like LDS, EHF-dielectrometry makes it possible to work with undamaged (intact) structures of the living cell. A cell's EHF-dielectrometry, specifically the correlations of free and bound water the cell contains, which is a functionally important index of its physiological state, is based on recording hydration changes. It is known [5] that water plays a part in the formation of protein structures, nucleic acids, supramolecular complexes, bio-layers of membranes, cytoskeletons, etc., and fuels the biological object's functions on different organizational levels of the life, depending on its functional activity. Hydration is a formative complex of atoms or molecules with water in a definite stoichiometric correlation in the shape of a steady crystalline hydrate or of a coat in the solution around the dissolved substance, whose properties differ from the surrounding water, and which are also called "volumetric". As a rule, the structure and the power and dynamic parameters of such a coat in macromolecules are heterogeneous, thus allowing judgments about their functional activity [6].

Dielectric penetrability was measured over the dispersion-of-free-water range under a wavelength of 7,56 mm. Changes in standing wave parameters, namely displacement of standing wave minimum (ΔL) and width of its doubled minimum (ΔX), were measured when a suspension of algae specimens was brought into the apparatus's wave guide. The parameters of the cultural medium solution were used as control parameters. In the range of specimen thickness used, it will be proportional to: ε' – real и ε'' – imaginary component of complex dielectric penetrability. Dielectric penetrability parameters are determined from nomograms, which are built on the basis of solving the transcendental equation for the evaluation of complex dielectric penetrability [7].

Experimental data testify that the dielectric indices of cells correlate with their photosynthetic activity. As shown on Figure 3, cells of algae belonging to different taxonomic groups, but grown under identical conditions and at identical growth phases, are characterized by a specific peculiarity of the cellular structures' hydrate medium. Changes in the cellular water state in the same species were also noted, depending on its growth phase and calendar age of culture. Intensively photosynthesizing cultures contain maximum amounts of free cellular water. While algal cells remain without reinoculation for a long time (in their stationary growth phase) and preserve the same level of hydrate medium within all structures, they differ by the state and volumes of free water.

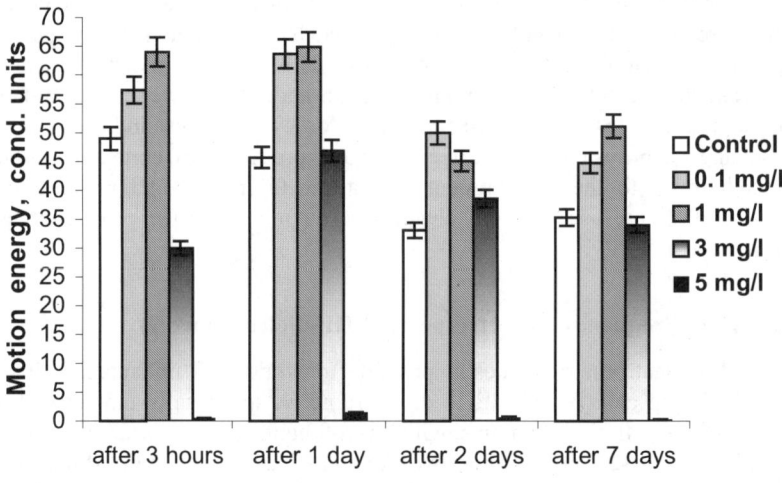

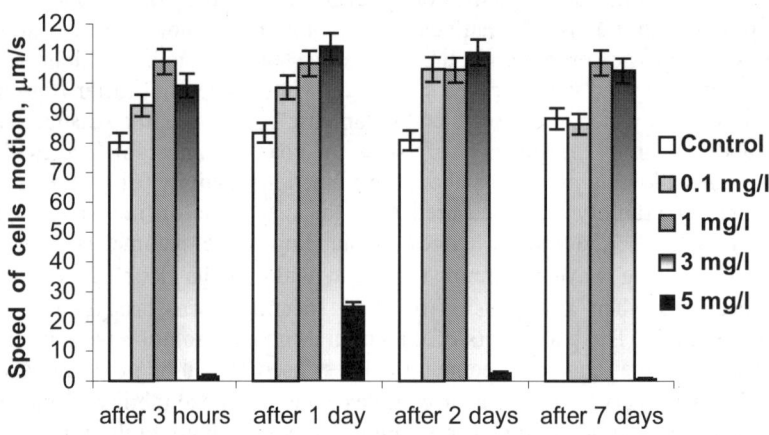

Figure 2: Changes of motion energy and speed of cells of *Chlamydomonas reinhardtii* under addition of catamine into cultural medium.

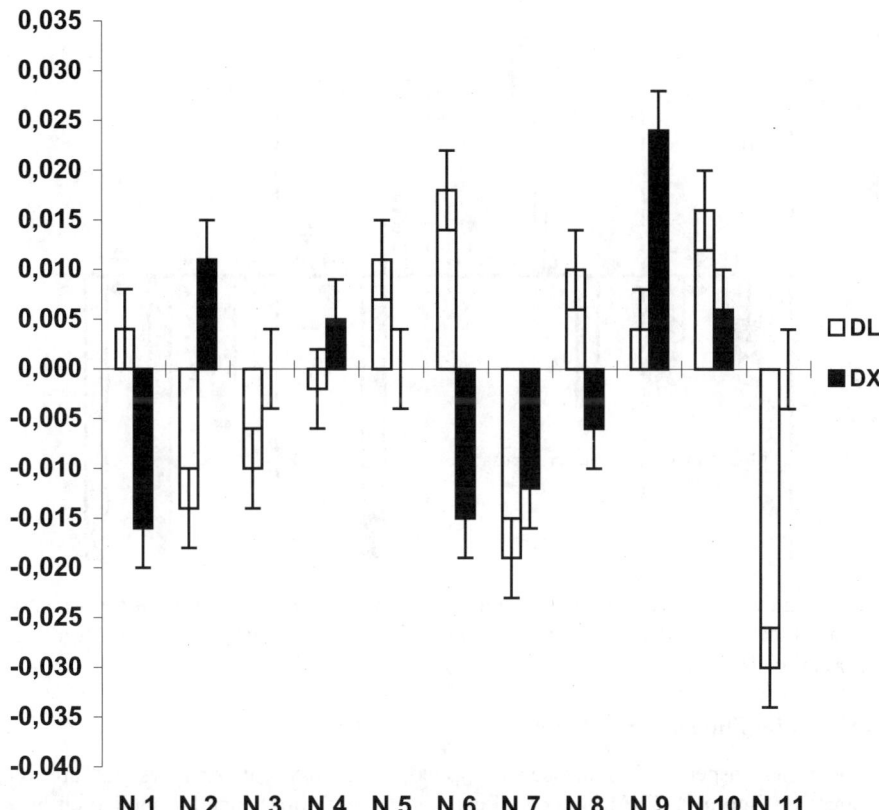

Figure 3: Dielectric parameters of water state for native cells of blue-green algae: ☐ – changes in hydrate surrounding of cells structures, ■ – amount of free water in respect of control. Symbols: 1 – *Microcystis aeruginosa* (log.), 2 – *M. aeruginosa* (stat.), 3 – *M. pulverea* (stat.), 4 – *Anabaena* PCC 7120 (log.), 5 – *A.* N 11 (mutant) (log.), 6 – *A.* N 12 (mutant) (log.), 7 – *A. flos-aquae* (log.), 8 – *A. hassalii* (stat.), 9 – *Spirulina platensis* (log.), 10 – *S. platensis* (stat.), 11 – *Nostoc punctiforme* (stat.).

The hydrate environment and the amount of free water of all cellular structures are very sensitive to the impact of negative factors on algal cells (Figure 4). When Chromium (VI) salt (potassium dichromate in concentration of 5 mg/l) was introduced into the algal culture, water regime indicators in algal cells underwent acute changes, which correlated with their deterioration and death. The method is immediate, the analysis can be carried out in only 2 to 3 minutes, it can be repeated and many samples can be processed. Therefore, this method may be recommended to control the physiological state and algae growth rates in cultures as well as in natural reservoirs.

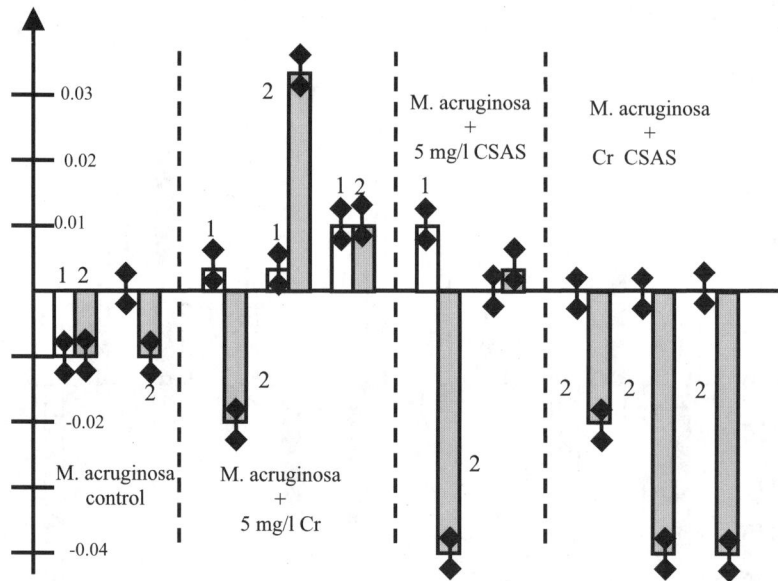

Figure 4: Indices of free cellular water in culture of *Microcystis aeruginosa* under effect of chromium and CS: 1 – changes in hydrate surrounding of cellular structures; 2 – amount of free water.

3. Differential Fluorimetry Method

The theoretical aspects of fluorescence applied to photosynthesizing systems are very complicated (Figure 5). The prospects of using variable fluorescence for applied purposes can be estimated thanks to this scheme. Most particularly, it was thus determined [8] that the ΔF index (distinctions in the fluorescence intensity of native algal cells before and after an electron transport disconnector (diuron or simazine, for example) was added) is only registered in the presence of the formed photosystem II (PS II), which launches the transport of electrons. Therefore, the variable fluorescence can be used as a good test for the presence of this photosystem and its activities, using the "yes-no" principle (living or dead cells depending on whether they photosynthesize normally or the processes are inhibited by something). In the course of work [9], it was established that biomass concentration (by chlorophyll), growth rate of a culture, degree of its viability and potential photosynthetic activity (index ΔF, Table 1) can be determined through the results of native algal suspensions, both in the norm and under the effects of various factors (changes of medium composition, introduction of various contaminants, different light intensity, pH, interfusion regime, etc).

Figure 6 shows that when a surfactant is added, the higher the concentration of CS and the longer the contact, the more chlorophyll concentration decreases. The algae's photosynthetic activity was also reduced under the effect of catamine. Exact concentration dependence was noted over 2 days of contact.

The Planctofluorimeter FL 003 M, developed by Krasnoyarsk University (Russia), was used. The different fluorimeters are currently used in laboratory practices and on board research boats.

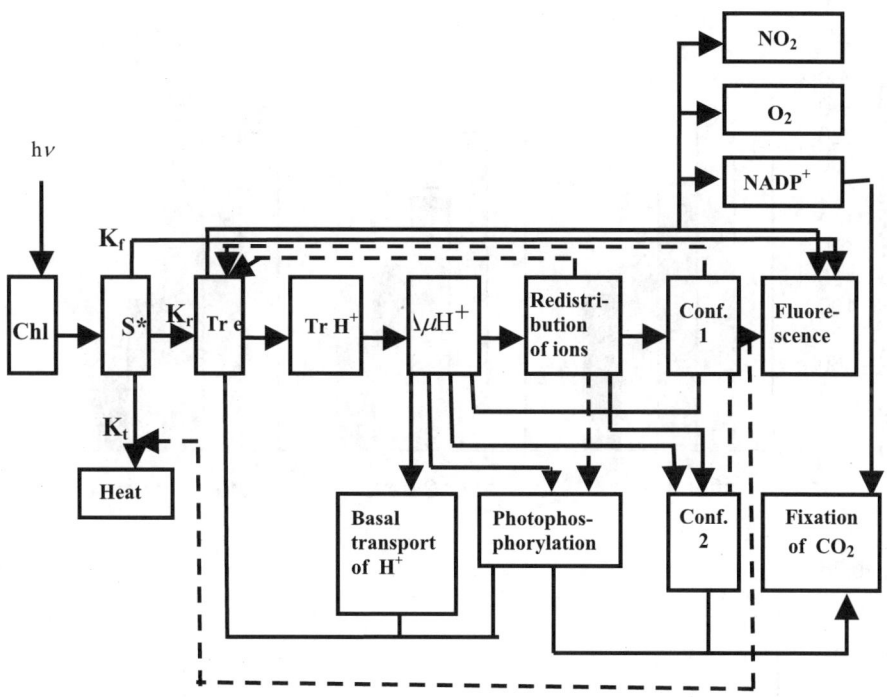

Figure 5: Diagram of links of fluorescence output with main stages of photosynthesis: S* – excited state of molecule; Kf, Kt, Kr – deactivation rate of excited molecules through the fluorescence channels, temperature deactivation and photochemical reactions; $\Delta\mu H^+$ – electrochemical potential of hydrogen ions; Tr e – transport of electrons; Tr H^+ – transport of hydrogen ions.

TABLE 1. ΔF indicators for algae cultures in different physiological states

ΔF index	Physiological state of culture
0.5-0.7*	Grows intensively and is characterized by maximum photosynthetic capacity
0.3-0.4	Growth rate is average, the photosynthesis is 50-70% of the maximum
0.1-0.2	The culture is weakened, growth speed is reduced, photosynthesis is no more than 30-40%
< 0.1	The algae are suppressed, most of them die, only separate cells photosynthesize

* These limits are typical for algal cells being to the maximum in optimized habitat and light conditions, nutrient medium and other factors

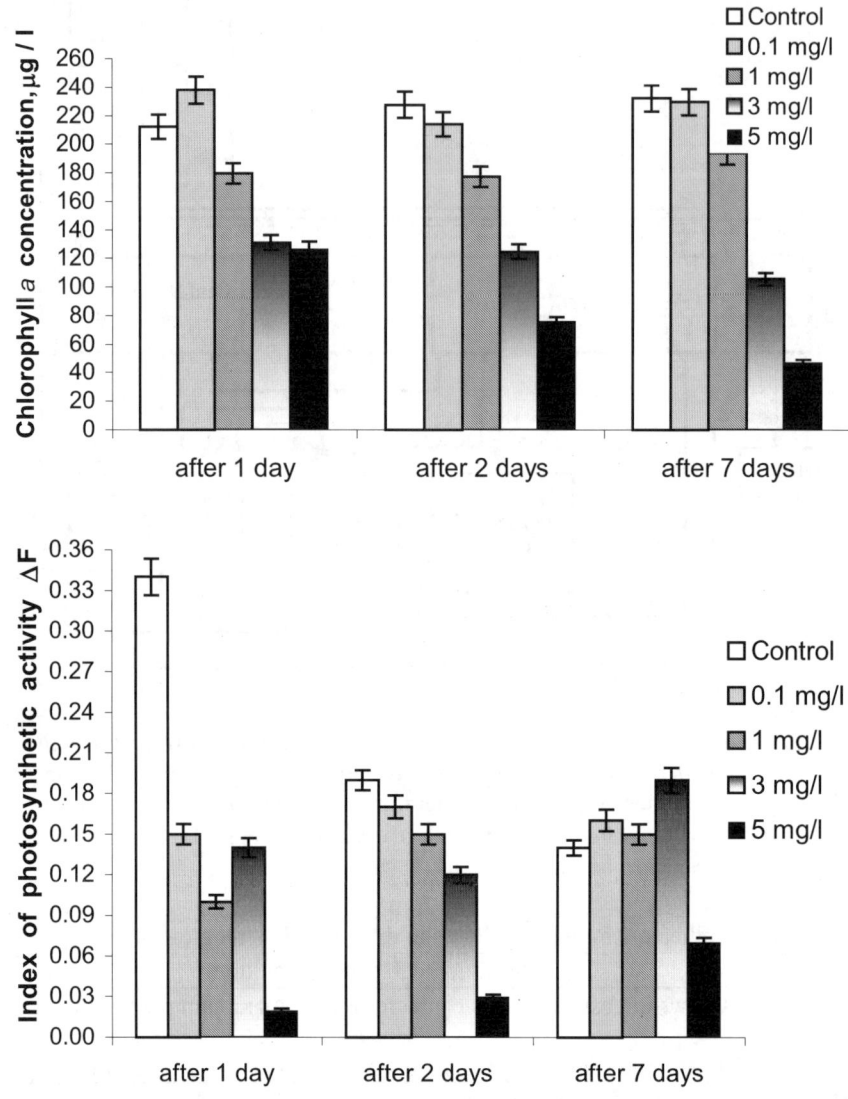

Figure 6. Effect of CS on dynamics of chlorophyll *a* content and potential photosynthetic activity in suspension of *Chlamydomonas reinhardtii* cells.

4. Chlorophyll photofading

Present material testifies that the process of chlorophyll photofading (its photodestruction) occurs as a result of the effect of light illumination and UV-stress. Both photodestruction and the quantity and state of water in the cells (which is conditioned by its functional deficit) play an important part in the development process, as do the spectral content of light, low temperature, the corresponding photophysical peculiarities of pigments, the presence of protective factors in the cell (for example, carotinoids, flavonoids) and the activity of some enzymes (for example 5-lipoxygenase). PS I is known to be less sensitive [10] to the damaging influence of UVs than PS II. When exposed to UVs with a wavelength of 280 nm, the long-wave forms (800 nm) of bacteriochlorophyll *a* were destroyed, and its photochemical activity decreased. When exposed to UVs with a wavelength of 365 nm, long-wave forms of chlorophyll (800 and 865 nm) were destroyed.

The experiments testify that using the blue-violet part of the spectrum to excite the photodestruction in the field of vision of a luminescence microscope instead of using traditional UVs is justified methodically. In this case the real term of loss for red fluorescence and the functional activity of chlorophyll was considerably greater. The registration of the process's dynamics was respectively more exact than under UV-irradiation. UVs also produced quicker deep destruction of the chlorophyll and loss of its functional activity.

Obtained data testifies that *Anabaena* cells were living and fluoresced intensively until the samples were exposed (Figure 7).

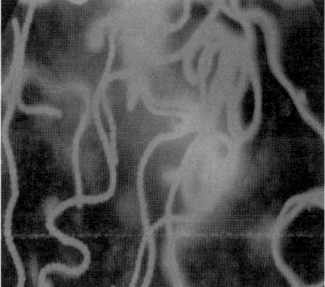

Figure 7: Mutant cells of *Anabaena PCC 7120* (control without surfactants). In original view under luminescent microscope all living cells are bright red.

In the process of exposure to light (2 min) the main part of chlorophyll lost its fluorescence and functional activity of pigments. This process becomes particularly intense in the presence of surfactants. The level of injury to photosynthetic organism (green color of cells under microscope) may be estimated by measuring the photofading speed. As shown on Figure 8, the photofading speed in different algae is significantly different. It is pertinent to note that increasing the acting concentration of cationic surfactants depresses the bonding strength of chlorophyll-protein-lipid complexes in both algae, especially at 3 mg/l. As a result, the concentration of unstable connections of chlorophylls in algae (relative to initial content) increases 3 to 6 times for 3 hours of contact. Eukaryotic cells (for example, *Chlorella*) are more stable to photofading in the presence of surfactants than prokaryotic (*Anabaena, Microcystis*).

This part of the experiments was provided jointly with Prof. A.V. Brayon from Kiev National University.

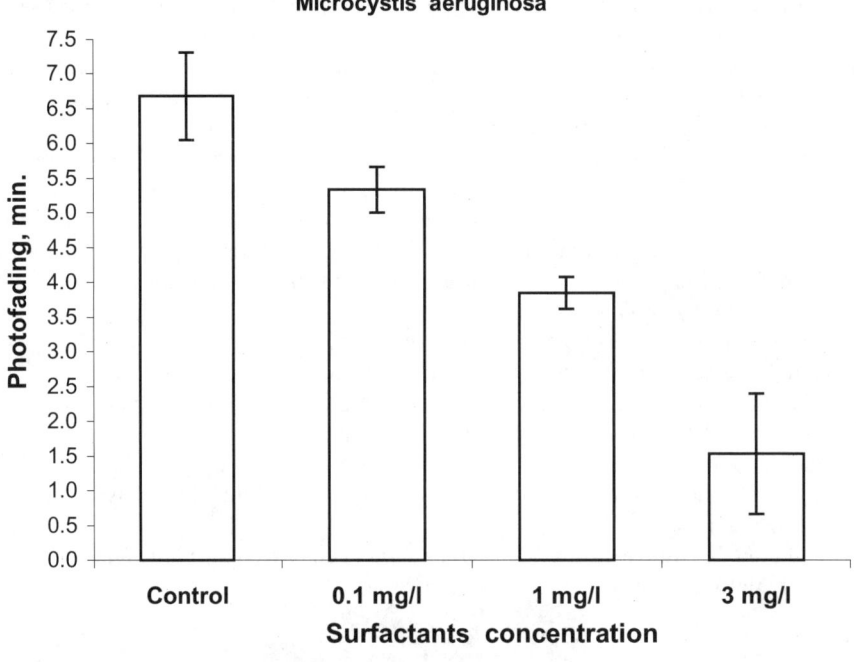

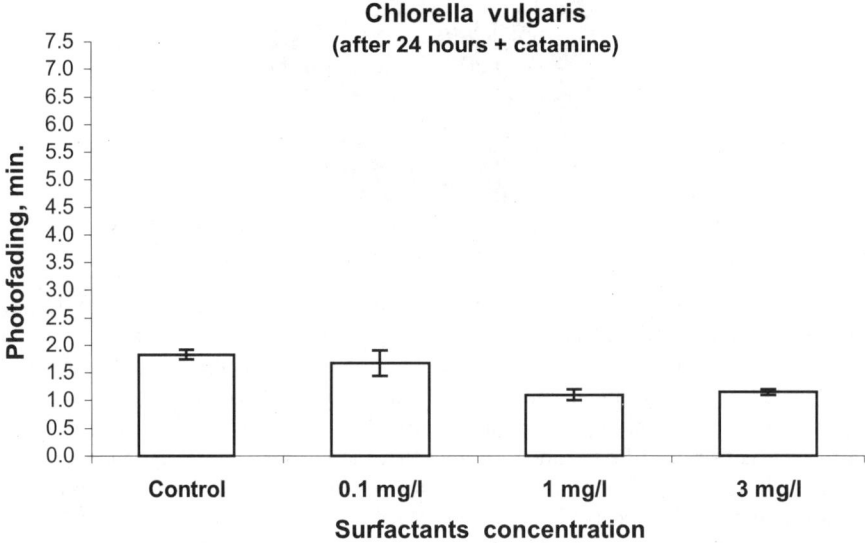

Figure 8: Speed of pigments photofading for algae under surfactants effect

Thus, one of these lifetime nondestructive methods of control may be used to carry out express-control and estimation of the physiological activity and state of microscopic algae (or fraction of macroalgae tallomes). Depending on the aims of the monitoring, it may be supplemented, if necessary, with visual microscopic or standard biochemical analysis.

Acknowledgements

The author is grateful to Prof. T.Y.Shchegoleva and Dr. S.P.Ponomarenko for their help and useful suggestions. This study was partially funded by a Grant from the Ukrainian State Committee on Science and Technology N 06.07/82.

References:

1. Parshikova, T.V. (2003) *The structure-functional markers of adaptation for microalgae under surfactant effect,* Thesis of dissertation for obtaining the scientific degree (doctor of biological sciences). Specialty 03.00.12 – Physiology of plants. Kyiv National University, Kyiv.
2. Vlasenko, V.V. et al. (1992) *Method of measurements for toxic influence of chemical compounds, which are present in water environment on culture of plankton hydrobiont,* Patent of Ukraine.
3. Ponomarenko, S., Vlasenko V. (2003) Usage of laser-doppler spectroscopy (LDS) method in biotechnology of moving algae. *Proc. of 5th European Workshop "Biotechnology of microalgae",* Bergholz-Rechbrucke, Germany, 69-71.
4. Parshikova, T.V. (2003) Participation of surfactants in regulation of microalgae development, *Hydrobiological* J. 1, 39, 64-70.
5. Schegoleva, T.Y. (1984) Hydrated medium of biopolymer macromolecules by EHF-dielectrometry data, *Biofizika* 29 (6), 935-939.
6. Aksenov, S.I. (1990) *Water and its role in regulation of biological processes,* Nauka, Moscow.
7. Shchegoleva, T.Yu., Kolesnikov, V.G., Vasilieva, E.V. et al. (2000) *Using of millimetric diapasons radiowaves in medicine,* ChIMB, Kharkov.
8. Gol'd, V.M., Gaevskiy, N.A., Grigorjev, Yu.S. et al. (1984) *Theoretical basis and methods of estimation for chlorophyll fluorescence,* Krasnojarsk University Publ., Krasnojarsk.
9. Parshikova, T.V., Sirenko, L.A., Shchegoleva, T.Yu., Kolesnikov, V.G. (2002) Express control of growth and physiological state of microalgae, *International J. on Algae* 4 (1), 106-117.
10. Kochubei, S.M. (2001) *Organization of photosynthetic apparatus of high plants,* Altpress, Kiev.

Chapter 2

Modeling Approaches and Mathematical Foundations of Environmental Management

STRATEGIC MANAGEMENT OF ECOLOGICAL SYSTEMS: A SUPPLY CHAIN PERSPECTIVE

E. LEVNER
Department of Computer Science, Holon Academic Institute of Technology, Holon, 58102, ISRAEL

J.-M. PROTH
INRIA, Metz, FRANCE

Abstract

The demand for new sophisticated methods for ensuring sustainable development of exploited ecosystems is increasing worldwide. Today, the natural resources, including water, air, soil, flora and fauna, etc., are significantly affected by disastrous pollution from industrial, agricultural, municipal, and other anthropogenic sources. Without careful global management and coordination of international anti-pollution efforts, the world's oceans are threatened with catastrophic changes; under an unprecedented high rate of natural resource degradation due to pollution and overexploitation, the whole marine ecosystems may lose its integrity and collapse. These worrying situations call for efficient approaches that may help biologists detect, analyze, assess and solve the problems that occur in ecosystems in general and marine ecosystems in particular. Solving these problems involves activities that should be implemented and managed. We consider this aspect in this paper.

1. **Tools for environmental integrity management.**

Several approaches are available to help scientists and engineers in charge of designing or improving, analyzing, assessing and managing complex systems. From now on, we focus on marine ecosystems.

Analyzing a marine ecosystem requires the competence of biologists. Only biologists are able to suggest rules that totally or partially explain the dynamics of the system. Rules connect decisions and / or external events to the dynamics of the system. Two kinds of rules may be suggested: (i) the so-called qualitative rules that express trends of quantitative parameters with terms like "increasing", "decreasing", or "vanishing", or change the values of qualitative parameters (colors, texture, etc.) which define the state of the system, and (ii) the quantitative rules that express a mathematical relationship between decisions and / or external events expressed in terms of quantitative variables, while the state of the system is also expressed in terms of quantitative variables.

Starting from these rules, it is possible to build models. The types of models that can be built and their usefulness are discussed hereafter.

Another important aspect is the management of marine ecosystems, the goal being to ensure sustainable development of exploited ecosystems or to correct the damage resulting from pollution. The management of marine ecosystems is based on the results of analysis and the knowledge provided by biologists. The efficiency of the decisions to be made to manage ecosystems depends on the quality of the analysis, the precision of the objectives and the scientific background available concerning the problem at hands. Assuming that we have a detailed analysis, a strong background and clear objectives, it is easy to define the decisions to be made. They are usually strategic decisions from which tactical, and then operational decisions are derived. We are at the level of planning and scheduling. It is at this level that the Supply Chain paradigm appears. It will also be developed hereafter.

1.1. Modeling marine ecosystem.

We do not discuss the models derived from qualitative rules since they only describe the ecosystem and allow neither interpolation nor extrapolation.

From now on, we consider only systems that can be described in terms of quantitative rules [14].

Let us first outline that clearly defining the limits of the system to be modeled is the first step of the modeling activity. This step is of utmost importance since any system is connected to the environment that influences it, and this influence must be analyzed. Another influence is the control that is intentionally introduced. The marine ecosystem evolves according to the relationships between the parameters that characterize its state and the relationships between these parameters and the parameters that characterize both the external influence and the control. The control is usually constrained.

Figure 1 represents this situation. $E(t)$ is the state of the system at time t.

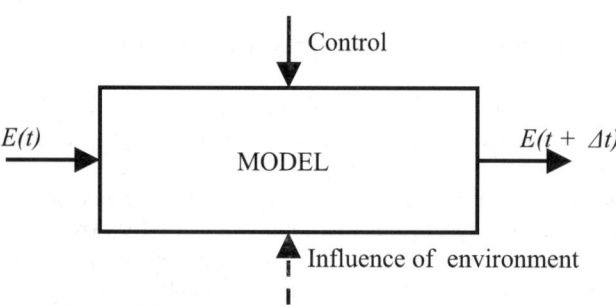

Figure 1: Evolution of the model.

The inputs of the model are the state of the ecosystem at time t and the values of the parameters that define the control and the influence of the environment. The output is the state of the ecosystem at time $t + \Delta t$, Δt being either a finite value or a value that is « as small as possible ».

Let c be the control, v the influence of the environment on period $[t, t + \Delta t]$ and F the model. Formally:

$$E(t + \Delta t) = F(E(t), c, v). \tag{1}$$

Usually, some constraints apply on the control.

If (1) concerns a period Δt that is « as small as possible », the model is said to be continuous. Otherwise, Δt is either constant and defined by the modeler, or variable and dependent on the events that arise in the system. In the first case, we say that the model is activity driven. In the second case, the model is said to be event driven.

If F is a set of mathematical equations, we may be in one of the following cases:

- Either equation (1) can be used to express the optimal control $c_0(E(T))$ to apply on period $[0, T]$ in order to reach the objective $E(T)$ at time T knowing the state of the system at time 0. Indeed, this implies that the influence of the environment, say $v(0-T)$, is known on period $[0, T]$. In this case:

$$c_t(E(T)) = G[E(T), v(0-T), E(T)] \tag{2}$$

If the influence of the environment is known only at the current time t for a short period Δt, the corresponding situation can be expressed as:

$$c_t(E(t + \Delta t)) = G[E(t), v(t), E(t)] \tag{3}$$

We say that such a model is analytical. Indeed, the situation expressed by (2) is by far the most interesting situation since it allows reaching the optimal solution just by using G to compute (2). A model of type (3) is more difficult to handle and it should be associated with a strategy that takes the state as close as possible to the objective $E(T)$ at time T.

- Or neither (2) nor (3) can be derived from (1). The only possibility, in this case, is to use (1) in order to build a simulation software that will compute the sequence $E(\Delta t), E(2.\Delta t), E(3.\Delta t), \ldots, E(n.\Delta t = T)$ starting from the initial state $E(0)$ and knowing the control $c(0), c(\Delta t), c(2.\Delta t), \ldots, c((n-1).\Delta t)$, and the parameters that characterize the external influence $v(0), v(\Delta t), v(2.\Delta t), \ldots, v((n-1).\Delta t)$ at times $0, \Delta t, 2.\Delta t \ldots, (n-1).\Delta t$. This kind of model does not lead to the optimal solution. Such a model only allows checking the consequences of scenarios under various hypotheses. This situation is the most common when ecosystems are concerned.

If the model is activity driven, the modeler provides the mathematical rules F that allow to derive the state of the system at time $t + \Delta t$ knowing the state of the system at time t. Δt is a constant given by the modeler. We should mention that in this case the rules are approximations integrating all the events that arise on each interval $[(i-1).\Delta t, i.\Delta t]$. Thus, the quality of an activity driven model depends on the quality of the approximations. For such a model, the sequence of states of the model diverges from the sequence of the corresponding ecosystem, which means that the time span which the model can cover efficiently is limited.

If the model is event driven, we compute the next state each time an event arises. Thus, Δt is variable and its value is defined by the system itself. Such an event can either be external or generated by the system itself (for instance, an event arises when a constraint that applies to the state of the system is saturated). If the rules are precisely defined, an event driven model is usually much more accurate than an activity driven model, but it is usually much more complex in terms of computation load, and this complexity is unpredictable.

What is the advantage of a model? It helps to understand the system by comparing the evolution of the marine ecosystem and the result provided by the model. When the model is adjusted, i.e. when the evolution of the ecosystem matches the results provided by the model, the model can be used to test various management approaches. A model is a tool that helps the manager designing an efficient management system based on the Supply Chain paradigm (see [6], [9] and [12]). For instance, Pitcher *et al.,* whose "Back-to-the-Future" approach is developed in this book, mention that their approach "employs recent developments in whole ecosystem simulation modeling". The importance of mathematics in modeling activities is widely discussed in the literature (see for instance [3], [5], [6], [10], [17] and [20]) and simulation models are pivotal in this field (see [4], [7] and [22]).

We will not go any further into the description of models. We just want to mention that a model encapsulates the knowledge held on the phenomenon under study, but a model *is not* the marine ecosystem under study. A model is always a simplified and incomplete representation of reality. As a consequence, several models may be designed based on a given ecosystem. Another observation is the importance of the precision of the rules for the quality of the model [15 – 16]. Note also that models may be biased by the tool (language or macro language) used to build it, since the structure of a language always constrains the model.

1.2. The Supply Chain paradigm.

In modern literature, the concept of a *supply chain* suffers from a confusion of meanings and the abundance of definitions. Here we use a definition of the supply chain given in [12], which encapsulates most of the known characteristics of this management concept. A *supply chain* is a global network of organizations that cooperate to improve the flows of material and information between suppliers and customers at the lowest cost, the highest speed and the best benefits. The ultimate objective of the supply chain is customer satisfaction. In other words, a *supply chain* is the network of physical, financial and information-processing activities that involve the movement of materials, funds, and related information through the full logistics process, starting with raw materials and terminating with finished products. The nodes of the supply chain network represent all vendors, service providers, intermediaries, and customers [12, 18]. In this paper, the terms "*supply chain management*" will be considered as synonymous to "*logistics*". According to Webster [24], logistics is the branch of military science and operations dealing with the procurement, supply, and maintenance of equipment, with the movement, evacuation, and hospitalization of personnel, with the provision of facilities, services, and with related matters. We define supply chain management (SCM) as the management of available material, financial and information facilities to design, procure, fabricate, produce, store, distribute, use, maintain, recycle and dispose of resources, goods and services in a partner-aware, customer-oriented and cost-effective manner [18].

We present the main components of the SCM according to [9]:
- demand forecasting and planning;
- materials requisition;
- production planning;
- manufacturing inventory;
- material handling;
- manufacturing;
- industrial packaging;
- finished goods inventory;
- distribution planning;
- order processing;
- transportation service;
- customer service.

Graphically, the SCM considers each agent in the system as a "node", with each "chain link" describing an interconnection transforming ideas into products and services. Thus, "supply chain" is a visual representation of the technological activities of the participants of the ecosystem. In its simplest form, a supply chain forms of a linear graph ("a chain").

Typical examples of a linear supply chain are:
- *Water source – Water transporting and distributing – Watering and irrigation – Water recycling – Repeated use - Effluents disposal.*
- Or: *Water source – Water transportation and distribution – Irrigation – Agricultural production – Product transportation and distribution - Storing - Marketing – Consumption – Maintenance- Disposal.*

More complicated supply chains may take the shape of a tree or a general graph. Several supply chains put together in a parallel or sequential manner can be graphically viewed as a general graph (network). This graphic representation is close to the idea of problem trees in target planning and strategic project management [12]. From the point of view of the graph-theoretic analysis, the three problems in the project's management: (i) the SCM time-compression problem, (ii) the target programming problem, and (iii) the problem of minimizing the project completion time, may be presented as equivalent mathematical programming problems. A simple SCM model, when formulated as the max-flow network programming problem, is dual to the PERT/CPM project management problem. The SCM is a new formal, computer-aided tool for coordinating the players's contradictitory economic, social, technological and environmental objectives and creating a mechanism for the fair distribution of resources among them. It is a new and effective tool for the strategic planning and management of ecosystems. The SCM methodology allows breaking down or smoothing out the barriers between the agents and processes in the cooperative ecological network.

The idea of introducing a common network that sends the customers' information simultaneously to all the partners according to their needs, and at the same time takes the main material and financial flows in the system into account, is a major characteristic of the supply chain management paradigm [12]. The use of the term "network" suggests that the companies involved in processes within the exploited ecosystem may not only perform complementary activities but also compete to perform the same activities. The definition also states that this network of organizations is considered globally, and that the partners cooperate. To obtain such an integrated system that is fair for each one of its participants, an internal policy that specifies the relationships between the participants should be defined and implemented. The goal of this internal policy is to make sure that workloads, benefits and losses are fairly shared among the participants. We call this internal policy a "sharing process". This sharing process based on ecological risk assessments is considered in the next section.

2. Environmental Integrity.

Today, natural resources, including water, air, soil, flora and fauna, are significantly affected by disastrous pollution from industrial, agricultural, municipal, and other anthropogenic sources. The demand for new sophisticated methods for ensuring sustainable development of exploited ecosystems is increasing worldwide. In particular, ecological problems caused by man's technological interference arise in the seas that surround the lands of the Middle East; they are especially acute in the Red Sea basin. The Red Sea is being dramatically polluted by municipal, maricultural, industrial, and agricultural sources, port and ballast waters, with severe negative consequences on the ecology and wildlife in the area, especially the coral reef. Without careful global management and coordination of international anti-pollution efforts the seas of the Middle East are threatened with catastrophic changes; under an unprecedented high rate of natural resource degradation due to pollution and overexploitation, the whole marine ecosystems may lose its integrity and collapse.

Supply Chain Management (SCM) opens up fresh opportunities for coordinating technological, economic and ecological contradictory demands and creating a mechanism for fair distribution of costs and benefits among the participants of the ecosystems. The CSM is the management of available resources to design, procure, fabricate, produce, store, distribute, use, maintain, recycle and dispose of goods and services in a partner-aware and cost effective manner. The SCM, when applied to coastal zone management, is an approach that focuses on integration and partnership in order to meet participants' needs on a timely basis, with high-quality marine aquacultural products being produced and ecological impacts being taken into account. An exploited ecosystem can be looked at as a system comprising several "supply chains" placed together and functioning in a parallel fashion. In the system, each chain comprises a number of "chain links", that is a number of interconnected components required to transform ideas into delivered products and services. Different activities comprising the supply chains can be graphically represented as a parallel-sequential network. The SCM directs the participants' various technological activities in the exploited ecosystems towards environment quality improvement, ecosystem integrity preservation and efficient utilization of environmental resources. The attractive, though ambiguous, concepts of ecosystem quality and ecosystem integrity hold great appeal for *ecologists, economists and managers, although they have different meanings for different people.*

Although it may not always be realistic, we accept the following descriptive definition of the ecosystem suggested by Odum [21]: "Any entity or natural unit that includes living and nonliving parts interacting to produce a stable system in which the exchange of materials between the living and nonliving parts follows circular paths is an ecological system or ecosystem. The ecosystem is the largest functional unit in ecology, since it includes both organisms (biotic communities) and the abiotic environment, each influencing the properties of the other and both necessary for maintenance of life as we have it on the earth.".

Within any corporate system, and particularly in large exploited coastal ecosystems consisting of industrial, agricultural, maricultural, municipal, shipping and other participating agents, there are organizational, social, psychological and other barriers between the agents [19]. They may be caused by contradicting criteria and demands preventing sustainable development and environmental integrity in a region. In recent years, it has been well recognized in the management science literature that many success stories of the SCM can be explained by breaking down or smoothing out the barriers in corporate systems - this is a general view pursued by the SCM methodology, and this view is being developed in the present paper.

Strategic planning of natural resources in general, and planning of water resources in particular, require an integrated, nation-wide approach. Integrated planning and management focuses not only on the performance of separate components of the ecosystem, but also on the performance of the entire ecosystem. Integration may be considered in at least three dimensions:

- Systematic analysis and balance of various categories of water: surface and groundwater, potable and technical water, etc.
- Introduction of integrated water quality and risk characteristics.
- Coordination of the actions and objectives favored by different players and agencies so as to achieve the best total result for the entire community, from the viewpoint of social, economic and environmental development.

Following the definition suggested by the USA EPA [23] we adopt the following, perhaps somewhat utopian, vision of the objectives of integrated water resource management:

- To balance competing uses of water and to efficiently allocate water resources through thorough coordination of social values and environmental costs and benefits.
- To coordinate and resolve conflicts by including all units of government, agencies and water stakeholders in the decision-making process.
- To promote water conservation, reuse, source protection, and enhance good water quality.
- To foster public health and safety.

We will extend this definition with one more point:

- To mitigate the environmental risks related to water pollution.

3. Case Study: Risk-Based Management of Coral Reef Deterioration in the Gulf of Aqaba-Eilat

The Gulf of Aqaba-Eilat is the northernmost coral reef ecosystem in the western Indo-Pacific region, a national wealth of four bordering countries – Israel, Egypt, Jordan and Saudi Arabia. The Gulf is 180 Km long, 16 Km wide on average and 900 meters deep on average. Israel's border spreads along a 14 Km coastal strip. The Gulf's water supports hundreds of species of corals, 1270 species of fish, and 1100 species of mollusks [1, 11].

It is well recognized that the fragile coral ecosystem of the Gulf of Eilat has been endangered in recent years:
- Lost diversity of corals (50%),
- Decrease in coral cover (50%),
- Low rates of coral-larval settlement and recruitment,
- Decreased rates of reef calcification,
- Coral mortality,
- Increase of macro-algal blooms during the spring.

Many reports claim that today about 70% of the corals in Eilat are dead and only 30% are alive, whereas in 1996 the proportions were reversed: about 70% were alive and 30% were dead ("Protecting the Gulf of Eilat/Aqaba". Israel Environment Bulletin, Summer 2002, v.25, no.3). The problem is to reduce flows of potential pollutants to Eilat's waters.

The major sources of risk (= starting points in supply chains) [1,2]

Main sources of risk	Size of pollution
Eilat port and phosphate terminals	One ton of dust per year
Eilat-Askelon oil pipeline	
Eilat municipal sewage	About 6 tons nitrogen annually
Eilat marina	About 5 tons nitrogen annually
Mariculture (fish cages)	260 T of N + 40 T of P per year
Oil spills from large and small vessels	
Industrial plants and hotels	
Ballast waters	
Agriculture	
Tourism and diving	
Siltation and sedimentation	
Pollution from port of Aqaba	
Groundwater inputs	

According to the report from the International Expert Team [1], the following activities and measures may be effective to decrease the pollution of the Gulf:
- Preventive equipment (pollution prevention station, oil-combat vessel, pumps, sorbents, etc.).
- 40-Km long Eilat sewage pipeline.

- The loading chute at Eilat port.
- Improving feeding technologies at fish farms.
- Decreasing the amount of feed from 4,150 T to 3,600 T per year.
- Decreasing protein contents from 45% to 40%.
- Reducing the maximum size of the fish grown.
- Shifting gradually to land fish pools.
- Construction of the artificial reef (absorbing C and N).
- Building within the limits of 200 M from the water line.
- On-shore waste-pumping at the marina.
- Controlling offshore and open-ocean ballast water exchange.
- Monitoring and economic assessments of coral reef deterioration at IOLR.

There are many formal and informal definitions of risk. Speaking informally, the ecological risk is a (quantitative) measurement of ecological hazards with their economic, ecological, social and related consequences being taken into account. Following the U.S. EPA definition of ecological risk assessment [23] we define environmental risk assessment as a quantitative appraisal of the actual or potential impact on humans, animals, plants and technological infrastructures of contaminants from a hazard. Another definition is given by the U.S. EPA: ecological risk assessment is the process that evaluates the likelihood that adverse ecological effects may occur or are occurring as a result of exposure to one or more stressors [23]

Most of the formal ones define risk R as the product of the likelihood P of a hazard (probability, frequency, expert estimation, etc.) and amount of damage Q (in a monetary or material form):

$$R = PQ$$

In a multi-dimensional case, numerous factors (n) related to risk are summed-up with weights (priorities) assigned to each risk factor:

$$R = wPQ = \Sigma_{i=1,...,n} w_i P_i Q_i$$

The US EPA [23] distinguishes four main types of ecological risks:
- Environmental risks;
- Health risks;
- Risks to public welfare/natural resources;
- Safety risks.

Safety risk caused by technological disasters:

$$R = P \bullet Q = \Sigma_{i=1,...,L} p_i(Q_i) Q_i,$$

where:
L: the number of catastrophic technological accidents;
$p_i = p_i(Q_i)$: the frequency (likelihood) of the i-th accident;
Q_i : the number of deaths caused by the i-th accident.

In the Netherlands, the admissible p_i is limited by the law:

$p_i(10) \leq 10^{-4}$ (during a year); $p_i(20) \leq 2.5 \bullet 10^{-5}$ (during a year) [13].

Health risk caused by toxicants in water, air and food:

$$R = P \bullet Q = \Sigma_{i=1,...,I} \Sigma_{j=1,...,J} p_{ij}(D_{ij}) Q_{ij},$$

Q_{ij}: number of people suffering from dose i of toxicant j, or the loss of life ex-pectancy, *LLE* [8]

$p_{ij} = p_{ij}(D_{ij})$: probability (= response factor) of occurring a heavy illness, cancer, or other heavy consequences caused by dose i of toxicant j;

D_j: dose of toxicant j;

I: the total number of dose levels;

J: the number of toxicants.

In the case of linear relation between the dose and the response factor, we have:

$$p_{ij} = p_{ij}(D_{ij}) = F_{ij} \bullet D_{ij} = F_{ij} \bullet c_j v_j t_j,$$

where F_{ij} is the risk factor (the "weight") of dose i of toxicant j;

c_j : concentration of toxicant j;

v_j: daily intake of toxicant j,(mg/day);

t_j : exposure duration for toxicant j.

Consider n scenarios of possible actions and measures to mitigate/avert risks.

Notation

W_i: costs to implement scenario i;

C_i and D_i: the capital and current expenses;

t: life cycle time of the environment protection project;

$$W_i = (1/t) \Sigma_{j=1,...,t} (C_j + t D_j)(1/(1+r_i))^j;$$

V_i: expected benefits (profit) from scenario i;

$E_i = V_i - W_i$: the netto profit (social-economical effect);

R_{ij}: the residual risk at year j for scenario i;

Y_i: the residual average social damage for scenario i;

α_j: the risk price, the coefficient defined by the relation

$Y = \alpha R$, where the social-economic damage incurred by risk R.

$$Y_i = (1/t) \Sigma_{j=1,...,t} \alpha_j R_{ij}(1/(1+r_i))^j;$$

Y_o: the initial social damage for scenario i; $E_i = V_i - W_i$

$\Delta Y_i = Y_o - Y_i$: the averted damage resulted by ith scenario;

$$\text{Benefit } V_i = \Delta Y_i;$$

Pure social-economic effect $E_i = V_i - W_i = Y_o - Y_i - W_i$.

$(Y_i + W_i)$ is called the total ecological expenses.

Conclusion I. (*Risk Optimization Principle I*): The problem of maximizing the pure social-economic effect E_i is equivalent to the problem of minimizing the total ecological expenses, $Y_i + W_i$.

Simple risk-management computer-aided model
- Input: C_{ij} is the annual expenses needed to introduce the preventive measure i (i=1,...,12) for decreasing the risk source j (j=1,..., 10), in its full extent;
- r_{ij}: the expected annual decrease of the risk j (j=1,..., 10), measured in probabilistic or monetary terms, when using the preventive measure i (i=1,...,12) in its full extent (i.e., when $x_{ij} = 1$).
- x_{ij}: 120 variables determining the rate (level) of using the preventive measure i (i=1,...,12) for decreasing the risk source j (j=1,..., 10).
- $\underline{R}_j$: the admissible level of risk of type j (j = 1,..., 10).
- R_j: the current level of risk of type j (j = 1,..., 10).

Mathematical programming model
Minimize $\Sigma_{ij} C_{ij} x_{ij}$
subject to

$\Sigma_{ij} r_{ij} x_{ij} \geq R_j - \underline{R}_j$, j=1,..., 10.

$0 \leq x_{ij} \leq 1$, i=1,..., 12; j=1,...,10

A Simple (One-dimensional) Linear Program

Minimize $\Sigma_i C_i x_i$
subject to $\Sigma_i r_i x_i \geq R_o - \underline{R}$,
$0 \leq x_i \leq 1$, i=1,..., n

Reciprocal mathematical programming model

Minimize $\Sigma_i r_i x_i$
subject to $\Sigma_i C_i x_i \leq C_o$,
$0 \leq x_j \leq 1$, i=1,..., n

The Reciprocity Theorem (Risk Optimality Principle II)

The problem of minimizing total ecological expenses is equivalent to the problem of minimizing the risk value (in a material form) subject to the expense constraint [18].

4. Conclusion

Human being never makes a decision starting from the system he / she is in charge of, but from the model he / she implicitly or explicitly derives from the system. The model may be altered by the preconceived idea of the model maker or by the tools used to build the model. It is why a strict approach is necessary to reach an adequate model. Due to the complexity of marine ecosystems, global mathematical models are not always encouraging. Mathematical models remain local, and global models are, at the best, simulation models build by adequately linking local mathematical models. The Supply Chain paradigm that advocates for a global approach of system management has become a powerful support for global management of marine ecosystems, assuming that a strict modelling approach is followed when building the model: it is what we tried to point out in this publication. We also insisted on the urgent need of efficient management in several domains such as environmental risks, health risks, risks to public welfare/natural resources, and safety risks.

Acknowledgements

We are very grateful to Dr. M. Steiner for the most fruitful discussion and critical review of the manuscript. This work was supported by the NATO Collaborative Research Grant and by the US NAS/NRC fellowship.

References.

1. M.J. Atkinson, Y. Birk, H. Rosenthal, Evaluation of Pollution in the Gulf of Eilat, *Report for the Ministries of Infrastructure, Environment and Agriculture of Israel*, December 2001.
2. D. Angel et al., *Review on the environmental influence of cultivating fish in the cages of Suf Fish in the Gulf of Eilat*, IOLP, 117 pp., 1990.
3. E. Beltrami, *Mathematical Models in the Social and Biological Sciences*. Jones & Bartlett, Boston, 197 p., 1993.
4. H. Bossel, *Modeling and Simulation*, Wellesley, MA: A. K. Peters, 484 p., 1994.
5. D. Brown and P. Rothery, *Models in Biology: mathematics, statistics and computing*. Wiley, 708 p., ISBN 0-471-93322-8, 1993.
6. C. W. Clark, *Mathematical Bioeconomics: The Optimal Management of Renewable Resources*, 2nd Ed., Wiley, 400p., ISBN 0-471-50883-7, 1990.
7. C. W. Clark and M. Mangel, *Dynamic State Variable Models in Ecology*, 2000.
8. B.L. Cohen, Catalog of risks extended and updated, *Health Physics* 1991, 61, 89-96
9. J. Coyle, E. Bardi, C. Langley, *The Management of Business Logistics*, West Publishing, Minneapolis, 1996.
10. L. Edelstein-Keshet, *Mathematical Models in Biology*, McGraw-Hill, 608 p., ISBN 0-07-554950-6, 1988.
11. S. Gabbay (ed.), *The Environment in Israel*, Ministry of the Environment, Jerusalem, 2002.
12. M. Govil, J.-M. Proth, *Supply Chain Design and Management: Strategic and Tactical Perspectives*, Academic Press, N.Y., ISBN: 0-12-294151-9, 2002.
13. W. E. Grant, E. K. Pedersen, and S. L. Marin, *Ecology and Natural Resource Management: Systems Analysis and Simulation*, Wiley, 373p., ISBN 0-471-13786-3, 1997.
14. B. Hannon and M. Ruth, *Dynamic Modeling*, Springer, 250p., 1994, 2nd ed: 2001.
15. B. Hannon and M. Ruth, *Modelling Dynamic Biological Systems*. Springer, 400p., 1997, 2nd printing 1999.
16. S. E. Jørgensen and G. Bendoricchio, *Fundamentals of Ecological Modelling*, Elsevier Science Publishers, 544p., 3nd edition, 2001.
17. S. A. Levin, T. G. Hallam, and L. J. Gross, editors, *Applied Mathematical Ecology*, volume 18 of *Biomathematics*, Springer, ISBN 3-540-19465-7, 1989.
18. E. Levner and J.-M. Proth, *Strategic Management of Marine Ecosystems*, A key lecture at the Computer-Aided Management of Ecosystems CAMES Workshop, Holon, Israel, April 2003.
19. E. Levner, D. Zuckerman and G. Meirovich, Total quality management of a production-maintenance system: A network approach, *IJPE*, pp. 407-421, 1998.

20. P.A. Leffelaar *ed., On Systems Analysis and Simulation of Ecological Processes: With Examples in Csmp, Fst, and Fortran: Product Features*, Kluwer Academic Publisher, ISBN: 0792355253, 318 pages, 1999.
21. E. P. Odum, *Fundamentals of Ecology,* W.B. Saunders Co, Philadelphia 1959.
22. S. Tuljapurkar and H. Caswell, *Structured-population models in marine, terrestrial and freshwater systems*, Kluwer Academic Publishers, 656p., ISBN 0-412-07271-8, 1997.
23. *U.S. EPA Ecological Risk Assessment Guidance for Superfund,* USA Environmental Protection Agency, PA, 1991.
24. *Webster's Encyclopedic Unabridged Dictionary of the English Language*, Gramercy Books, New York, 1999.

MODELLING THE ENVIRONMENTAL IMPACTS OF MARINE AQUACULTURE

William SILVERT
INIAP – IPIMAR, Avenida de Brasília s/n, 1449-006 Lisboa, Portugal

Abstract

In order to understand, manage, and regulate the environmental impacts of marine aquaculture (or any other form of ecosystem utilisation) we need to be able to predict its effects. Prediction is based on modelling, and reliable models of how activities like aquaculture affect the marine environment are essential if such activities are to develop in a way that is environmentally friendly and compatible with other uses of marine ecosystems. Modelling must be supplemented with sophisticated and comprehensive data management, and the results of model runs must be presented to stakeholders in a clear and transparent format.

While this paper focuses on modelling the environmental impacts of marine aquaculture, most of the issues are relevant to many aspects of management of marine ecosystems, and the material is presented with this generality in mind.

1. Introduction

The modelling of aquaculture impacts has much in common with the modelling of other activities that affect the marine environment, and indeed with ecological modelling in general, so the development of such models should be undertaken in the general context of modelling and should not be considered a narrow speciality. For this reason it is desirable to take a broad view of modelling and to treat aquaculture impact modelling as a special case. Several recent publications dealing in greater depths with technical aspects of modelling marine aquaculture impacts provide many of the details that cannot be covered here [19, 20].

WHAT IS MODELLING?

Modelling is a universal activity carried out by all scientists, all human beings, and all organisms with any capacity for learning. A model is any representation of a system and any kind of model can be used to tell us something about the properties of the system – of course the model may be misleading, but it is still a model [6, 16].

Models are therefore not only mathematical abstractions or computer programs, but also include physical models as well as non-quantitative concepts (often referred to as "conceptual models"). The use of the term "model" to represent mannequins, or fashion models, is not a misnomer – when a woman looks at a fashion model wearing a gown she tries to imagine what she would look like in the same gown. Of course this kind of modelling is not very reliable – a gown that looks good on a fashion model may look ridiculous on a real person – but it is still modelling.

Occasionally one hears a scientist claim that he (or she) doesn't use models but only works with real data. This is impossible. One cannot design and carry out any experiment without the use of models. Consider for example one of the basic experiments in marine ecology, the measurement of zooplankton displacement volume. Typically one hauls a net through the water, measures the amount of zooplankton in the net (Z), and divides by the volume of water filtered, which is the area of the mouth of the net (A) multiplied by the length of the tow (L).

Clearly this model for the zooplankton concentration $C = Z / AL$ is a mathematical model, even though a very simple one. The mathematical complexity of a model is really not important, but the validity of the model is. We need to look at this model more critically, and in particular identify and understand the underlying assumptions.

ASSUMPTIONS

Every model is based on a set of assumptions, and the validity of these assumptions are absolutely critical in determining whether the model gives reliable results. This model of zooplankton concentration is based on many different assumptions, and the degree to which these assumptions are violated limits the validity of the model on which the experiment is based, and consequently the correctness of the data. Some of the questions we can raise about these assumptions are:

1. Can we neglect the bow wave? When we haul the net through the water it pushes some water ahead of it, and the actual volume of water filtered is therefore less than the geometrical volume AL. Experimentalists are of course aware of this and try to minimize the effect of the bow wave by pulling the net slowly.

2. Can zooplankton avoid the net? They can sense the net and will try to avoid it, with varying degrees of success. This will bias the sampling, and will probably lead to an underestimation of Z. The slower we haul the net (to minimise the bow wave), the easier it is for zooplankters to avoid it, so hauling the net slowly enough to minimise the bow wave may lessen the capture rate.

3. Do all the zooplankton that enter the net stay in it? Small zooplankton that are of a size comparable with the mesh of the net may be extruded through it, as are gelatinous zooplankton which are fragile and easily broken into small fragments. The faster the net is hauled, the greater the possibility of extrusion.

4. Can we identify all the zooplankton that get caught? Live carnivores in the cod-end of a net (the collector at the bottom) may keep feeding and consume some of the other zooplankters, which will not be identified and measured, especially if they are partially excreted before sampling.

These are just some of the issues that can be raised about this simple experiment, and similar issues arise for any type of experiment. Denying that data are based on models is not only a distortion of reality, it is a dangerous practice that leads scientists to ignore the assumptions underlying their experiments and thus can produce seriously biased "data".

Frequently one hears the counter-argument that one doesn't need models to analyse data, one can simply use statistics. Statistics is based on models, but these are so standard that many users are unaware of them. Most of these statistical models are based on the assumption that the errors in the data are normally distributed, which is very often not true, and application of such models, or of linear models to nonlinear data, often leads to nonsensical, or at least incorrect, results [16].

OBJECTIVES

Although it is common for projects to specify the development of an ecological model as an objective, modelling is more a means of achieving objectives than an objective in itself. Models are used to answer questions – in modelling the environmental impacts of aquaculture we use models to tell us what probable consequences of fish farming are expected, or what the risks are. It is essential to identify the kinds of questions that area likely to be posed in designing the model, since otherwise there is a risk that the model may not be able to answer those questions.

It must be kept in mind that models are simplified representations of the original system, and in the process of simplification some information is discarded. If this information contains the answers to relevant questions, then the model will not be able to answer those questions. It is therefore essential to identify at least some of the ways in which the model will be used and the kinds of questions it will be expected to answer before actual modelling begins.

It is tempting to build a model and then run it to see what the results are. A far better strategy is to decide what kind of information one wants the model to provide and to design the model with this in mind.

ISSUES OF SCALE AND RESOLUTION

One of the ways in which models are simplified is by designing them with specific scales of time and space, which reflect both the basic dynamics of the system and the types of questions which the models are designed to answer. One of the greatest challenges in modelling marine ecosystems is determining the appropriate scale for the model, since the marine environment is characterised by a multitude of processes which operate on dramatically different space and time scales. Bacterial growth occurs in minutes or hours and can be affected by the presence of bubbles and particulates only a few millimetres in diameter. Phytoplankton blooms occur in patches that may be only a few meters across and come and go in a matter of days to weeks. As we move up the food chain and size scale we find organisms that migrate tens of thousands of kilometres every year and have lifetimes measured in decades. Models cannot capture this enormous range of scales, and attempts to do so require so many parameters and such detailed information on all of the processes that link different scales that they are unworkable, to say nothing of the practical problems of programming models with millions of grid cells and time steps.

These problems are less severe when dealing with aquaculture, but still cannot be ignored. The spatial scales are determined by the dimensions of the fish farms and of the inlets where they are located, and thus run from several meters to several kilometres in most cases. Time scales run from days or weeks to years, although for some impacts, such as oxygen minima in tidal systems, occur over intervals measured in minutes [7, 9].

2. Types of Models

There are many different types and subtypes of models, three of which will be discussed here – steady-state models, analytic mathematical models, and numerical models.

STEADY-STATE MODELS

There is an important difference that needs to be stressed between steady-state systems and those at equilibrium. Equilibrium refers to a state at which all of the forces acting on a system are in balance and there is no net flux, while steady state is a condition in which the forces are constant and generate constant fluxes. While equilibrium is a special case of the steady state, systems in a steady state can be far from equilibrium.

A bowl of water provides a good physical illustration of these concepts. When the water is at rest in the bowl it is at equilibrium, but if we stir the water at a constant speed there is a net force (actually a torque) which sets up a constant circular current in the bowl – this is a steady state.

The closest an ecosystem comes to equilibrium is when all of the organisms in it are dead, and clearly a functioning ecosystem with trophic flows and reproduction is far from equilibrium, no matter how constant the populations may be. Although there are many useful theorems in thermodynamics that apply to systems close to equilibrium, such as the Onsager relations which explain phenomena like the piezoelectric effect, efforts to apply these to ecosystems have not been very productive and are probably misguided for this reason.

Steady-state is almost always an approximation which involves averaging over a finite time interval. For example, a population that remains "constant" in an unchanging environment appears to be in a steady state, but in fact is constantly changing due to birth and death processes as well as individual growth. Often the averaging time is one year, and steady-state models are commonly expressed in terms of annual means. Such models are often of great value – for example, if we want to calculate the role of phytoplankton in the global carbon dioxide cycle we are more likely to want to know the effect over a period of a year than in the seasonal or daily fluctuations – but the limitations of these models must be recognised.

ANALYTIC MODELS

Although analytic models are little used in applied ecological modelling, they have important theoretical value and merit discussion. An analytic model is one that can be solved by strictly mathematical means without the need for numerical computation. The best known example is probably the Lotka-Volterra model and its many variants, based on equations of the form

$$dx/dt = ax + bxy$$

where x and y are two populations and a and b are constants which can be either positive or negative. The term ax represents the net population change due to birth and death processes (excluding birth by predation) while bxy represents the predator-prey interaction between x and y. The solutions of this equation are perfectly cyclic when only two species are present, and the Lotka-Volterra model is often used to describe cycles in predator-prey systems. Lotka-Volterra models with more species are often used, but they have some odd properties – for example, if there is an odd number of species the model is always unstable [3].

Another extremely common analytic model is the uptake-clearance equation,

$$dC/dt = a - bC$$

where C is a concentration, a is a constant input rate, and b a constant relative loss rate. This equation describes many biological phenomena, such as the transfer of contaminants between physiological compartments and the spread of nutrients through a water body, in addition to the applications to ecological phenomena like plankton concentrations. The solutions approach a steady-state condition given by the relationship

$$dC/dt = 0, \quad a = bC, \quad C = a/b$$

which shows the power of analytical methods to provide answers in a simple and easily understood way.

Unfortunately the analytic nature of these equations depends on the constancy of the parameters a and b, and if these vary, as they usually do in a real ecosystem, or if additional terms need to be included, simple mathematical solutions may not exist. For example, if we include density-dependence in the Lotka-Volterra model by adding a term in x^2,

$$dx/dt = ax + bxy - dx^2$$

then the model no longer has simple analytic solutions and most of the beautiful mathematical elegance of the model is lost. In the uptake-clearance model we often find that the input term, a, is variable and this makes it impossible to obtain analytic solutions.

NUMERICAL MODELS

Models which cannot be solved analytically can usually be solved in other ways, most often by numerical evaluation on a computer. Differential equations of the type shown above are ideally suited for numerical solution. Hydrodynamic models are almost always computer-based, although physical scale models of complex inlets can sometimes provide more realistic results.

In practice, virtually all dynamic models of ecological phenomena (meaning models which represent the change of systems over time) are computer-based, and the following discussions will deal only with computer models.

AGGREGATION

One of the most important considerations in building a model is deciding how much to simplify it. Every model is an approximation of the system modelled, and when the original ecosystem consists of millions of unique individuals from many different species, it is essential to group some of them together. The degree to which individual organisms are lumped together defines the degree of aggregation in the model, and often reflects both ecological considerations and human objectives. For example, an ecological model of a fishery might have a variable representing the number of 2-year-old herring, and another variable representing the biomass of copepods, which is much more aggregated from a taxonomic point of view, or even the total displacement volume of all zooplankton, which covers several phyla.

Aggregation is the process of grouping different variables and this can be done in different ways. Taxonomic aggregation, such as the grouping of all copepods described above, is common but not always desirable, since related species may play very different roles in an ecosystem. Functional aggregation based on what different organisms actually do can be more useful, even though it may involve lumping together very different organisms – for example, carnivorous zooplankton include crustaceans, cnidarians, coelenterates and ctenophores from different phyla. Other forms of aggregation are less obvious, such as aggregation by size; this works remarkably well in aquatic ecosystems.

Determining the appropriate level of aggregation is difficult, as too much aggregation may produce an over-simplified model that works very well but does not answer the questions of interest. It is an essential part of the modelling process to decide what needs to be included in the model. The number of measurements that can be made on any complex system tends to be far too much to be included in a model, and deciding what needs to be included and what can safely be excluded is difficult but critically important [6]. It is often argued that because ecosystems are very complex, models of ecosystems must also be complex and therefore the simplification achieved by aggregation is not of value. An alternate viewpoint is that models can be thought of as data-processing channels and thus the complexity of models should reflect not the underlying structure of the system but rather the amount of information available, and since ecological data are usually very uncertain this suggests that simple models are most appropriate [15]. In any case, several extremely complicated ecological models have been developed over past decades, few if any of which have stood the test of time.

PREDICTABILITY

In developing a model it is essential to recognise that not everything that we want to predict can actually be predicted. This is well recognised in some fields, such as meteorology, but is not always understood in ecological applications. For example, if a couple is trying to set a date for an outdoor wedding several months in advance, they recognise that it is not possible to predict reliably whether or not it will rain on a specific target date, even though the probabilities of precipitation are probably well known.

Often there is a gap between our ability to make ecological predictions and the needs of management. Classic examples of this are the observation of Dickie [2] that the total landings of cod and haddock on the Scotian Shelf was relatively constant, and the work of Sutcliffe *et al. [22]* on relations between environmental factors and total fish landings for the Gulf of Maine which make quite good predictions, but do not provide desired information about individual fish stocks because they are too aggregated. Unfortunately it is often the case that the best predictions can only be made at a level of aggregation that is poorly suited for management.

Often the burden is placed on ecologists to refine models to answer the questions that managers and stakeholders are asking, but this is not always realistic. There are times when the end users of models have to come up with better ways of using the information that can be provided rather than insisting that modellers do the impossible. A trivial example is the development of new technologies for catching pelagic schooling fish – models can predict the catchable biomass, but cannot predict where a school will be at any particular time, so the ability of a small boat to find fish is limited. By using search methods that cover large areas, such as high-speed boats with electronic fish-finding gear, or even airplane spotters, the schools can be tracked and exploited wherever they are.

3. Model Development

The task of building models is complicated by the enormous variety of ways in which fish farms and other marine activities interact with their surroundings. There are of course structural impacts in the way that farm sites and other physical installations can interfere with navigation, recreation and other uses of marine systems. There are chemical interactions through the release of nutrients into the water column and carbon into the sediments. Farmed fish can escape and mingle with wild populations, causing genetic changes that are not widely considered desirable. Use of antibiotics and other therapeutants can have serious consequences for natural ecosystems. Both physical structures and chemicals affect the binding of the sediments and their susceptibility to transport and turbation. All of these and more factors can have far-reaching ecological consequences in addition to the scope for conflict with other coastal zone uses.

We also need to consider the complication that different types of aquaculture have different kinds of environmental effects which require different modelling approaches. In light of all these differences it might seem that every situation is unique and requires an independent modelling effort, but fortunately this is not the case. We can tackle the problem in stages and break the models into systems of submodels, or modules, which can be assembled to mimic the variety of the real world [10].

BIOLOGICAL ASPECTS

We begin by modelling the individual organisms that are being cultured. These may be finfish, shellfish, algae – they may differ in other respects too, as the culture of larvae is very different from the culture of adult fish, and many innovative approaches are being explored in the field of aquaculture. However, one point that these models have in common is that they include a nutrient budget describing the relationship between what the organisms consume and what they excrete. Consumption may relate to naturally occurring resources or to introduced food – shellfish are normally cultured in waters where they can extract plankton from the water, while finfish are almost always grown with artificially feed. On the other hand, both shellfish and finfish excrete waste products which can have an impact on both the water column and the seabed.

These models of individual organisms can be combined to represent the total nutrient budget of a farm. If all the animals being farmed are equivalent, we simply need to multiply the budget of a single organism by the total number, but we can also add together the contribution of different cohorts and even of different species. In this way we are able to construct a model of a farm by combining modules representing individual fish [10].

Once the effluents from a farm have been modelled, there are usually several different types of environmental impact to be considered, which can be treated within a single model but may more easily be analysed with a set of different models which reflect the scale and type of different effects. Benthic impacts are due mainly to carbon loading from particulate effluents like faeces and unconsumed feed and mostly occur in the "footprint" of the cage array, which is determined by looking at the sedimentation trajectories of particles at different tide stages [4]. The spatial scale of benthic impacts is usually on the order of several hundred metres, although resuspension and other effects may produce noticeable impacts up to several kilometres from the source. On the other hand, nutrification by soluble effluents or contamination caused by traces of therapeutants easily mix with the entire water column and can affect an entire inlet in very short time.

Many impacts seem simple to model but turn out to involve unanticipated processes which complicate the situation and may completely upset the modelling process. As pointed out above, we usually model benthic impacts by calculating the trajectories of falling particulates and identifying a "footprint" where they land, but if the particulates are resuspended they may end up far from where they first hit the bottom. This may however be a misleading picture, since the faeces can also be mucoid strings of uncertain density which are likely to be trapped on the structure of the pens. Much of the impacts may also be due to fouling organisms growing on the pens which fall to the bottom during episodic events like storms or pen cleaning.

Some environmental effects are so complex that very little work has gone into modelling them. The spread of disease and parasitic organisms between farms and to wild populations is one of the most serious environmental impacts of aquaculture, but calculation of the probability of transmission depends on so many factors – intensity of infection, persistence of disease organisms in the water column, virulence – that theoretical calculations are almost impossible to make with any degree of reliability.

Since shellfish consume plankton which are naturally present, the production of a farm is limited by the quantity of plankton available, and this limit is referred to as the <u>carrying capacity</u>. The plankton may grow *in situ*, or they may be advected in by tidal action. This limitation does not apply to finfish, since they receive artificial feed in quantities which are in principle unlimited. However, attempts to grow too much fish in a given area are inevitably disastrous, and the size of finfish farms is limited by the <u>holding capacity</u>, which depends on a number of limiting factors. Some of these are analogous to carrying capacity, since, for example, if the respiratory requirements of the fish exceed the amount of oxygen in the water the result can be stress or worse. Other limitations on the holding capacity can arise from the accumulation of waste products and on the ability of the environment to remove or assimilate wastes. There are also intrinsic limitations on the size of fish farms due to increased susceptibility to disease and other consequences of crowding.

Shellfish also of course are subject to holding capacity limitations since too many shellfish can certainly excrete too many wastes for the environment to assimilate – however, it seems that the holding capacity is always much larger than the carrying capacity, so this is only a theoretical limitation. This seems reasonable, since the areas where shellfish farms are introduced are presumably ones with a functioning ecosystem and the farmed shellfish simply replace natural grazers.

PHYSICAL ASPECTS

There is considerable need for physical modelling in understanding and predicting the interactions of marine aquaculture, especially for shellfish where it is necessary to model the primary production that determines the carrying capacity of a farm site. Other types of physical modelling are needed for evaluation of the impacts, and impacts on different scales require different types of models [7, 9].

Oxygen depletion is a very localised impact that usually occurs only in the immediate vicinity of a fish farm at slack tide when the current speeds fall close to zero. This is relatively easy to model, and often just a series of tidal current measurements is enough information to use in the biological modelling. Benthic impacts on the other hand require more extensive physical modelling, not only to calculate the tidal ellipse in order to estimate the "footprint", but also information about benthic currents as well as sedimentology to determine whether resuspension is a significant factor.

The flushing of dissolved nutrients poses a particular problem, since even though all that is needed for calculation of the steady-state nutrient levels is the flushing rate (or the residence time, which is its inverse), this is difficult to determine. Physical oceanographers usually focus on calculation of the currents, but integrating the currents in an inlet to determine the flushing rate is usually difficult.

There are also important physical processes which are not predictable in detail but which can have important consequences, such as upwelling events and storms, or river runoff [11]. These require a different type of physical modelling, but such events should not be ignored.

THRESHOLD VALUES

Calculating the impacts as described above is only part of the problem of modelling the environmental impacts of marine aquaculture, since we also need to assess the severity of these impacts and determine what levels are acceptable. In some cases this is relatively simple – for example, we can calculate how much dissolved oxygen levels are depressed at slack tide, and how long the condition lasts, which can be used to determine the stress on the fish and other nearby organisms.

It is harder to estimate critical levels for benthic carbon loading or nutrification, and the threshold values are likely to be site-specific. The ability of local fauna to assimilate fish farm effluents varies with the type of organisms present and their degree of adaptation to enhanced levels of nutrients and carbon. Nutrification of coastal waters may increase primary productivity if nutrients are limiting, but production may also be reduced by the release of particulate matter which increases turbidity. In either case there is likely to be an optimal level of primary production, and too much production can lead to harmful blooms which are either toxic or which are more than resident herbivores can graze. The importance of arriving at reasonable target levels cannot be ignored [7].

RECOVERY PROCESSES

The severity of environmental impacts depends not only on the immediate degree of impact but also on its persistence. For this reason it is important to understand how undesirable impacts can be mitigated and how long it takes an impacted site to recover.

Although there have been experimental observations of how the seabed recovers after a fish farm has been removed, there have been few modelling studies. Sowles *et al.* *[21]* described an exponential decline in measures of impact, although Angel *et al.* *[1]* found evidence for biphasic recovery. This is an important area for future study.

OVERVIEW

Although there have been many projects where an effort has been made to develop a comprehensive models of a specific site, there are obvious advantages to developing generic models which can be customised for different sites. Aside from the reduction in modelling and programming effort which can be achieved by using standardised model components many times, this practice also ensures a degree of uniformity which helps maintain a good standard of quality in the models, since it means that every model includes the conceptual background of many different modellers, and is thus less likely to be based on questionable assumptions or to overlook important factors. This type of approach can best be implemented with the above-mentioned modular structure, since in different environments one can include different modules to deal with factors of local importance. For example, in highly turbid macrotidal estuaries like those found in the Bay of Fundy it is not necessary to model light limitation, but in clear bodies of water like the Scottish lochs the attenuation of light by particulate effluents has a significant impact on primary production which should be incorporated in impact models.

PROGRAMMING CONSIDERATIONS

The choice of programming languages is largely a matter of taste, since virtually all general-purpose languages are adequate for computer models of aquaculture impacts. All of the languages currently in use support some sort of modular programming approach, although some approaches work better than others – for example, object-oriented languages have some advantages, since many of the components of the models like *fish* and *pens* represent objects with great similarities but important differences [8].

The manner in which the models will be used is probably more important – will the models be used only by scientists, or will they be released as desk-top management tools? Do they need to interface with a Geographical Information System? If a model is designed to be accessed and run on the Web, Java is probably a good programming choice.

Some otherwise promising computer languages have short-comings which may not be immediately visible. Valuable "packaged" routines that come with mathematical libraries are available only for some of the more widely used scientific programming languages. Graphical programming software can be very easy to use, but programming certain ecological functions becomes very difficult if they do not conform to the preconceptions of the software developers.

In programming, as in other aspects of modelling, one should always identify how the model will be used and formulate objectives before starting to write code. Choices may be constrained by the need to interface the model with a GIS or other database program.

Increasingly, special software packages are being used for ecological modelling, to the extent that many marine scientists think that this is the only way to do modelling. Unfortunately there is a trade-off between ease of use and flexibility, and often the more user-friendly and accessible to non-programmers the package is, the more limited it is and the harder it can be to represent processes that were not designed into the program. It is always important to start by clearly specifying the modelling task before deciding how to build the model – if the model needs to describe things that are not easily described in the programming language or software package, or if the software requires data that are different from what is available, it may be far better off to switch to another approach.

Above all, the relation between available data and the modelling approach must be taken into account in deciding how to implement the model. This is particularly true when using software packages which require certain specific data in a rigidly defined format and may therefore require that the user provide only these data, regardless of their quality or availability, while ignoring other existing data that may be better suited for a different kind of modelling approach. It is essential to recognise that all models are based on data, and the choice of model and its design and implementation must take into account the availability of data and their compatibility with the needs of the model for initialisation and parametrisation.

DATA MANAGEMENT

The potential applicability of modelling is vast, but the process requires that the modelling work be supplemented by extensive data collection, since running models requires that relevant variables be known and entered into the model. Two kinds of data are required for coastal zone modelling. First of all, it is essential to have geographically indexed data for each site under consideration – for modelling aquaculture impacts these include bathymetry, sediment characteristics, currents, temperature ranges and information on ecological community structure, as well as general coastal zone information such as maps of competing usages like transport and recreation. Second, each proposal for use of the coastal zone must include specific project details, ranging from the biological (biomass and production levels, age groupings) to the physical (cage types, mooring requirements). It is increasingly popular to store the geographically indexed data in a Geographic Information System (GIS) and these are widely used for all aspects of coastal zone management. A GIS is ideal for storing spatial data and in fact much useful scientific data is already archived in GIS's. It is also possible to import other types of data, such as digital bathymetric charts, into a GIS, although this is not always as easy as it should be (for that matter, transferring data from one type of GIS to another is not always a simple task). With a GIS it is possible to visualise spatial data and spot potential conflicts by the use of overlays, where, for example, one projects a chart of swimming areas or crustacean fishing sites on top of a map showing a proposed farm site to see whether they are too close, or if perhaps a sensitive site is located downstream from a source of effluents.

Within the context of geographic data one can then introduce the second type of data relevant to the specific proposal under evaluation. This includes information about the size and scope of the activity, and operational considerations. Integration of these two types of data within the model can lead to two types of model output – either the proposal can be evaluated on a pass-fail basis, or a more flexible type of evaluation which considers alternatives may be generated. For a fish farm the first of these approaches would result simply in acceptance or denial of the application. The second approach might lead to suggestions that the size of the farm be reduced, or that the farm be relocated where the water is deeper or the current speeds higher. The former approach implies a very literal type of licensing where applications are either accepted or rejected, while the second is more of a planning tool and combines regulation with positive steps to develop the coastal environment in a way that is compatible with alternate uses as well as meeting environmental standards.

4. Producing Meaningful Results

It is important that the output of models be expressed in terms that are meaningful not only to the scientists who design the models, but also to the managers and other stakeholders. The most important consideration is usually incorporation of reference points that make quantitative results understandable. It may be sufficient for a benthic ecologist to know that the loading from a farm site or sewage outlet is 2.4 g-C/m^2/d, but even if a non-scientist knows what the number means, that is not enough – it has to be characterised as low, high, excessive, or otherwise expressed in meaningful terms if it is to be understood and acted on.

One way of translating complex scientific output into something meaningful to managers and other stakeholders is through the concept of triage. Triage, which was originally developed as a means to allocate limited medical attention to injured soldiers in Napoleon's army [5], is based on a simple classification scheme in which impacts are classified as acceptable, marginal, and unacceptable. Models are of course not totally reliable, so the usual application of these to categories to environmental impacts is to define the three categories as ones requiring minimal, moderate, and extensive evaluation and monitoring.

One method that is increasingly popular for expressing environmental impact and other measures of system status is by the use of "traffic lights", namely coloured indicators similar to the red-yellow-green pattern of ordinary traffic lights. A green symbol indicates an acceptable effect, red one that is unacceptable, and a yellow light represents a zone where caution is indicated, either because the effect is close to a known critical value or because the threshold level is not known precisely – this is clearly one way of implementing the triage concept. A further refinement is to use "fuzzy traffic lights" which retain the informational clarity of the more traditional traffic light representation while permitting more finely structured indication of the level of the effect. Instead of a sharp transition from green to yellow and from yellow to red, a fuzzy traffic light permits a gradual admixture of different colours to show how acceptable or unacceptable the effect is.

It is important to anticipate how modelling will fit into a management scheme. Silvert [17, 18] and Silvert and Cromey [19] describe a "3M" framework involving Modelling, Monitoring and Mitigation, with emphasis on the requirement that modelling must be followed up with a monitoring program to ensure either that the predictions of the model are valid or that the actual operation of aquaculture sites is consistent with the data fed into the model. Whether the model is wrong or the data are incorrect (which can occur through undocumented changes in the way a site is managed), it is essential to have contingency plans for mitigation of any undesirable impacts.

THE USER INTERFACE

Models are of little value unless they are used, and designing models with the needs and capabilities of the end user in mind is an essential part of the modelling process. Good user interfaces are just as important as proper model design, and far more important than the usual preoccupations of modellers, namely accuracy and efficiency. A very simple model which can be understood and manipulated by stakeholders is far more useful than a very sophisticated one which can only be run by skilled professionals and can only be interpreted by knowledgeable scientists, no matter how reliable its output may be. This is especially true with models of fish farm impact and other aspects of coastal zone management where most of the stakeholders have little or no experience with models, but where there is often heated controversy about impacts. In situations where there is a lack of confidence between antagonists, credibility is generally the most important characteristic for a model. If the workings of a model are transparent to the users, and if it can be manipulated so that the users can test it against their own conceptions of the system, it is much more likely to be accepted than if the stakeholders are told to accept the authoritative word of experts, no matter how well qualified those experts may be.

This is reflected in the trend to present models less as mathematical abstractions which simply generate outputs in response to whatever the modellers have specified as inputs ("black boxes"), and more as expert systems with which users can interact in much the same way as they communicate with real experts – asking questions, running different scenarios, perhaps testing the limits of validity. While the comparison of expert systems to experts is inexact, since one can talk to a person, even a scientist, in more flexible language than that required to interact with a computer program, the user interface is of over-riding importance to the acceptance of the model.

This of course has its bad side, since a poor model with a well-developed interface may receive greater user acceptance than a much better but less well presented one. This is a general problem not restricted to coastal zone modelling – a beautiful car may be poorly constructed, an elegant restaurant may serve inferior food, the salesman in the fine silk suit may be selling shoddy goods. Almost all areas of human interaction involve a balance between intrinsic quality and polished presentation. This tends to be overlooked in scientific work since scientists are used to working with systems that are designed solely for functionality, as a visit to almost any laboratory will attest, and it requires a great deal of conscious effort to develop an interface that communicates well with members of the general public.

One aspect of the human interface that is commonly overlooked is that a model must not only inspire confidence so that clients will accept the validity of its outputs, but in many areas of human activity – and this is very much true of aquaculture – it will not be possible to obtain reliable input data unless the people who are supposed to provide it have confidence that the information that they provide will be used appropriately and in their best interests. If proprietary data are leaked to competitors or otherwise mishandled, no regulatory agency is likely to receive reliable data from the same source in the future.

Even a degree of diplomacy can play an important role in the development and implementation of models. If a fish farmer provides data that indicate the use of excess amounts of feed, and on the basis of this information the model produces a recommendation that the farm be closed down (excess feed generally being environmentally harmful), there is likely to be a general reaction among all fish farmers against the model, the modellers, and the agency that uses the model. If on the other hand the model output results in confidential advice to the farmer on how to run his operation more efficiently and economically, the acceptability of the model will be enhanced.

Many expert systems are classified as decision support systems (DSS) since they provide expert input to the decision-making process. This is very much the case in coastal zone management, since models are generally used to resolve conflicts or evaluate applications for the use of certain areas for mariculture or similar purposes, and thus they contribute to the formulation of decisions. In aquaculture licensing procedures it is increasingly common to use models to evaluate the viability of a site and its probable environmental impacts. Running these models in the framework of an expert system has clear benefits, since in this way evaluations can be carried out without the direct involvement of scientific specialists, so that the decision-making process can be both rapid and decentralised. This means that local offices may be able to make sophisticated decisions about site applications without having to bring in scientific experts or sending masses of local data to a central location for analysis.

The design of DSS for modelling and evaluating aquaculture impacts has been extensively discussed elsewhere [12, 13, 14] and will not be dealt with here. However, the importance of designing models that feed directly into the decision-making process cannot be overstated.

5. Summary

There are three major aspects to modelling the environmental impacts of marine aquaculture – developing the model, managing the data on which the model is based, and presenting the results in a useful form. Generally models are developed by scientists with little direct connection with the ultimate users of the models, so it is important to ensure that there is good communication between the developers and their clients.

Because of the many different environmental situations in which aquaculture can be considered, a modular approach to model-building which lets the modeller include those aspects relevant to a particular site and omit those of limited significance has emerged as the most effective modelling strategy. This approach permits a unified methodology for dealing with both shellfish and finfish aquaculture, and often modules developed for the study of aquaculture impacts can be incorporated in models of other types of coastal zone impacts, such as those associated with sewage and factory effluents.

Data management is a major concern in all aspects of coastal zone management and relies heavily on Geographical Information Systems (GIS). These play an increasingly important role in aquaculture, and often provide the best way of addressing all the data needs of management, not just those of environmental impact. For example, a GIS could be of great value in predicting the path of an epizootic or analysing vulnerability to oil spills.

The user interface is an important aspect of the modelling procedure, since without a good interface it is difficult both to incorporate the kind of data required for modelling and to present results in a manner meaningful to the stakeholder community. Until a model has been implemented and accepted by its intended clients, the work cannot be considered complete.

Models have an essential role to play in all aspects of coastal zone management, including but not limited to the planning, development and regulation of all forms of aquaculture. The development of suitable models is only part of modelling however – we need to manage the data they use effectively, and the outputs of models have to be understood and accepted by all of the people affected. This is a major challenge, and one which requires a commitment of effort and resources, but it is a commitment well worth making.

References

1. Angel, D., Krost, P., and Silvert, W. (1998) Describing benthic impacts of fish farming with fuzzy sets: theoretical background and analytical methods. *J. Appl. Ichthyology 14*, 1-8.

2. Dickie, L. M. (1965) Difficulties in interpreting trends in cod and haddock landings from the Eastern Scotian Shelf. *ICNAF Res. Bull. 2*, 80-82.

3. Goel, N. S., Maitra, S. C. and Montroll, E. W. (1971) On the Volterra and other nonlinear models of interacting populations. *Rev. Mod. Phys. 43*, 231-276.

4. Gowen, R. J., Smyth, D., and Silvert, W. (1994) Modelling the spatial distribution and loading of organic fish farm waste to the seabed, in B. T. Hargrave (ed.), *Modelling benthic impacts of organic enrichment from marine aquaculture*, Can. Tech. Rep. Fish. Aquat. Sci. 1949. xi+125 p., p. 19-30.

5. Larrey, D.-J. (1832) *Surgical Memoirs of the Campaigns in Russia, Germany, and France (trans. John C. Mercer)*. Carey and Lea, Philadelphia.

6. Silvert, W. (1981) Principles of ecosystem modeling, in A. R. Longhurst (ed.) *Analysis of Marine Ecosystems*, Academic Press, London., p. 651-676.

7. Silvert, W. (1992) Assessing environmental impacts of finfish aquaculture in marine waters. *Aquaculture 107*, 67-71.

8. Silvert, W. (1993) Object-oriented ecosystem modelling. *Ecological Modelling 68*, 91-118.

9. Silvert, W. (1994) Modelling environmental aspects of mariculture: problems of scale and communication. *Fisken og Havet 13*, 61-68.

10. Silvert, W. (1994) Simulation models of finfish farms. *J. Appl. Ichthyol. 10*, 349-352.

11. Silvert, W. (1994) Modelling benthic deposition and impacts of organic matter loading, in B. T. Hargrave (ed.), *Modelling benthic impacts of organic enrichment from marine aquaculture*, Can. Tech. Rep. Fish. Aquat. Sci. 1949. xi+125 p., p. 1-18

12. Silvert, W. (1994) Decision support systems for aquaculture licensing. *J. Appl. Ichthyol. 10,* 307-311.
13. Silvert, W. (1994) Putting management models on the manager's desktop. *J. Biol. Systems 2,* 519-527.
14. Silvert, W. (1994) A decision support system for regulating finfish aquaculture. *Ecological Modelling 75/76,* 609-615.
15. Silvert, W. (1996) Complexity. *J. Biol. Syst. 4,* 585-591.
16. Silvert, W. (2001) Modelling as a Discipline. *Int. J. General Systems 30,* 261-282.
17. Silvert, W. (2001) Impact on habitats: determining what is acceptable, in M. F. Tlusty, D. A. Bengston, H. O. Halvorson, S. D. Oktay, J. B. Pearce and R. B. Rheault (eds.), *Marine Aquaculture and the Environment: A Meeting for Stakeholders in the Northeast,* Cape Cod Press, Falmouth, Massachusetts, p. 16-40.
18. Silvert, W. (2001) Scientific Overview of the ESSA Project, in B. T. Hargrave and G. A. Phillips (eds.). *Environmental Studies for Sustainable Aquaculture (ESSA): 2001 Workshop Report.* Can Tech. Rep. Fish. Aquat. Sci. 2352: viii + 73 pp. p. 60-68.
19. Silvert, W., and Cromey, C. (2001) Modelling Impacts, in *Environmental Impacts of Aquaculture,* K. D. Black (ed.), Sheffield Academic Press, Sheffield, p. 154-181.
20. Silvert, W. and Sowles, J. W. (1996) Modelling environmental impacts of marine finfish aquaculture. *J. Appl. Ichthyology 12,* 75-81.
21. Sowles, J. W., Churchill, L., and Silvert, W. (1994) The effect of benthic carbon loading on the degradation of bottom conditions under farm sites. in B. T. Hargrave (ed.), *Modelling benthic impacts of organic enrichment from marine aquaculture,* Can. Tech. Rep. Fish. Aquat. Sci. 1949. xi+125 p., p. 31-46.
22. Sutcliffe, W. H. Jr, Drinkwater, K.F., and Muir, B.S. (1977) Correlations of fish catch and environmental factors in the Gulf of Maine. *J. Fish Res. Bd. Can. 34,* 19–30.

ADDRESSING UNCERTAINTY IN MARINE ECOSYSTEMS MODELLING

Lyne MORISSETTE
Fisheries Centre, The University of British Columbia, Vancouver, BC, CANADA

Abstract

Ecosystem modelling has become a very important way to study marine ecosystems processes. A valuable tool for model development is the use of the software package *Ecopath with Ecosim*, which enables the construction of foodwebs and their simulation over time and space according to different scenarios. An important part of the process of ecosystem modelling is to compare results from the model with those from observations, followed by an analysis of the remaining sources of error. However, few of the currently developed *Ecopath* models have gone so far as to examine the uncertainty in analyses. Thus, it would be useful to address this problem, to clearly define the type of uncertainty that may be encountered in ecosystem modelling, and the means by which it may handled. Sensitivity analyses represent one solution by which one might address uncertainty in *Ecopath with Ecosim*. This approach functions by examining the sensitive elements as revealed in model results with differing scenarios of model-building and construction. In addition, other tools can also be used to perform uncertainty analysis routines. Examples are the *Pedigree*, *Ecoranger* and *Autobalance* tools, all of which are included the *Ecopath* software package. Furthermore, it is possible to combine these approaches with other modelling techniques in order to get an even stronger analysis of uncertainty. The purpose of this paper is to examine the strengths and weaknesses of these different approaches in addressing uncertainty.

1. Introduction

When we want to address issues at the ecosystem level, one of the options is to create models that will represent these ecosystems and simulate different scenarios to see how they will react to diverse situations. Ecosystem modelling has become very popular in applied ecology, and this is even more truth for marine ecosystems, where the habitats are harder to investigate directly. Even if this approach is more and more used, only a minority of studies address the uncertainty related to the model results. For the marine environment, a very popular tool to create ecosystem models is called *Ecopath*. With this approach, many options exist to address uncertainty. However, many sources of uncertainty exist, and they are virtually infinite in ecosystems modelling. It is thus important to identify the various sources of uncertainty and to recognize how to deal with each of them. This paper will try to review the different options available for uncertainty analyses in marine ecosystem modelling, and propose some alternatives for a better investigation of uncertainty.

2. *Ecopath* ecosystem modeling and its sources of uncertainty

Ecopath is a modelling approach that creates a simple static model to describe the average interactions of the populations within an ecosystem during a certain period. The model assumes mass-balance, i.e., that the system is the same at the end as it was at the start of the period. Hence, its parameters can change. Such an approach is judicious, precise and much less complex than the others attempts to model whole ecosystems, such as MSVPA [1] for which an enormous quantity of catch-at-age data, and stomach contents analyses is required [2]. The principal advantage with Ecopath is that the input values (mainly total mortality, consumption and diet composition) are often already available for several species or groups in the ecosystem, and that they can easily be placed in an ecological model [3]. Ecopath is thus an approach allowing the construction and the fast checking of balanced ecosystem models [3]. Different to the traditional approaches, *Ecopath* models consider the ecosystem in its whole rather than at individual components. On the other hand, because the information at the ecosystem level is never complete, there is no single solution for a specific region or period of time. The main advantage of this model is that it makes it possible to insure that available data for an ecosystem will be completely used and put in an ecosystem context [4]. During the last decades, *Ecopath* models were constructed for more than 150 ecosystems, and more than 60 others are at present in construction (L. Morissette, unpublished data). Models were published for ecosystems as various as Peru upwelling system [5], coral reefs in the Philippines [6], the Gulf of Mexico [7], Antarctica [8], Lake Victoria (Kenya) [9], etc. This type of modeling was also applied to various uses (comparison of the structure of estuaries [10], estimate of the trophic levels of the species [11] or the modeling of the inundated rice fields in the Philippines [12].

Most *Ecopath* models constructed so far have been based on a single set of input parameters representing the mean of the model period, typically for a given year [13]. The way we reach a balanced solution from the input datasets consists mainly in modifying manually the parameters so as to obtain mass balance and the outcome represent one of the many possible representations of how the trophic structure of the ecosystem may have been during the period covered. Obtaining a balanced network with the *Ecopath* approach was mostly left to trial and error, either through user intervention or Monte-Carlo simulations.

In its simplest form, the master equation of Ecopath defines the mass-balance between consumption, production, and net system exports over a given time period for each functional group (i) in an ecosystem [14]:

$$B_i \left(\frac{P}{B}\right)_i EE = Y_i + \sum_j B_j \left(\frac{Q}{B}\right)_j DC_j$$

where B_i and B_j are biomasses (the latter pertaining to j, the consumers of i); P/B_i is the ratio of production to biomass, equivalent to total mortality under most circumstances [15]; EE_i is the ecotrophic efficiency which is the fraction of production ($P = B(P/B)$) that is consumed within, or caught from the system (by definition between 0 and 1); Y_i is equal to the fisheries catch (i.e., $Y = FB$); Q/B_j is the food consumption per unit of biomass of j; and DC_{ji} is the contribution of (i) to the diet of (j), and the sum is over al predators (j). Biomass accumulation and migration can also be added to the right hand side of the equation. Each model can deal with an unknown parameter (B or EE; P/B or Q/B) that can be estimated by the model if no data is available. Most often, when the datasets are relatively complete, the EE is left unknown and is then used as a verification parameter to see which compartment of the model does not meet mass balance constraints.

The problem when we build such a model is that the estimates of biomass, production (P/B), consumption (Q/B) and diet composition do not necessarily result in an ecotrophic efficiency (EE) between 0 and 1, as required by mass-balance constraints. Having a EE higher than 1 for a tropic group means that the predation and/or the catch on this group is exceeding its biological production.

Traditionally, to reach a balanced solution when building an Ecopath model, its designer generally had to modify the diet composition of the major predators of species for which we had an excess of EE. The decision process was mainly based on ecological knowledge of the modeler, but presented the risk of modifying high-quality (=reliable) estimates to balance some that were of lower-quality.

It is crucially important in ecosystem modeling to compare model results with observations and to analyze remaining errors. However, it is difficult to distinguish between errors which are related with the model structure and those which are due to the improper choice of parameter values [16]. Sources of uncertainty are virtually infinite in ecosystem modeling. However, the more we learn about species that are parts of marine ecosystems, the more we can address uncertainty in these systems.

The degree of predictability of ecosystem models is statistically uncertain by itself [17]. However, there is many other types of uncertainty that we need to take into account in ecosystem modelling. One of them can be called "predictable uncertainty" [18], which arises from the known stochastic nature of the environment (e.g., climate fluctuation that follows a historical pattern). These fluctuations in the environment, such as El Niño, can be incorporated in the model as a known fluctuation that would affect some species groups (for example, the primary production).

A more fundamental source of uncertainty (and one much more difficult to take into account) is called "structural uncertainty" [18]. Our lack of knowledge on marine ecosystems and fishery is a good example of that [19]. In Canada, for example, there is no consensus on the causes of the collapse of cod stocks. Some authors argue that it was due to intensive exploitation combined with a period of reduced productivity of cod stocks (poor condition and growth, and increased natural mortality) [20-21].

Others believe that the collapse of cod stocks could be attributed solely to overexploitation [22-23]. However, a large part of cod mortality remains unexplained in the Gulf of St. Lawrence's *Ecopath* models. Savenkoff *et al.* [24, submitted] concluded that much of the unexplained or other mortality in the NGSL ecosystem in the 1980s resulted from the under-reporting and discarding of catches. As a result, fishing mortality was substantially underestimated in the mid 1980s, just before the demise of a cod stock that historically was the second largest in the Northwest Atlantic [25, submitted].

Sources of uncertainty in fisheries models are presented in a long list by Seijo *et al.* [26] for the FAO, in an attempt to incorporate risk and uncertainty in bioeconomic modeling and to address some alternative ways of contending with it in a precautionary fishery management context. According to these authors, uncertainty can come from as many sources as abundance estimates, model structure, model parameters, future environment conditions, behavior of resource users, future management objectives, and economic, political and social future conditions.

Conceptually, it is easy to agree that ecosystem models are designed to do a more complete -and hence more valid- task than simpler models. However, in practice they may or may not perform better, because we have too many functional relationships whose true functional forms are poorly known (or not known at all) and too many ecosystem components that we cannot parameterize. On a case-by-case basis, improvements have to be demonstrated, not just asserted as occurring.

Development and evaluation of new ecosystem models have to seek for optimal parameters which remain constant in time, as it is usually assumed (but not always true), in order to fully address uncertainty.

3. Sensitivity analyses as a tool to address uncertainty

In ecosystem modeling, there are different approaches to reach a balanced scenario. As a result, it is very important to examine how sensitive are the results (or outputs) of the model to changes in the way it was constructed and balanced. Each model constructed must carry on such sensitivity analyses in order to observe if the decisions taken when trying to get a balanced solution for an ecosystem model might conceivably have affected the results. Unfortunately, in Ecopath modeling, not all authors are using this approach. Indeed, only 20% of published models explicitly present the sensitivity analysis that has been done with input parameters (L. Morissette, unpublished data).

Inspired by Alderson and colleagues [27] and their study on review methods for healthcare research, we can describe the types of decisions and assumptions that might be examined in sensitivity analyses, including:
- changing the input parameters to reach the mass-balance constraints of the model;
- compare the strength of the model with other studies of ecosystem models;
- reanalyzing the data using a reasonable range of possible values as new inputs;
- reanalyzing the data imputing a reasonable range of values for missing data
- reanalyzing the data using different statistical approaches (e.g. using a random effects model instead of a fixed effect model, or *vice versa*).

If the sensitivity analyses that are done do not significantly change the results, this strengthens the confidence that can be placed in the results of the model. If the results change in a way that might lead to different conclusions, this indicates a need for greater caution in interpreting the results and drawing conclusions

Sensibility analyses are the key to address the consequences of uncertainty, and this is particularly true for ecosystem models.

According to Silvert [18], there is a general tendency to put too much confidence in ecosystem models, *a priori*. This is why the use of sensibility analyses becomes such an important part of the modeling approach.

The main idea when we use sensitivity analysis in an ecosystem modeling approach is to test if the results are robust or if they are very sensible to small changes in the way the models are constructed, or changes in the value of input parameters, in a way that an trivial change could radically affect the results. The approach that allows testing the performance of the models is called "sensitivity analysis".

After perturbations on the input data within their range of uncertainty, the derived probability distributions are likely to be narrower than the original distributions indicating that we have gained information in the process of checking for mass balance constraints, and eliminating parameter combinations that violate thermodynamics constraints [29].

Sensibility can also be a biological phenomenon. Indeed, if we perturb an ecosystem, we cannot be entirely certain of how it will respond, but there are some common ecological principles that allow us to do general predictions on what could happen [18]. For example, if a fish population decrease in abundance, there is a high probability that the ecological niche left by this population will be soon re-occupied by other species. This happened in the Gulf of St. Lawrence, where the quasi-disappearance of Atlantic cod (*Gadus morhua*) after an intense period of fishing in the 1980s, left an empty ecological niche that was quickly filled by forage species such as capelin (*Mallotus villosus*), herring (*Clupea harengus*), sandlance (*Ammodytes dubius*) and Arctic cod (*Boreogadus saida*) [25, submitted].

4. Sensitivity analysis in *Ecopath*

A simple sensitivity routine is included in *Ecopath*, to allow users to explore the effects of uncertainty on the model results. The method is quite simple, and consists in plotting relative output changes against relative changes in the inputs. However, this does not allow testing of the "structural sensibility" discussed above.

The routine varies all basic input parameters (biomass [B], production / biomass ratio [P/B], consumption / biomass ratio [Q/B], ecotrophic efficiency [EE]) in steps from -50% to +50% for each species or trophic group of the model, and then checks what effect each of these steps has for each of the input parameters on all of the "missing" basic parameters for each group in the system [13]. The output is then given as the proportion of the difference between the estimated and original parameter to the original parameter, and converted to a percentage [13].

Unfortunately, this method only re-estimate the parameters for which no data was available, and that were left to be estimated by the model, using the mass-balance constraints. If a model if fully constructed and has no parameter but EE left to be estimated by the model, it is usually EE that is left to be estimated by the model. As a consequence, this becomes the only parameter to be taken into account in terms of impacts of the variation of input parameters on outputs values. The other outputs provided by Ecopath (diet composition, catch, mortalities, system's emergent properties, etc.) can not be analyzed by this method.

The *Ecosim* part of the software (used for temporal simulations) does not include any formal sensitivity analysis. However, Morissette and colleagues developed a methodology with the CDEENA program (see L. Morissette's section in Pitcher *et al.*, this issue) to address uncertainty through temporal simulations. The idea was to construct a model for a given period of time (in this case, the Gulf of St. Lawrence in the mid-1980s) and another one for the exact same ecosystem, but for a later period (ten years later, in the mid-1990s). The "real" 1990s model was then compared with a 1990s model obtained after a 10-year simulation of the 1980s model, including functions such as change if the fishing effort, or environmental factors (changes in water temperature, ice coverage, El Niño effect, etc.).

Moreover, it is a common misconception to believe that because we have complete databases for some ecosystem models, we can necessarily construct models that are reliable. Indeed, all the input parameters used in models can have very wide ranges of variation. Even when the data are well known, there can still be a lot of uncertainty related to the inputs, and thus, the outputs of the models.

5. *Pedigree* and *Ecoranger* routines

When we compare the different research fields in marine ecology, we quickly realize that charismatic species such as seals or whales are much more studied than invertebrates or parasites. However, when we analyze the ecosystem as a whole, the latter species can become very important in the global response of the system to its environment.

The interactions between the different species and the fishery in a marine ecosystem are complex, generally not well understood, and can become a major source of uncertainty. The interactions that are generally better understood are predator-prey relationships, primarily based on stomach contents analyses. On the other hand, information on production, consumption or mortality sources for each species of the system is usually less understood and thus represents a considerable part of the variability of input values in the marine ecosystems models that we construct.

The *pedigree* is a coded statement that quantifies the uncertainty related to each input value in Ecopath models. For each input that we use in a given model, a choice can be made to describe the kind of data used, and thus the confidence we can have in these data. The routine uses percent ranges of uncertainty based on a set of qualitative choices relative to the origin of biomass, P/B, Q/B, catch and diet input or model estimates (model estimates have a high range of uncertainty) (Table 1). When these choices are made for each single input values, an overall *pedigree* of the model is calculated as the average of the individual pedigree values. This overall *pedigree* is then very useful for comparisons with other models [29], because comparing models with a different amount of trophic compartments (and thus with different amount of input values and individual *pedigrees*) would not be rigorous. The overall *pedigree* is calculated as:

$$\tau = \sum_{i=1}^{n} \frac{\tau_{i,p}}{n}$$

where $\tau_{i,p}$ is the pedigree index value for group i and input parameter p for each of the n living groups in the ecosystem. Parameters (p) can be B, P/B, Q/B, DC or catch data [29].

Table 1. Default options for the *pedigree* routine, for each input parameter used in *Ecopath* models. Defaults (Percentage CI) are means based on actual estimates of CI in various studies. Modified from Christensen et al. [13].

Parameter	Pedigree index	Default CI (± %)
Biomass		
Sampling based, high precision	1.0	10
Sampling based, low precision	0.7	40
Approximate or indirect method	0.4	50-80
Guesstimate	0.0	80
From other model	0.0	80
Estimated by Ecopath	0.0	n.a.
P/B and Q/B ratios		
Same group/species, same system	1.0	10
Same group/species, similar system	0.8	20
Similar group/species, same system	0.7	30
Similar group/species, similar system	0.6	40
Empirical relationship	0.5	50
From other model	0.2	80
Guesstimate	0.1	90
Estimated by Ecopath	0.0	n.a.
Diet compositions		
Quantitative, detailed, diet composition study	1.0	30
Quantitative but limited diet composition study	0.7	40
Qualitative diet composition study	0.5	50
General knowledge for same group/species	0.2	80
From other model	0.0	80
General knowledge of related group/species	0.0	80
Catches		
Local study, high precision/complete	1.0	10
Local study, low precision/incomplete	0.7	30
National statistics	0.5	50
FAO statistics	0.2	80
From other model	0.0	>80
Guesstimates	0.0	>80

The confidence intervals associated to each quality parameter attributed in the pedigree table can be defined by the constructor of the model or else left to default values (Table 1). Specifying the pedigree of data used to generate Ecopath input make users aware of the danger of constructing an Ecopath model mainly from input taken from other models, but also provides defaults for the *Ecoranger* routine of *Ecopath* (see below) [13]. When the pedigree table is complete, models are then implemented with this "quality footprint" that will be unique and make comparisons between models possible, based on single parameters pedigree, or overall pedigree indices (see an example for the northern Gulf of St. Lawrence in Table 2)

Table 2. *Pedigree* of biomass (B), production (P/B), consumption (Q/B), diet and catch inputs for the fish components of the northern Gulf of St. Lawrence ecosystem model constructed by Morissette et al. [30] with *Ecopath*. The overall pedigree of this model was 0.651.

Species or group	B	P/B	Pedigree Q/B	Diet	Catch
Large cod	1.0	0.5	0.8	1.0	0.7
Small cod	1.0	0.5	0.8	1.0	0.7
L. Greenland halibut	1.0	0.5	0.8	1.0	0.5
S. Greenland halibut	1.0	0.5	0.8	1.0	-
American plaice	1.0	0.5	0.8	0.7	0.7
Flounders	1.0	0.5	0.8	0.7	0.5
Skates	1.0	0.5	0.8	0.7	0.5
Redfish	1.0	0.5	0.8	0.7	0.7
L. demersals	0.7	0.5	0.7	0.7	0.5
S. demersals	0.7	0.5	0.7	0.0	-
Capelin	0.0	0.8	0.8	0.0	-
Sand lance	0.0	0.2	0.8	0.7	-
Arctic cod	0.0	0.2	0.8	0.7	-
L. pelagics	0.7	0.5	0.7	0.7	0.5
S. pisciv. pelagics	0.7	0.5	0.7	0.7	0.5
S. plankt. pelagics	0.7	0.2	0.7	0.0	0.7

However, even when we have a parameter for which a high pedigree score is assigned, this doesn't necessarily mean that the range of uncertainty associated to this parameter is small. In the Gulf of St. Lawrence, for example, the biomass estimates come from very important sampling survey (such as DFO groundfish survey database that was used for many species of the Gulf of St. Lawrence models, see Morissette et al.[30] for more information) and still, theses have quite large confidence intervals. This is also true for diet composition studies. For example, in the Gulf of St. Lawrence, some inshore and offshore diets had to be combined, or we had to assume that the diet for the 1980s was the same as during the 1990s, or else that the diet for a key species was representative of the functional group [30-32]. When aware of such cases, users should change/overwrite the default values in Table 1.

Ecoranger is a resampling routine based on input probability distributions for B, P/B, Q/B, EE, DC and Catches that uses a Monte Carlo approach. The distribution ranges of each parameters can be entered explicitly for each input or Ecoranger can pick up the confidence intervals from the pedigree tables and use these as prior probability distribution for all input data. After perturbing the input data, the routine attempt to reach a balanced solution that solves the physiological and mass balance constraints [29]. Most of the time, a balanced solution is not found, and Ecoranger propose a "best unabalanced model" (or BUM) that can then be worked on manually. However, when such a situation happens, it does not necessarily means that there is no balanced solution available for the input ranges of possible values. After running the same inputs through other perturbation methods (see inverse approach, in a section below), many models ended up having more than one possible balanced solution, that were just not found by the Ecoranger processes (L. Morissette, unpublished data for the northern Gulf of St. Lawrence model).

The main advantage of the *Ecoranger* routine is that it will change all parameters at once within the confidence interval limits defined by the constructor of the model. However, this method is not used a lot, and many users have encounter troubles using this routine (A. Bundy, Department of Fisheries and Oceans, Bedford Institute of Oceanography; J.J. Heymans, University of British Columbia, Fisheries Centre; pers. comm.). Some authors mention that, indeed, the results of this procedure can be confusing and reverted to standard manual methods for most of their analyses [18].

6. Autobalance

An important step was done addressing uncertainty in *Ecopath* models with the *Comparative Dynamics of Exploited Ecosystems in the Northwest Atlantic* (CDEENA) program in 2004. Within this program, a total of 10 *Ecopath* models were constructed for 2 periods (the 1980s, prior to the collapse of groundfish species; and the 1990s, after that decline) and 5 different ecosystems: the grand banks of Newfoundland (NAFO zones 2J3KLNO), the eastern and western Scotian shelves (NAFO zones 4VsW and 4X), and the northern and southern Gulf of St. Lawrence (NAFO zones 4RS and 4T) (Fig. 1)

The "autobalance" routine of *Ecopath* is a new parameter optimization for ecosystem models. This tool is used mainly to obtain reproducible mass-balanced models from an unbalanced state [14], but can also be used (as it is the case here) for perturbation analyses in order to assess uncertainty associated with certain models. The CDEENA program represents the first case-study where such an approach was used in an attempt to fully address uncertainty in *Ecopath* modeling.

Autobalance is an algorithm that allows reaching mass-balance constraints of Ecopath in a quantitative way. This routine should be used jointly to the traditional manual balance process which is way more informative (and can take ecological knowledge of the modeler into account) when used alone.

Autobalance is a structured way (as opposed to the traditional manual approach) to use input data and get to a balanced solution for an ecosystem model, based on clear assumptions. The routine uses the uncertainty definitions associated to data provided in Ecopath, which are described by the *pedigree* (see section above on *pedigree*), and then apply perturbations (within these uncertainty ranges) to the data based on the degree of confidence we have in them. Perturbations on parameters with better *pedigree* (that we know more about) would be less important than the ones on parameters with lower *pedigree*, assuming that if we know less about these parameters, there is a better chance they can have a different value than the one provided in the model.

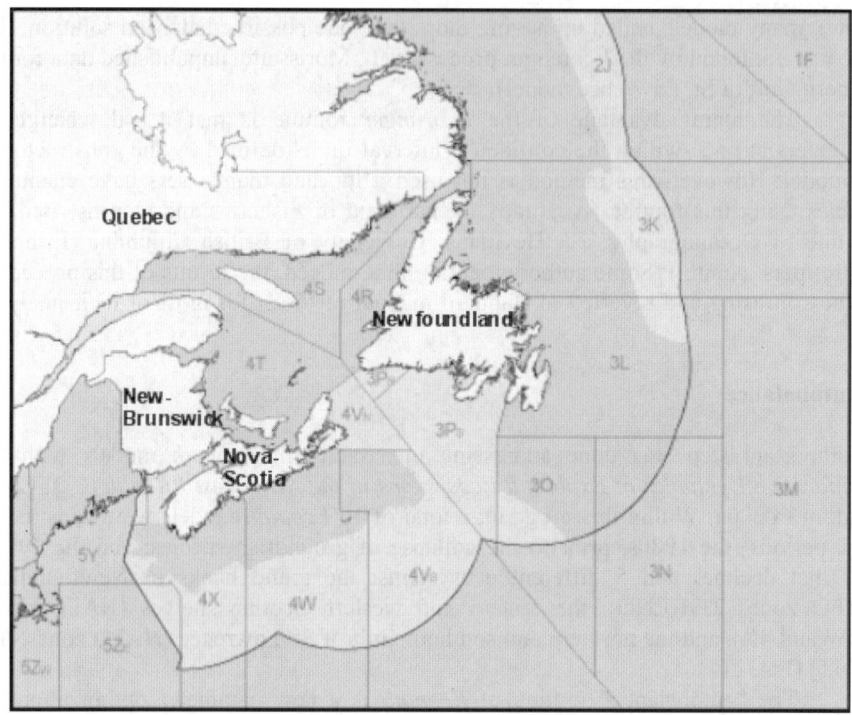

Fig. 1. The Nortwest Atlantic areas studied by the CDEENA program for ecosystem modeling

The main objective of the autobalance process is to obtain a model for which the *EE* are all below 1, but without exceeding the confidence intervals on diet compositions or biomass. The method also allows to choose the magnitude of perturbations according to the amount of reduction needed to reach EE < 1 [14]. Unfortunately, only *DC* and *B* parameters are presently modified with this routine. The authors who developed this method [14] explain that they allow only *DC* and *B* to be varied as these parameters are generally the most uncertain. However, this is not always the case. There are a lot of models where the quality of information (i.e. the *pedigree*) is way better for *B* or *DC* than it is for parameters such as *P/B*, *Q/B* or catch (L. Morissette, unpublished data). This is in fact the case for many models, such as a Caribbean coral reef [33], a lake of Sri Lanka [34], the West Florida shelf [35], or the coast of Guinea [36]. However, even when the *B* or *DC* are well-known, they are still more uncertain than Q/B and P/B, because they can take a wide range of values (which *Q/B* or *P/B* cannot).

When used as an uncertainty analysis, the autobalance routine presents some disadvantages. Firstly, *EE* < 1 is the only hard constraint to achieve mass-balance in the autobalance process. Moreover, the approach only deals with parameters affecting species that are unbalanced. This means that not all parameters are perturbed in such an uncertainty analysis and not all species as well.

For the CDEENA models, 30 alternate models were created with the autobalance by perturbing the input B and DC within their *pedigree* confidence intervals, and then re-balanced by the automated process. Ranges of possible values used for *pedigree* information are fully described in Morissette [30], Heymans [37], Bundy [38], and Savenkoff *et al.* [31-32].

The re-balance process was not checked for ecological logic, as a complete model construction and balance would be. Thus, the 30 alternate solutions represent balanced scenarios, but may be incorrect. Since the new solution is within the confidence intervals of all parameters entered, the solution can be as logical as a manually fixed model. However, the confidence intervals used for the perturbation of input parameters are not related to the range of possible values of these parameters, so some solutions can be totally erroneous ecologically.

Each run started with a different set of conditions, and the routine searched for the combination that will produce a balanced model. The Autobalance routine was programmed to run for 10,000 runs in order to reach this target [38]. The thirty solutions were used to define 95 % confidence intervals for the model estimates, giving an idea of the uncertainty associated with the model output.

The uncertainty analysis consisted in comparing the two models (for the mid-180s vs the mid-1990s), with confidence intervals generated from the 30 autobalance runs used to determine whether differences between models were real, or an artifact created by the uncertainty of the input parameters.

Even if this methods has some important weaknesses, this was at least the first time an attempt to address uncertainty in a structured way was done for Ecopath ecosystem models comparisons. Thus, incorporating the autobalance into an ecosystem comparison process is already a huge step forward to understand changes between models. Part of the weaknesses of the approach presented in this section were however addressed in the new method used by Morissette *et al.* [25] for the Gulf of St. Lawrence, part of the same *CDEENA* program.

7. A step further: combining *Ecopath* with other modeling approaches

There is a true advantage to use different approaches on the same data to ascertain the robustness of inferred differences between periods and among ecosystems, and this is what was done by Morissette and colleagues [25] for the CDEENA program on ecosystem modelling. To obtain a balanced solution as well as to test the sensibility of the models, *Ecopath* was coupled with the inverse approach for the analysis of four Gulf of St. Lawrence models (one model for the northern Gulf, one model for the southern Gulf, both for the mid-1980s and the mid-1990s).

The use of inverse modeling to find balanced food-web solutions provided a substantial improvement in objectivity and quantitative rigor compared to previous ecosystem modeling approaches using only *Ecopath*, with or without its *autobalance* routine. This method solved the flows of the different mass-balance equation by minimizing the imbalances between inputs and outputs. This inverse approach provided a global criterion for an optimal (balanced) solution [39-43].

In addition to the basic steady-state constraints of ecosystem models, additional constraints had to be added to obtain a meaningful solution. Each flow was taken to be non-negative and the flows and ratios of flows (metabolic efficiencies) were assumed to fall within certain ranges to satisfy basic metabolic requirements. Gross growth efficiency (GE) is the ratio of production to consumption and for most groups should have values between 0.1 and 0.3 [44]. Exceptions were top predators, e.g., marine mammals and seabirds, which can have lower GE, and small, fast growing fish larvae or nauplii or bacteria, which can have higher GE [44]. Following Winberg [45], 80% of the consumption was assumed to be physiologically useful for carnivorous fish groups while the non-assimilated food (20% consisting of urine and feces) was directed to the detritus. For herbivores, the proportion not assimilated could be considerably higher, e.g., up to 40% in zooplankton [44]. Assimilation efficiency (AE) was also constrained to fall between 70 and 90% for all the groups except for large and small zooplankton (between 50 and 90%) [46, submitted].

Certain flows have a minimal and maximal value imposed (export for detritus, production, consumption, diet composition, etc.). The production and consumption values that were not estimated from local field studies were used as constraints. To avoid a model that was too severely constrained (constraints on production, consumption, and growth efficiency), we constrained growth efficiency and either production or consumption depending on data availability (e.g., confidence level and local sampling). Diets with reasonable estimates of uncertainty (SD greater than 0.6%) were also specified as constraints. To facilitate comparisons with other *Ecopath* models, constraints were also added on the *EE* [46, submitted].

When the system of equations was strongly underdetermined, additional constraints (inequality relations) were added to constrain the range of possible solutions and thus to obtain a meaningful solution. Each flow had to be non-negative. The mass-balance equations and the additional constraints reduced the potential range of flux values, and trophic flows were estimated using an objective least-squares criterion for an optimal (balanced) solution (sum of flows in the system is as small as possible). The solution process thus generated the simplest flow network that satisfied both the mass conservation and constraints. The best solution was the model that produced the smallest sums of squared residuals for the compartmental mass balances. The solution minimized the imbalances between inputs and outputs. The mass balance was closed by residuals (inputs-outputs) instead of ecotrophic efficiencies as in the *Ecopath* approach [32, 47].

The equations solution process is fully described in Savenkoff *et al.* work [32, 46-47]. This generated the simplest flow network that satisfied both the mass conservation and biological constraints. These operations were done with a Matlab® program (Optimization Toolbox). The best solutions were those that produced the smallest sums of squared residuals for the compartmental mass balances (e.g., the solution thus minimizes the imbalances between inputs and outputs.).

To assess the solution's robustness to variations in the data, random perturbations were applied to both input data and right-hand sides of the mass balance equations. A total of 31 balanced solutions corresponding to 31 random perturbations (including a response without perturbation) on each model input to a maximum of its standard deviation was used for each model. The inverse approach was useful to obtain a first balanced solution by finding the solution that minimized both the sum of squared flows and the sum of squared residual consistent with the constraints. A more complete description of these balanced scenarios is given in Savenkoff *et al.* [24] for the northern Gulf of St. Lawrence and in Savenkoff *et al.* [47] for the southern Gulf of St. Lawrence models. Each of the 31 balanced inverse solutions were then transposed into *Ecopath* to obtain the on fishing mortality, predatory mortality, and other mortality, as well as the basic emergent estimates and network analysis indices of the two time periods, combined with their associated uncertainties. The confidence intervals generated by the 31 balanced solutions allowed a strong comparison of many indices between the two ecosystems (northern and southern Gulf) and/or time periods (mid-1980s or mid-1990s) [25].

Using conjointly inverse and *Ecopath* modeling approaches represent a great strategy in ecosystem modeling, each tool supplying the other with optimized solutions. The inverse model is very useful to obtain a first balanced solution and to supply *Ecopath* with first-cut diet compositions and efficiencies (metabolic and ecotrophic) using an objective least-squares criterion. This approach also generates complete perturbations on all input parameters for sensibility analyses. The *Ecopath* model is then used for its strengths to estimate biomass of each groups, generate global ecosystem indices such as network analyses or emergent properties of ecological systems. An approach combining the two modelling methods thus gains robustness and represent an important step further in the comparison of marine ecosystem models through time or space.

8. Conclusions

Uncertainty and variability are inherent in the very nature of ecosystem modeling. Therefore we need to use appropriate tools to define, represent and analyze this uncertainty. *Ecopath* is only one of the many approaches that are used worldwide for ecosystem modeling. Within this sole approach, many tools has been created and used –with more or less success– to address uncertainty. The best solution seems to be the combination with other modeling approaches, in order to use the strengths of each approaches and thus gain more robustness.

9. Acknowledgements

The author would like to thank Dr. C. Savenkoff for his important collaboration with the inverse modeling methods. LM is also very grateful to Drs. A. Bundy and J.J. Heymans for the most fruitful discussion and development of the uncertainty analysis using autobalance. Finally, special thanks are dedicated to Drs. D. Pauly and V. Christensen for their critical review of this manuscript and their precious advices on ecosystem modeling during the past three years. This work was supported by the Sea Around Us Project and by the Comparative Dynamics of Exploited Ecosystems in the Northwest Atlantic program.

10. References:

1. Sparre, P. 1991. Introduction to multispecies virtual population analysis. ICES Marine Science Symposia 193: 12-21.
2. Morissette, L. 2001. Moddisation écosystémique du nord du Golfe du Saint-Laurent. M. Sc. thesis, Université du Québec, Rimouski, QC, Canada.
3. Christensen, V. and Pauly, D. (1992) 'ECOPATH II - A system for balancing steady-state ecosystem models and calculating network characteristics." *Ecological Modelling*. 61, 169-185.
4. Christensen, V. 1991. On ECOPATH, Fishbyte and fisheries management. Fishbyte 9(2): 62-66.
5. Jarre, A., Muck, P. and Pauly, D. (1991) 'Two appr oaches for modelling fish stock interactions in the Peruvian upwelling ecosystem."ICES Marine Science Symposia. 193, 171-184.
6. Alino, P.M., McManus, L.T., McManus, J.W., Nanola, C.L. Jr., Fortes, M.D. Trono, G.C. Jr. and Jacinto, G.S. (1993) 'Initial parameter estimations of a coral reef flat ecosystem in Bolinao, Pangasinan, northwestern Philippines." In: Christensen, V. a nd Pauly, D. (eds.). Trophic Models in Aquatic Ecosystems. ICLARM Conference proceedings 26. p. 252-258.
7. Arreguin-Sanchez, F., E. Valero-Pacheco and E.A. Chavez. 1993. A trophic box model of the coastal fish communities of the southwestern Gulf of Mexico. p. 197-205 In Christensen, V. and D. Pauly (eds.). Trophic Models in Aquatic Ecosystems. ICLARM Conference Proceedings 26.
8. Schalk, P.H., T. Brey, U. Bathman, W. Arntz, D. Gerdes, G. Dieckmann, W. Ekau, R. Gradinger, J. Plotz, E. Nothig, S.B. Schnack-Schiel, V. Siegel, V.S. Smetacek and J.A. VanFraneker. 1993. Towards a conceptual model of a Weddell Sea ecosystem, Antarctica. p. 323-337 In Christensen, V. and D. Pauly (eds.). Trophic Models in Aquatic Ecosystems. ICLARM Conference Proceedings 26.
9. Moreau, J., W. Ligtvoet and M.L.D. Palomares. 1993. Trophic relationship in the fish community of Lake Victoria, Kenya, with emphasis on the impact of Nile perch (Lates niloticus). p. 144-152 In Christensen, V. and D. Pauly (eds.). Trophic Models in Aquatic Ecosystems. ICLARM Conference Proceedings 26.
10. Monaco, M.E. and R.E. Ulanowicz. 1997. Comparative ecosystem trophic structure of three U.S. Mid-Atlantic estuaries. Marine Ecology Progress Series 161: 239-254.
11. Pauly, D., A. Trites, E. Capuli and V. Christensen. 1995. Diet composition and trophic levels of marine mammals. ICES Council Meeting Papers 1995/N:13. 22p.
12. Lightfoot, C., P.A. Roger, A.G. Cagauan and C.R. Dela Cruz. 1993. Preliminary steady-state nitrogen models of a wetland ricefield ecosystem with and without fish. p. 56-64 In Christensen, V. and D. Pauly (eds.). Trophic Models in Aquatic Ecosystems. ICLARM Conference Proceedings 26.
13. Christensen, V. Walters, C.J., and Pauly, D. 2000. *Ecopath* with Ecosim: a User's Guide. October 2000 Edition, Fisheries Centre, The University of British Columbia, Vancouver, B.C. and ICLARM, Penang, Malaysia.
14. Kavanagh, P., Newlands, N., Christensen, V., and Pauly, D. 2004. Automated parameter optimization for Ecopath ecosystem models. Ecological Modelling 172: 141-149.
15. Allen, K.R. 1971. Relation Between Production and Biomass. Journal of Fisheries Research Board of Canada 28: 1573-1581.
16. Schartau, M., A. Oschlies and J. Willebrand (2001). Parameter estimates of a zero-dimensional ecosystem model applying the adjoint method. Deep Sea Research II (48), 1773-1802.
17. Hilborn, R. 1987. Living with uncertainty in resource management. North American Journal of Fisheries Management 7: 1-5.
18. Silvert, W. 2004. Managing uncertainty in ecosystem Model Dynamics and the Implications and Feasibility of Spceific Management Scenarios. EFEP Work Package 5, Final Report (Deliverable 5) 65 p.
19. Gomes, M. do Carmo. 1993. Predictions under Uncertainty: Fish Assemblages and Food Webs on the Grand Banks of Newfoundland. ISER, Memorial University, St. John's, Newfoundland. 205 p.
20. Dutil, J.-D., Castonguay, M., Gilbert, D., and Gascon, D. 1999. Growth, condition, and environmental relationships in Atlantic cod (Gadus morhua) in the northern Gulf of St. Lawrence and implications for management strategies in the Northwest Atlantic. Can. J. Fish. Aquat. Sci. 56: 1818-1831.
21. Dutil, J.-D., and Lambert, Y. 2000. Natural mortality from poor condition in Atlantic cod (Gadus morhua). Can. J. Fish. Aquat. Sci. 57: 826-836.

22. Hutchings, J.A., and Myers, R.A. 1994. What can be learned from the collapse of a renewable resource? Atlantic cod, Gadus morhua, of Newfoundland and Labrador. Can. J. Fish. Aquat. Sci. 51: 2126-2146.
23. Hutchings, J.A. 1996. Spatial and temporal variation in the density of northern cod and a review of hypotheses for the stock's collapse. Can. J. Fish. Aquat. Sci. 53: 943-962.
24. Savenkoff, C., Castonguay, M., Chabot, D., Bourdages, H., Morissette, L., and Hammill, M.O. Effects of fishing and predation in a heavily exploited ecosystem I: Comparing pre- and post-groundfish collapse periods in the northern Gulf of St. Lawrence (Canada). Can. J. Fish. Aquat. Sci. Submitted
25. Morissette, L., Savenkoff, C., Castonguay, M., Swain, D.P., Chabot, D., Bourdages, H., Hammill, M.O., and Hanson, J.M. Contrasting changes between the northern and southern Gulf of St. Lawrence ecosystems associated with the collapse of groundfish stocks. Canadian Journal of Fisheries and Aquatic Sciences, submitted.
26. Seijo, J.C., O. Defeo, and S. Salas. 1998. Fisheries bioeconomics. Theory, modelling and management. FAO Fisheries Technical Paper. No.368. Rome, FAO. 108p.
27. Alderson P, Green S, Higgins JPT, editors. Formulating the problem. Cochrane
28. Reviewers' Handbook 4.2.1 [updated December 2003]; Section 8.10 Sensitivity analyses. In: The Cochrane Library, Issue 1, 2004. Chichester, UK: John Wiley & Sons, Ltd.
29. Christensen, V., and Walters, C.J. 2004. Ecopath with Ecosim: methods, capabilities and limitations. Ecological Modelling: 109-139.
30. Morissette, L., Despatie, S.-P., Savenkoff, C., Hammill, M.O., Bourdages, H., and Chabot, D. 2003. Data gathering and input parameters to construct ecosystem models for the northern Gulf of St. Lawrence (mid-1980s). Can. Tech. Rep. Fish. Aquat. Sci. No. 2497.
31. Savenkoff, C., Bourdages, H., Castonguay, M., Morissette, L., Chabot, D., and Hammill, M.O. 2004a. Input data and parameter estimates for ecosystem models of the northern Gulf of St. Lawrence (mid-1990s). Can. Tech. Rep. Fish. Aquat. Sci. No. 2531.
32. Savenkoff, C., Bourdages, H., Swain, D.P., Despatie, S.-P., Hanson, J.M., Méhot, R., Morissette, L., and Hammill, M.O. 2004b. Input data and parameter estimates for ecosystem models of the southern Gulf of St. Lawrence (mid-1980s and mid-1990s). Can. Tech. Rep. Fish. Aquat. Sci. No. 2529.
33. Trophic interactions in Caribbean coral reefs. S. Opitz. 1996. ICLARM Tech. Rep. 43, 341 p.
34. Moreau, J. MC Villanueva, U.S. Amarasinghe and F.Schiemer. 2001. Trophic relationships and possible evolution of the production under various fisheries management strategies in a Sri Lankan reservoir. p. 201-214 in S.S. de Silva (ed) Lakes and Reservoir fisheries management in South East Asia ACIAR publ. 97.
35. Okey, T.A., G.A. Vargo, S. Mackinson, M. Vasconcellos, B. Mahmoudi, C.A. Meyer. 2004. Simulating community effects of sea floor shading by plankton blooms over the West Florida shelf. Ecological Modelling 172(2-4): 345-365.
36. Guénette, S., and Diallo, I., 2003. Modèles préimin aires de la côte guinéenne pour les années 1985 et 1998. Ed. by M. L. D. Palomares, J. M. Vakily, and D. Pauly, Fisheries Centre Research Report, UBC, in press, Vancouver, BC,
37. Heymans, J. J., 2003a. Revised models for Newfoundland (2J3KLNO) for the time periods 1985-87 and 1995-97. In: Ecosystem models of Newfoundland and Southeastern Labrador (2J3KLNO): Additional Information and Analyses for "Back to the Future". Fisheries Centre Research Reports, Vol. 11(5), in press.
38. Bundy, A. In press. Mass balance models of the eastern Scotian Shelf before and after the cod collapse and other ecosystem changes. Can. Tech. Rep. Fish. Aquat. Sci.
39. Parker, R.L. 1977. Understanding inverse theory. Annual Review of Earth and Planetary Sciences 5: 35-64.
40. Enting, I.G. 1985. A classification of some inverse problems in geochemical modelling. Tellus 37B: 216-229.
41. Vézina, A. F. and T. Platt (1988). Food web dynamics in the oceans. O. Best0estimates of flow networks using inverse methods. Marine Ecology Progress Series 42: 269-287.
42. Vézina, A. F., C. Savenkoff, S. Roy, B. Klein, R. Rivkin, J. C. Therriault and L. Legendre (2000). Export of biogenic carbon and structure and dynamics of the pelagic food web in the Gulf of St. Lawrence Part 2. Inverse analysis. Deep-Sea Research 47(3-4): 609-635.
43. Savenkoff, C., A. F. Vézina and A. Bundy (2001). Inverse analysis of the structure and dynamics of the whole Newfoundland-Labrador Shelf ecosystem. Canadian Technical Report of Fisheries and Aquatic Sciences 2354: 56.

44. Christensen, V., and Pauly, D. 1992. ECOPATH II-a software for balancing steady-state ecosystem models and calculating network characteristics. Ecol. Model. 61: 169-185.
45. Winberg, G.G. 1956. Rate of metabolism and food requirements of fish. Fish. Res. Board Can. Trans. Ser. 253.
46. Savenkoff, C., Castonguay, M. Despatie, S.-P., Chabot, D. and Morissette, L. Inverse modelling of trophic flows through an entire ecosystem: the northern Gulf of St. Lawrence in the mid-1980s. Canadian Journal of Fisheries and Aquatic Sciences, submitted.
47. Savenkoff, C., Swain, D.P., Hanson, J.M., Castonguay, M., Hammill, M.O., Bourdages, H., Morissette, L., and Chabot, D. Effects of fishing and predation in a heavily exploited ecosystem II: Comparing pre- and post-groundfish collapse periods in the southern Gulf of St. Lawrence (Canada). Can. J. Fish. Aquat. Sci. Submitted b.

ENVIRONMENTAL GAMES AND QUEUE MODELS

Charles S. TAPIERO
Ecole Supérieure des Sciences Economiques et Commerciales
tapiero@essec.fr

Abstract

This paper considers a pollution and control game which uses a queuing framework. This framework allows to account for pollution events, environmental pollution quality and the application of controls to maintain a desirable quality of the environment. A number of examples are used to highlight the approach and demonstrate both its theoretical and practical usefulness.

Key words: Environment, Control, Quality, Queuing

1. Introduction

Conventional wisdom states that a society has to maintain the quality of its environment to be sustainable. Environmental quality, however, may mean different things in various circumstances and to several actors, each responding to their specific needs. ISO, for example, defines quality as all the features and characteristics of a product or service that bear on its ability to satisfy stated or implied needs (Tapiero, 1996, Kuhre, 1998). By contrast, environmental quality can have several attributes that have various, potentially contradictory meanings for different groups, agreeing or disagreeing on the measurements applied to specify quality. There may also be objective measure-based and subjective attributes, expressing both tangible and intangible characteristics of environmental quality. For these reasons, environmental quality and its management also involve complex issues which are often very difficult to resolve efficiently. Rather, models and analyses based on environmental models can at best provide satisfying solutions. These issues are assuming an accrued importance. For example, Le Monde (October 2, 2003, page 12), in a recent article reports that for the first time French Authorities have taken a sharp view of ships unloading fuel oil at sea and have established laws extending their control to a 200 mile radius around French coasts. The ship captain guilty of such an act was fined 600,000 Euros, compared to symbolic fines in previous cases. This polluting event was detected by a helicopter pilot regularly watching out for such violations. City mayors of adjoining beaches have also taken part in the court hearings raising the environmental issues associated to oil spills, as well as cleaning costs and effects on the quality of the environment and of beaches, an essential source of the livelihood of these communities.

The purpose of this paper is to suggest a queue based process for environmental quality assessment and management (Harris and Gross, 1985, Chaudhry and Templeton, 1983). We describe some environmental situations in terms of the elements that make up a queue and thereby use the many theoretical and empirical results available in queuing theory to provide explicit theoretical results applicable to specific environmental problems. A number of examples are treated providing a pedagogical background to using queue related models in ecological and environmental problems. Finally, the problems we consider are also framed in an environmental game framework, providing an approach to environmental policy formulation in a conflict based set up consisting of polluting firms and an environmental agency which fight over environmental quality control. The modeling approach we present is consistent with the growing concern for environmental issues in Operations Research and Operations Management (for example, see Angel and Klassen, 1999, Bloemhof-Ruwaard et al., 1995 and Revelle, 2000).

2. An Environmental Queue Model and Gaming

The mathematical theory of queues has its origin in the study of telephone systems initiated by Erlang, a mathematician with the phone company in Copenhagen in 1917. Subsequently numerous applications were initiated in the 50's in many fields spanning information technologies, industrial processes, pharmacology, population studies, biology etc. The mathematical theory of queues is thus an important subfield of discrete events stochastic processes (for example, see Harris and Gross, 1985).
Queue models consist essentially of three components:
(1) An input process expressing an arrival process in the queue. For example, such an input might be a polluting event occurring in a random manner at specific instants called epochs. The pollution process might depend on a number of variables, due to multiple polluting firms and what not. Furthermore, preventive efforts made by polluting firms might be applied to reduce the probabilities that such events occur. For example, investments in pollution abatement technologies might reduce the rate at which polluting events occur. In contrast, increased economic activity and production can increase the probability of such polluting events.
(2) The service queue process expresses the amount of time that an incoming arrival remains within the queue. For example, given a polluting event, it may take a certain amount of time for the polluting event to dissipate itself. This may be a natural time or the time needed to clean the polluting event once it has been detected. It may de deterministic, stochastic and depend on the efforts applied to clean the environment.
(3) The queue's attributes and discipline. In a queue model, the discipline describes the behavior of incoming events which are "blocked", that is, joining a queue. In our case, "waiting queues" make little sense since pollution events, once they have occurred, do not "wait". This situation fortunately corresponds to a class of queuing models called infinite servers models where there is no waiting and each incoming event is "serviced".

A simple queue is represented graphically in Figure 1 below.

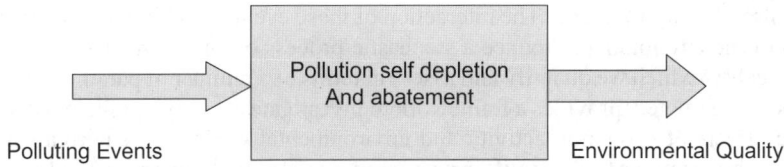

Figure 1: An Environmental Process Queue

Once a queue is defined, mathematical analysis is applied to determine its theoretical properties. These properties might include the number of events in a queue at a given time as well as the probability distribution of such events in steady state. For example, if arrivals are polluting events, and assuming that a polluting abatement technology is applied combined with controls and efforts to clean the environment, the probability distribution of effective polluting events over time and in steady state might provide a measurement of environmental quality. The time in the system of a specific event, the throughput rate at which polluting events are cleaned and their like are additional measurements that can be calculated theoretically.

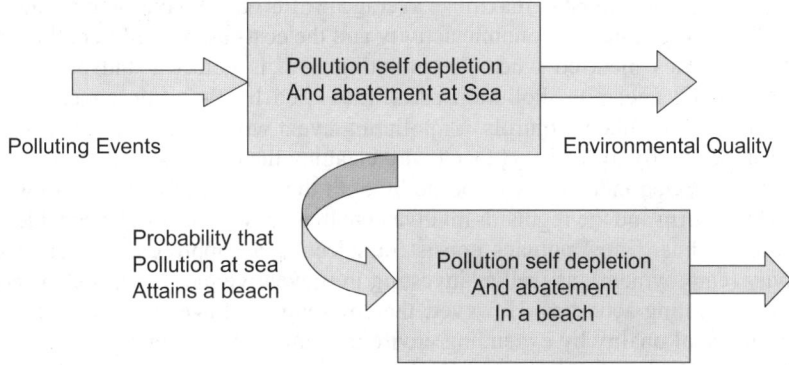

Figure 2: An Environmental Queue Network

A queuing system may also consist of interacting queues that can be defined by a network of queue. For example, let us assume that a pollution occurring at sea defines one environmental queue. This pollution can pollute a nearby beach as well, thereby creating a network of two queues where pollution movement from one queue to the other occurs with known (or estimated) probability. Figure 2 above highlights such a network.

Similarly, we can construct series of queues, interacting queues, queues with feedback etc. to represent the dependent complexity of geographical environmental pollution.

In this paper we shall assume explicitly that polluting events occur in a random manner but they may be cleaned, either naturally or through interventions by the polluting firms themselves or by the authorities that regulate and control the polluting firms and thereby control the environment. The interaction of these events combining regulation and clean up of the environment produce a stochastic process of environmental quality (or rather unquality) which we quantify and assess in terms of a number of parameters. Our model, once constructed, provides a framework to investigate a number of factors that determine the effects of economic activity and environmental control on environmental quality. Initially, a number of simplifying assumptions are made in order to obtain analytical results. Subsequently some of these assumptions are released to make the process more realistic. For complex situations, however, simulation can be used based on the queue framework which is deemed to model the environmental problem at hand.

The environmental problem we consider also involves an environmental agency-- "the regulator" --and potentially "polluting firms", each with varied motivations and thereby leading to a game played between the agency and the firms (Nash, 1950, Reyniers and Tapiero, 1995a, 1995b, Tapiero, 1995, Tapiero, 2001). For our purposes and for simplification we assume that the firm uses a pollution technology determined by the quantity of products (or employment) it produces as well as by the preventive and pollution abatement technology it applies to its industrial and production processes. The firm's motivation will be to maximize average profits once it takes into account both the payoff resulting from its economic activity and the costs associated to pollution (as well as the penalties incurred when the polluting firm is detected and penalized by the environmental agency). Pollution risks, measured by their consequences, however, depend *on* the regulator controls. A polluting event which is not detected is costless to the firm but costly to "society" faced with cleaning the environment. A polluting event which is detected induces a cost borne by both the firm and "society". Environmental costs to the firm and the regulator involve penalties as well and can be considered shared (or not). Thus, firms' policies consist in selecting an appropriate level of industrial activity (employment), as well as investing in preventive measures, such as controlling ex-post polluting activities. However, the environmental agency will seek to optimize environmental quality by expending environmental control efforts which are subject to numerous constraints (such as budget, employment requirements and the like). This results in an environmental game which is used to draw some conclusions regarding the process of investment in pollution abatement technologies and the preventive efforts and controls to be exercised by polluting firms while at the same time, determining the control effort that environmental regulators ought to apply.

In an environmental game framework, both the firm and the environmental regulator-managers must become aware of the mutual relationships and inter-dependencies of investments in pollution (production) abatement technologies, in the control effort they exercise on the processes under their control and the control regulators might exercise. For example, demanding a zero-pollution technology might lead to excessive costs and thereby to a firm's demise, resulting in a loss of jobs, tax income and the like, which are needed for a collective and population survival. Over-protecting fish or animals while the population faces starving may not be realistic. By the same token, an oil tanker with a propensity to pollute that does not take effective preventive measures ought to be penalized if a polluting event takes place. The same rationale can also be applied to oil tankers that produce pollution at sea when they believe they may not be caught (see also Reyniers and Tapiero, 1995a, 1995b and Tapiero 1995). So, the problems that both the firm and the regulator are faced with are two-fold: (1) Given a polluting technology and a shared penalty cost for a polluting event, what the control efforts should be exercised by the firm and what control and preventive efforts should be exercised by the firm and (2) What are the effects of the technology choice and penalty-cost sharing parameters on the firm and society's payoffs.

3. The Pollution Process and Environmental Quality

For simplicity and expository purposes, we shall assume first that pollution events by a firm (and generally a number of M potentially polluting firms) are known to occur as a Poisson process. This implies that: (1) Polluting events are independent of one another; (2) The probability that any one time a polluting event occurs is known and proportional to the interval of time considered; (3) Polluting events occur one at a time. This assumptions as sufficient to justify our use of the Poisson counting process, determining the probability distribution of the number of polluting events within a given time period.

Explicitly, say an individual firm, j, is engaged in an economic-industrial activity a_j which generates a return denoted by $\pi(a_j)$. This activity affects the likelihood of a pollution event occuring in a small time interval, and it is assumed to be given by $q_j(a_j(1-u_j))dt$, where u_j represents the preventive actions taken by the firm. Note that $0 \leq u_j \leq 1$ and let the cost of prevention be given by a function $C_{pr,j}(a_j u_j)$ with $C_{pr,j}(a_j) = \infty$, $C'_{pr,j}(\ ) > 0$. Further, we also assume that $q_j' > 0, q_j'' < 0$, with $q_j(0) = 0$ (i.e. a firm which environmentally controls its economic output fully will not pollute at all) and the maximum probability of pollution is evidently $q_j(a_j)dt$.

Assuming that firms' polluting events are statistically independent, the number of pollution events has also a Poisson distribution as with mean pollution rate given by:

(1) $$\lambda = \sum_{j=1}^{M} q_j(a_j(1-u_j))$$

Furthermore, due to the Poisson assumption, given that a polluting event has occurred, the probability that it is due to a specific firm j is given by:

(2) $$\theta_j = \frac{q_j(a_j(1-u_j))}{\sum_{j=1}^{M} q_j(a_j(1-u_j))}$$

Each polluting event by firm j produces a random damage expressed in terms of both time and money. "Time" is measured by the amount of time needed for polluting events to be cleaned naturally. For example, some organic pollution may be self deteriorating and thereby eventually self-dissipated. Alternatively, when a pollution event is detected by a regulator, special actions might be taken to help "Mother Nature" in negating the consequences of such an event. Let the time for a pollution by firm j to be cleaned be given by a random variable $\tilde{T}_j$ with $B(.)$ its cumulative density functions. This density function is of course a function of the firms and the agency efforts applied to clean the environment. We let (v_j, α_j) be these respective variables.. If we let $b_j(.)$ be the probability distribution of $\tilde{T}_j$ and $b_j^*(.)$ be its generating function, then since polluting firms are independent, the probability distribution that a polluting time event—any event, has a generating function given by:

$$(3) \quad b^*(s) = \prod_{j=1}^{M} b_j^*(s) \text{ with distribution function } b(.)$$

This model is therefore equivalent to an infinite servers queue process with a Poisson "arrival rate" given by equation (1) and "service time" given by the arbitrary distribution function $b(.)$. In other words, the environmental quality process can be considered as an M/G/∞ queue with parameters $(\lambda, b(.))$ which we can analyze with the standard techniques in the queuing theory (for example, see Harris and Gross, 1985). In this model, the number of active polluting events at any one time and the time needed to clean them can therefore be used as a measurement of environmental quality. In this framework, environmental quality is determined by the firms' economic activity and their preventive and control measures, and of course by the environmental agency's controls. In this context, a number of properties can be determined directly. A first proposition provides the probability that a polluting event has not been cleaned by time t. Proof of this proposition is a standard result in the queuing theory (Harris and Gross, 1985).

Proposition 1:
The probability that a polluting event in $(0,t]$, Q(t), is cleaned before or at time t is given by:

$$(4) \quad Q(t) = \frac{1}{t}\int_0^t B(z)dz, \ B(z) = \int_0^z b(x)dx = P(\tau \leq z)$$

While 1-Q(t) is the probability that a polluting event has not been cleaned by time t.

Further, due to the Poisson property, if there are exactly $N(t) > 0$ polluting events in $(0,t]$, the probability that there are exactly $0 \leq k \leq N(t)$ polluting events still un-cleaned is given by the binomial distribution :

$$(5) \quad P(K(t) = k) = \binom{N(t)}{k}(1-Q(t))^k (Q(t))^{N(t)-k}, 0 \leq k \leq N(t)$$

However, since the number of polluting events is given by the Poisson distribution, we have the following result as well:

Proposition 2:
The probability distribution of the number of still un-cleaned polluting events is:

(6) $P(K(t) = k) = e^{-\lambda\Phi(t)} \dfrac{(\lambda\Phi(t))^k}{k!}, k = 0,1,2,3,...; \Phi(t) = \int_0^t [1 - B(z)]dz$

Where λ is given by equation (1), $\lambda = \sum_{j=1}^{M} q_j (a_j(1-u_j))$,

and $0 \leq u_j \leq 1$. $B(.)$ is the time to clean up of a polluting event (any event and by any firm) density function with first two moments given by:

The probability distribution given by proposition 2 is a non-homogenous Poisson process with parameter $\lambda\Phi(t)$ which is also equal to the mean number of polluting events that have not yet been cleaned, again providing an assessment of environmental quality. At the limit, in steady state, the number of un-cleaned pollution events is therefore given by

(7) $E(S) = \int_0^\infty [1 - B(z)]dz$

and the number of polluting events is a Poisson distribution given by:

(8) $P(K = k) = e^{-\lambda E(S)} \dfrac{(\lambda E(S))^k}{k!}, k = 0,1,2,3,...$

and determined as a function of the economic activity of the firm, environmental controls and preventive measures. An example to this effect will highlight these relationships.

Example:
Say that pollution events occur at the Poisson rate and let the time a pollution event is active be exponential. We set by $P(n,t)$, the probability that at time t there are n active pollution events. In this case, it is simple to show that the pollution counting process has a Poisson distribution with mean λ/μ:

$P_n = \dfrac{e^{\lambda/\mu}(\lambda/\mu)^n}{n!}$, $E(n) = (\lambda/\mu) = \dfrac{qa(1-u)}{\mu}$ $n = 0,1,2,...$

The evolution over time can be calculated as well by noting that:

$\dfrac{dE(n(t))}{dt} = -\mu E(n(t)) + \lambda;\ n(0) = n_0$

Interestingly, the pollution counting process remains the same even if the polluting time of any pollution event has any other distribution. For example, if there are M firms each with its own polluting time (t_j), then the probability distribution of a polluting event is:

$$\tilde{t} = \begin{cases} \tilde{t}_1 & w.p \quad \theta_1 \\ \cdot \\ \cdot \\ \tilde{t}_M & w.p \quad \theta_M \end{cases}$$

Or,

$$\tilde{t} = \sum_{j=1}^{M} \tilde{t}_j \theta_j, \quad \theta_j = \frac{q_j(a_j(1-u_j))}{\left[\sum_{j=1}^{M} q_j(a_j(1-u_j))\right]}$$

Assuming that firms pollute independently, we also have:

$$E(\tilde{t}) = \sum E(\tilde{t}_j)\theta_j, \quad \mathrm{var}(\tilde{t}) = \sum \theta_j^2 \mathrm{var}(\tilde{t}_j)$$

An approximation might be the Weibull distribution given by (see also Heo, Salas and Kim, 2001, for estimation of this distribution using environmental data)::

$$b(\tilde{t}) = \frac{\delta}{\upsilon}\left(\frac{\tilde{t}}{\upsilon}\right)^{\delta-1} e^{-\left(\frac{\tilde{t}}{\upsilon}\right)^{\delta}}, \quad B(\tilde{t}) = 1 - e^{-\left(\frac{\tilde{t}}{\upsilon}\right)^{\delta}}, \quad \tilde{t} \geq 0$$

where the mean and the variance are:

$$E(\tilde{t}) = \upsilon\Gamma(1+1/\delta) = \sum E(\tilde{t}_j)\theta_j, \quad \mathrm{var}(\tilde{t}) = \upsilon^2\left[\Gamma(1+2/\delta) - \Gamma^2(1+1/\delta)\right]$$

and therefore using the first two moments fit, we can calculate the parameters (υ,δ) and calculate:

$$E(S) = \int_0^\infty [1 - B(z)]dz = \int_0^\infty e^{-\left(\frac{z}{\upsilon}\right)^{\delta}} dz$$

The pollution counting process is again Poisson with mean $\lambda E(S)$ which we can write explicitly by:

$$\lambda E(S) = \left[\sum_{j=1}^{M} q_j(a_j(1-u_j))\right] \int_0^\infty e^{-\left(\frac{z}{\upsilon}\right)^{\delta}} dz$$

Where the parameters (υ,δ) are determined by the mean and variance of the pollution time calculated above. By the same token,

$$Q(t) = \frac{1}{t}\int_0^t B(z)dz = 1 - \frac{1}{t}\int_0^t e^{-\left(\frac{z}{\upsilon}\right)^{\delta}} dz, \quad t \geq 0$$

which can be written in term of incomplete gamma integrals. Over time, the mean number of pollution events can thus be written as an evolution over time given explicitly by:

$$\frac{dE(n(t))}{dt} = -\frac{1}{E(S)} E(n(t)) + \left[\sum_{j=1}^{M} q_j(a_j(1-u_j))\right]; \quad n(0) = n_0$$

This equation can of course be calculated numerically.

4. The Firm and Environmental Agency's Management Policies

We consider next the management problems of both a controlling environmental agency and the firms. For simplicity however we consider only a representative individual firm and its environmental game with the environmental agency. Furthermore, we shall also define long run average objectives calculated with average cycle costs justified by application of a renewal theorem. Explicitly, we shall define a cycle time as the time between two detection events when the environmental agency detects a polluting event by the firm, in which case the polluting firm will be penalized. In this case, the firm long run average profit is given by:

$$\text{Average Profits} = \frac{E(\text{Profits Less Costs})}{E(\text{Cycle Time})}$$

The Cycle Time
Assume that en environmental agency applies a control effort which consists in effecting a control with probability θ while the probability that the firm generates a polluting even at this time is $\lambda = qa(1-u)$. As a result, the probability that a polluting event is detected is defined in the following proposition. This proposition specifies as well the number of undetected polluting events within a cycle which is mostly borne by society in a poorer environmental quality.

Proposition 3:
Let a detection (renewal) cycle be defined by the inter-event of two controls applied and detecting pollution events. Let environmental controls be applied by the environmental agency with a probability θ and let $\lambda = qa(1-u)$ be the firm pollution rate. Then, the joint probability distribution that such a cycle is of length K with i undetected polluting events within such a cycle is given by:

(9) $F(i,K) = \binom{K-1}{i-1} \lambda\theta\left[(1-\theta)\lambda\right]^{i-1} [1-\lambda]^{K-i}$; $i = 1,2,...,K-1$; $K = 1,2,...$

The marginal distributions are given by:

(10) $g(K) = \lambda\theta\left[(1-\lambda)(1+(1-\theta)\lambda)\right]^{K-1}$

(11) $h(i) = \theta\lambda\left((1-\theta)\lambda\right)^{i-1} \sum_{j=0}^{\infty} \binom{j+i-1}{i-1}(1-\lambda)^j$

With means

(12) $E(K) = \dfrac{\theta\left[(1-\lambda)(1+(1-\theta)\lambda)\right]}{\lambda(1-(1-\theta)(1-\lambda))^2}$

(13) $E(i) = \dfrac{1-\theta}{\theta} + 1 - \lambda(1-\theta)$

As a result, the probability of detecting a polluting event is:

$$(14) \quad \alpha = \frac{1}{E(K)} = \frac{\lambda}{(1-\lambda)} \frac{\left[\theta+(1-\theta)\lambda\right]^2}{\theta\left[1+(1-\theta)\lambda\right]}$$

While the firm propensity to pollute has a cumulative density function given by:

$$(15) \quad P\left(\frac{i}{K} \leq \xi\right) = \sum_{K=\left[\frac{1}{\xi}\right]}^{\infty} (\lambda\theta)(1-\lambda)^{K-1} \sum_{i=1}^{\lfloor \xi K \rfloor} \binom{K-1}{i-1} z^{i-1}; \quad z = \frac{\lambda(1-\theta)}{1-\lambda}$$

Proof: See Appendix

Proposition 3 has a number of implications that are worth mentioning. First, the average number of polluting events that are not detected is defined by the renewal theorem:

$$(16) \quad \overline{i} = \frac{E(i)}{E(K)} = \frac{\lambda\left(1-(1-\theta)(1-\lambda)\right)^2 \left\{[1-\lambda(1-\theta)]+\frac{1-\theta}{\theta}\right\}}{\theta\left[(1-\lambda)(1+(1-\theta)\lambda)\right]}$$

where E(K) is the cycle time (see also Figure 3) below.

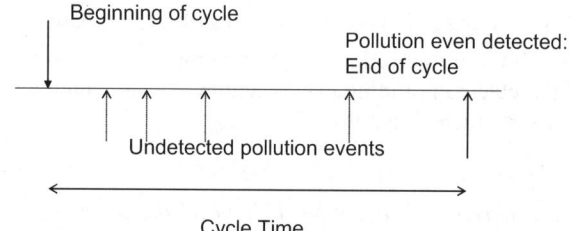

Figure 3: Detection Cycle

Thus, if the expected cost of an undetected polluting event equals C_i, the average cost of non-detected pollution events is $\overline{i}C_i$. If there are M firms, the total average cleaning cost is equal to $\sum_{j=1}^{M} \overline{i}_j C_{i,j}$. At the same time, the expected length of time during which pollution events are active is given by:

$$(17) \quad \overline{i}_{Time} = \frac{E(i)\Lambda(K)}{E(K)}$$

which can be used as a measure of environmental quality. These elements can be used next to calculate both the firm and the agency's long run average objectives.

The Firm Long Run Average Profit

Within a cycle the firm profit is assumed to be given by:

$$(18) \quad C_F = \left[K\pi(a) - KC_{pr}(au)\right](1-\kappa) - C_3 - C_i iv$$

where $\pi(a)$ is the firm period profit due to its economic activity $a \geq 0$, $C_{pr}(au)$ defines a period prevention cost of environmental pollution. When $u = 1$, $C(a) = \infty$

while $u = 0$, $C_{pr}(0) = 0$. As a result, the firm prevention cost is denoted by $0 \leq u < 1$. The firm tax rate is assumed to be given by κ while the penalty cost sustained by the firm if it is detected in a polluting act is C_3. Finally, if a polluting event occurs and is undetected, the firm may choose to attend to it. In this case, the cleaning cost over the cycle is $C_1 i v$, $0 \leq 0 \leq K$ where v is the probability that it does so. For simplicity, however, we shall assume that $v = 0$ and the undetected pollution events are borne only by society.

As a result, the firm long run average profit is given by:

(19) $\underset{a,u}{\text{Max}} \; A_F = \dfrac{E(C_F)}{E(K)} = \left[\pi(a) - C_{pr}(au)\right](1-\kappa) - \dfrac{C_3}{E(K)}$

With:

(20) $E(K) = \dfrac{\theta(1-\lambda)(1+(1-\theta)\lambda)}{\lambda(1-(1-\theta)(1-\lambda))^2}$, $E(i) = \dfrac{1-\theta}{\theta} + 1 - \lambda(1-\theta)$

and $\partial \pi(a)/\partial a > 0, \partial^2 \pi(a)/\partial a^2 \leq 0$, $\partial C_{pr}(x)/\partial x > 0$, $\partial^2 C_{pr}(x)/\partial x^2 > 0$ and of course the pollution rate $\lambda = qa(1-u)$. In this expression, note that the firm is oblivious to undetected polluting events. Furthermore, the firm's policy variables are determined by its economic activity and its investment in pollution preventive efforts only. However, these policy variables will necessarily be a function of the agency's propensity to implement environmental controls. An analysis of the firm's objective leads to the following proposition:

Proposition 4.
1. For a given environmental control policy, and assuming interior solutions, the firm's marginal revenue equals the marginal cost of environmental prevention,

(21) $\dfrac{\partial \pi(a)}{\partial a} = \dfrac{\partial C_{pr}(x)}{\partial x}; x = au$

2. Let $\alpha = 1/E(K)$ be the probability of detecting a polluting event and consider the marginal effect of a change in pollution rate occurrences and the probability of detection. This marginal effect is then proportional to the cost of prevention and given by:

(22) $\dfrac{\partial \alpha}{\partial \lambda} = \dfrac{(1-\kappa)}{qC_3} C_{pr}(x)$

Proof:
Optimization of the average cost with respect to a yields,

(23) $\left[\dfrac{\partial \pi(a)}{\partial a} - u \dfrac{\partial C_{pr}(x)}{\partial x}\right](1-\kappa) - C_3 \dfrac{\partial \alpha}{\partial a} = 0$ where $\dfrac{\partial \alpha}{\partial a} = \dfrac{\partial \alpha}{\partial \lambda} q(1-u)$

Similarly, in the case of optimization with respect to u, the preventive effort of the firm yields result 2. of the proposition:

(24) $\dfrac{\partial A_f}{\partial u} = 0$, leading to: $qC_3 \dfrac{\partial \alpha}{\partial \lambda} = \dfrac{\partial C_{pr}(x)}{\partial x}(1-\kappa)$

Combining these two equations we obtain the first result of the proposition.

Q.E.D.

The implications of these results are revealing. The larger the pollution occurrence rate, the larger the probability of detection. The larger this term, the larger u--the pollution prevention effort. Similarly, the larger the penalty cost and the larger the tax rate, the greater effort u is. In addition, if the firm is oblivious to environmental cost, then $C_{pr}(0) = 0$ and therefore, $\dfrac{\partial \pi(a)}{\partial a} = 0$. As a result, investing in pollution abatement is likely to reduce the level of economic activity of the firm. The reduction in this economic activity depends on the marginal profits and marginal costs of prevention. Using implicit differentiation, we have:

(25) $\dfrac{da}{du} = -\dfrac{\partial \Phi(a,u)/\partial u}{\partial \Phi(a,u)/\partial a} = \dfrac{a \partial^2 C_{pr}(x)/\partial x^2}{\dfrac{\partial^2 \pi(a)}{\partial a^2} - u \dfrac{\partial^2 C_{pr}(x)}{\partial x^2}} < 0$

Note $\dfrac{\partial^2 \pi(a)}{\partial a^2} \leq 0, \dfrac{da}{du} < 0$. However, if $\dfrac{\partial^2 \pi(a)}{\partial a^2} > 0, \dfrac{da}{du} < 0$ if $\dfrac{\partial^2 \pi(a)}{\partial a^2} \leq u \dfrac{\partial^2 C_{pr}(x)}{\partial x^2}$. If this is not the case, then we have $\dfrac{da}{du} > 0$ implying that pollution preventive efforts can increase the level of economic activity.

The Environmental Agency's Problem

The environmental agency long run average objective will be assumed to be given by an environmental quality objective while at the same time recognizing the constraints (budgets and otherwise) it is subjected to. As a result, we shall assume for simplicity that the agency's policy problem consists in selecting a control strategy which minimizes the long run average environmental non-quality subject to a set of constraints stated below:

Environmental Non-Quality: the expected average length of time during which pollution events are active within an inspection cycle

(26) $\underset{0 \leq \theta \leq 1}{\text{Min}} \; \bar{i}_{Time} = \dfrac{E(i)E(\Lambda(K))}{E(K)}$, $E(\Lambda(K)) = qa(1-u)E\left(\int_0^K [1-B(z)]dz\right)$

Subject to the Constraint Cost:

(27) $\theta C_\theta \leq B_A + \alpha C_3, \; a \geq a_{\min}$

and

$$(28) \quad \frac{1}{\alpha} = E(K) = \frac{\theta(1-\lambda)(1+(1-\theta)\lambda)}{\lambda(1-(1-\theta)(1-\lambda))^2}, \quad E(i) = \frac{1-\theta}{\theta} + 1 - \lambda(1-\theta)$$

Note here that the penalty to the firm C_3 is a revenue to the agency which is added to its budget B_A while θC_θ is the average environmental control costs borne by the agency. Furthermore, note that undetected polluting events are not attended to either by the firm or the environmental agency (although our analysis can be easily extended to deal with such an issue). In addition, while the environmental agency spends funds to control potentially polluting firms and clean the environment when necessary, it also collects money from government allotted budgets and penalties imposed on firms. At the same time, however, it has contradictory objectives, seeking a greater economic activity (given by a constraint for minimum economic activity a_{min}), augmenting the tax base to finance government budgets and at the same time seeking to increase the quality of the environment. These results in the environmental games which the firm and the agency are involved in. Assuming interior solutions only, the resolution of the agency's problem combined with the firm's optimization conditions provides a pure Nash equilibrium. However, assuming that the agency is the leader in a Stackleberg game (Stackleberg, 1934) with the firm, the optimal solution of such a game is an optimization problem which is defined by the following:

$$(29) \quad \text{Min } \bar{i} = \frac{E(i)E(\Lambda(K))}{E(K)}, \quad E(\Lambda(K)) = qa(1-u)E\left(\int[1-B(z)]dz\right)$$

Subject to the Constraints

$$(30) \quad \theta C_\theta \leq B_A + \alpha C_3; \quad a \geq a_{min}; \quad \text{and} \quad \frac{\partial \pi(a)}{\partial a} = \frac{\partial C_{pr}(x)}{\partial x}; \quad \frac{\partial \alpha}{\partial \lambda} = \frac{(1-\kappa)}{qC_3} C_{pr}(x), \quad x = au$$

This is of course a nonlinear optimization problem which can be dealt with by the usual Kuhn-Tucker conditions. Finally, a generalization to the environmental control of M firms is straightforward and is left here for further study and application.

5. Extensions and Discussion

This introduction to queue models of environmental games and environmental quality control have provided an approach which can be used extensively in the modeling of such of such problems as well as in providing a statistically based approach for predicting and estimating environmental pollution effects. The combination of such models, their validity testing, parameter estimation and optimization (or simulation) can then be used as an environmental management tool. This paper has explicitly considered simple models and examples to demonstrate the possibility of obtaining results. Extensions and generalizations, some of which are straightforward, are of course possible. For example, pollution events, once they occur, can have different magnitudes. In this case, Bulk queues with infinite servers might be used to predict and estimate the effects of pollution events. In other cases, events might occur continuously in time rather than at discrete epochs. This might be the case of lake pollutions. In such cases, diffusion approximation to infinite server queues might be used when modeling continuous pollution processes. As stated in the introduction of this paper, dependent pollution events can also be construed as networks of queues, representing the causal (albeit probabilistic) relationship between pollution events. A similar approach might be used to model pollution and development of interacting species. In this context, a broad number of feedback phenomena, multiple pollution sources etc. might be used.

References

1. Angell, L.C., Klassen, R.D., 1999. Integrating environmental issues into the mainstream: an agenda for research in operations management. *Journal of Operations Management* 17, 575–598.
2. Bloemhof-Ruwaard, J.M., van Beek, P., Hordijk, L., Van Wassenhove, L.N., 1995. Interactions between operational research and environmental management. *European Journal of Operational Research* 85, 229–243.
3. Chaudhry M.L. and J.G.C. Templeton, A *First Course in Bulk Queues*, Wiley, New York, 1983
4. Gross D. and C.M. Harris, 1985, *Fundamentals of Queueing Theory*, (2nd Edition), New York, Wiley
5. Heo J.H., J.D. Salas and K.D. Kim, 2001, Esimation of confidence intervals of quantikles of the Weibull distribution, *Stochastic Environmental Research and Risk Assessment*, 15, 284-309.
6. Kuhre, W.L., 1998. ISO 14031: *Environmental Performance Evaluation*. Prentice-Hall, Englewood Cliffs, NJ.
7. Nash, J., 1950, Equilibrium points in N-person games, *Proceedings of the National Academy of Sciences*, 36, 48-9.
8. Reyniers D.J. and C.S. Tapiero, 1995a, 1995a, The supply and the control of quality in supplier-producer contracts, *Management Science*, October-November, 1995, vol 41, no.10, pp. 1581-1589
9. Reyniers D.J. and C.S. Tapiero, 1995b, Contract design and the control of quality in a conflictual environment, *Euro. J. of Operations Research*, 82, 2, 1995, pp.373-382
10. ReVelle, C., 2000. Research challenges in environmental management. *European Journal of Operational Research* 121 (1), 218–231.
11. Stackleberg von, 1934, H., *Marktform and Gleichgweicht*, Vienna, Springer Verlag
12. Tapiero C.S., 1995, Acceptance sampling in a producer-supplier conflicting environment: Risk neutral case, *Applied Stochastic Models and Data Analysis*, vol. 11, 3-12
13. Tapiero C.S., *1996, The Management of Quality and Its Control*, Chapman and Hall, London, May.
14. Tapiero, C.S., 2001, Yield and and Control in a Supplier-Customer Relationship, *International Journal of Production Research*, vol 39, no.7, May, pp. 1505-1515

Appendix: Proof of Proposition 3

The proof below has the distinct advantage of being both general and adaptable to other approaches and control-detection cycles. The joint distribution $F(i,K)$ satisfies a system of recursive equations together with a stopping boundary when the cycle is ended and a pollution event is detected.. Namely, we have:

(A1) $\quad F(0,j) = (1-\lambda)F(0,j-1), j = 0,1,2,3,...K-1$
$\quad F(i,j) = (1-\lambda)F(i,j-1) + \lambda(1-\theta)F(i-1,j-1), i = 1,2,...j; j = 0,1,2,3,...K-1$
$\quad F(i,K) = \lambda\theta F(i-1,K-1)$

A solution by induction yields the solution stated in the proposition. The first expression calculates the probability that in j periods no pollution occurs leading to $(1-\lambda)^j$. The second equation calculates the probability that i polluting events occur when j

periods have passed prior to the detection of a polluting event. Finally, the third equation is a stopping condition since a polluting event both occurs and is detected. Using joint distribution we also find the following marginal distributions:

(A2)
$$g(K) = \sum_{i=1}^{K-1} F(i,k) = \sum_{i=1}^{K-1} \binom{K-1}{i-1} \lambda\theta\left[(1-\theta)\lambda\right]^{i-1}[1-\lambda]^{K-1} =$$
$$= \lambda\theta[1-\lambda]^{K-1}\sum_{j=0}^{K-1}\binom{K-1}{j}\left[(1-\theta)\lambda\right]^{j} = \lambda\theta[1-\lambda]^{K-1}\left[(1-\theta)\lambda+1\right]^{K-1} =$$
$$= \lambda\theta\left[(1-\lambda)(1+(1-\theta)\lambda)\right]^{K-1} = \lambda\theta v^{K-1}$$

where $v = \left[(1-\lambda)(1+(1-\theta)\lambda)\right]$ which conforms to the proposition result. The mean is in this case given by:

$$\text{(A3)}\quad E(K) = \lambda\theta\sum_{K=1}^{\infty} K v^{K-1} = \lambda\theta v \frac{d}{dv}\sum_{j=1}^{\infty} v^{j} = \lambda\theta v \frac{d}{dv}\left[\frac{1}{1-v}\right] = \frac{\lambda\theta v}{(1-v)^{2}}$$

And therefore,

$$\text{(A4)}\quad E(K) = \frac{\theta\left[(1-\lambda)(1+(1-\theta)\lambda)\right]}{\lambda(1-(1-\theta)(1-\lambda))^{2}}$$

As a result, the probability of detection is:

$$\text{(A5)}\quad \alpha = \frac{1}{E(K)} = \frac{\lambda}{(1-\lambda)}\frac{\left[\theta+(1-\theta)\lambda\right]^{2}}{\theta\left[1+(1-\theta)\lambda\right]}$$

It is obvious that we have a renewal cycle since each of the cycles during which the pollution event is detected for the first time are independent.

By the same token, the marginal distribution $h(i)$ is:

(A6)
$$h(i) = \sum_{K=1}^{\infty} F(i,k) = \sum_{K=1}^{\infty}\binom{K-1}{i-1}\theta\lambda\left((1-\theta)\lambda\right)^{i-1}(1-\lambda)^{K-i} =$$
$$= \theta\lambda\left((1-\theta)\lambda\right)^{i-1}\sum_{j=0}^{\infty}\binom{j+i-1}{i-1}(1-\lambda)^{j}$$

Of course, the means and other moments can be computed using these distributions. For example, the expected value of i for a given K, the number of defectives equals one plus a random variable given by the binomial distribution with parameters (K,ϑ) where ϑ has a mixture distribution with parameter $1-\theta$ and therefore its mean is $\lambda(1-\theta)E(K-1)$. Thus,

$$\text{(A7)}\quad E(i) = 1+\lambda(1-\theta)(1/\lambda\theta-1) = [1-\lambda(1-\theta)]+\frac{1-\theta}{\theta}$$

If $\theta = 0$ and the agency performs no environmental controls, then the number of polluting events is infinite while for full control, $\theta = 1$, we have $E(i) = 1$ as expected.

Q.E.D.

COMPUTATIONAL COMPLEXITY OF MODELING ECOSYSTEMS

Vladimir NAIDENKO
Institute of Mathematics, NASB
Surganov Str., 11, 220012 Minsk, BELARUS
naidenko@im.bas-net.by

Inna BOURIAKO
Institute of Microbiology, NASB
Kuprevich Str., 2, 220141 Minsk, BELARUS
bia@mbio.bas-net.by

Jean.-Marie PROTH
INRIA / SAGEP, UFR Scientifique
Universite de Metz, Ile du Saulcy, 57012 Metz cedex 01, FRANCE
Jean-Marie.Proth@loria.fr

Abstract

Algorithmic undecidability of the problem of determining computational complexity of models describing various ecosystems has been proved from a formal language theory point of view

Key Words: Ecosystems, Models, Computational complexity, NP-completeness, Formal languages, Context-free grammars

1. Introduction

The need to be able to measure the complexity of a problem, algorithm or structure, and to obtain bounds and quantitative relations for complexity arises in more and more scientific fields besides computer science. The traditional branches of mathematics, statistical physics, biology, medicine, social sciences and engineering are also confronted with this problem more and more frequently. In the approach taken by computer science, complexity is measured by the quantity of computational resources used up by a particular task.

When formulating models, including models of ecosystems, the sometimes neglected or underestimated notion of complexity should be taken into account. For instance, in some cases it is proposed to estimate the complexity of a model in its informational-theoretical aspect as the amount of information on the system, and, according to the author [Silvert 1996]: "This demand for computing power, which at least intuitively seems to bear some relationship to the complexity of the model, clearly does not reflect any great intrinsic complexity in either the model or the system, and it cannot be attributed to the complexity of the algorithms which are used...". Yet, from the standpoint of the theory of computational complexity, even a simply formulated model containing a relatively small amount of information on the system, cannot always be processed by a computer due to its computational intractability.

2. Computational complexity

Computation theory can basically be divided into three parts of different character [Lovasz et al 1999]. First, the exact notions of algorithm, time, storage capacity, etc. must be introduced. For this, different mathematical machine models must be defined, and the time and storage needs of the computations performed on these must be clarified (this is generally measured as a function of the size of input). When the available resources are limited, the range of solvable problems gets narrower; this explains how different complexity classes are obtained. The most fundamental complexity classes provide an important classification of problems arising in practice, but also of those arising in classical areas of mathematics; this classification reflects the problems' practical and theoretical difficulty quite well. The relationship between different machine models also belongs to this first part of computation theory. Secondly, one must determine the resource need of the most important algorithms in various areas of mathematics, and give efficient algorithms to prove that certain important problems belong to certain complexity classes. In these notes, we do not strive for completeness in the investigation of concrete algorithms and problems; this is a task for the corresponding fields of mathematics (combinatorics, operations research, numerical analysis, number theory). Thirdly, one must find methods to prove negative results, i.e. to prove that some problems are actually unsolvable under certain resource restrictions. Often, these questions can be formulated by asking whether certain complexity classes are different or empty. This problem area includes the question of whether a problem is algorithmically solvable at all; this question can be considered classical today, and there are many important results concerning it; in particular, the decidability or undecidability of most concrete problems of interest is known. The majority of algorithmic problems occurring in practice is such, however, that algorithmic solvability itself is not in question; the question is only what resources must be used for the solution. Such investigations, addressed to lower bounds, are very difficult and are still in their infancy.

We will treat the concept of computation or algorithm. This concept is fundamental for our subject, but we will not define it formally. Rather, we consider it an intuitive notion, which is amenable to various kinds of formalization (and thus, investigation from a mathematical point of view). An algorithm means a mathematical procedure serving for a computation or construction (the computation of some function), and which can be carried out mechanically, without thinking. This agreement is often formulated as Church's thesis. A program in the Pascal (or any other) programming language is a good example of an algorithm specification. Since the mechanical nature of an algorithm is its most important feature, we will introduce various concepts of a mathematical machine instead of the notion of algorithm. Mathematical machines compute some output from some input. The input and output can be a word (finite sequence) over a fixed alphabet. Mathematical machines are very much like the real computers the reader knows but somewhat idealized: we omit some inessential features (e.g. hardware bugs), and add an infinitely expandable memory.

In general, a typical algorithmic problem has an infinite number of instances, with an arbitrarily large size. Therefore we must consider either an infinite family of finite computers of growing size, or some idealized infinite computer. The latter approach has the advantage of avoiding the questions of what infinite families are allowed. Historically, the first pure infinite model of computation was the Turing machine, introduced by the English mathematician Turing in 1936, before the invention of programmable computers. The essence of this model is a central part that is bounded (with a structure independent from the input) and an infinite storage (memory). More exactly, the memory is an infinite one-dimensional array of cells. The control is a finite automaton capable of making arbitrary local changes to the scanned memory cell and of gradually changing the scanned position. All the computations that could ever be carried out on any other mathematical machine-models can be carried out on Turing machines. This machine notion is used mainly in theoretical investigations. The algorithmic solvability of some problems can be very far from their practical solvability. There are algorithmically solvable problems that cannot be solved, for an input of a given size, in fewer than exponential or doubly exponential steps. The complexity theory, a major branch of the theory of algorithms, investigates the solvability of individual problems under certain resource restrictions. The most important resource is time. We define these notions in terms of the Turing machine model of computation. This definition is suitable for theoretical study. Generally, it does not matter which machine model is used in the definition. This leads us to the definition of various complexity classes: classes of problems solvable within given time bounds, depending on the size of the input. Every positive function of the input size defines such a class, but some of them are particularly important. The most central complexity class is polynomial time. Many algorithms which are important in practice run in polynomial time (in short, are polynomial). Polynomial algorithms are often very interesting mathematically, since they are built on deeper insight into the mathematical structure of the problems, and often use strong mathematical tools. We mainly restrict the computational tasks to "yes-or-no" problems; this is not too much of a restriction, and pays off in what we gain in simplicity of presentation. Note that the task of computing any output can be broken down into computing its words in any reasonable alphabet.

A finite set of symbols will be called an alphabet. A finite sequence formed from some elements of an alphabet Σ is called a word. The empty word will also be considered a word, and will be denoted by $\varnothing$. The length of a word is the number of symbols in the word. The set of all words (including the empty word) over Σ is denoted by Σ^*. A subset of Σ^*, i.e., an arbitrary set of words, is called a language. Note that the empty language is also denoted by $\varnothing$; but it is different from the language $\{\varnothing\}$ containing only the empty word.

Informally, a Turing machine is a finite automaton equipped with an unbounded memory. This memory is given in the form of one or more tapes, which are infinite in both directions. The tapes are divided into an infinite number of cells in both directions. Every tape has a distinguished starting cell. On every cell of every tape, a symbol from a finite alphabet Σ can be written. With the exception of a finite number of cells which denote words, this symbol must be a special symbol "*" of the alphabet, denoting the empty cell. To access the information on the tapes, we supply each tape with a read-write head. At every step, this sits on a cell of the tape. The read-write heads are connected to a control unit, which is a finite automaton. Its possible states form a finite set. There is a distinguished starting state "START" and a halting state "STOP". Initially, the control unit is in the "START" state, and the heads sit on the starting cells of the tapes. In every step, each head reads the symbol in the given cell of the tape, and sends it to the control unit. Depending on these symbols and on its own state, the control unit carries out three things:

1) it sends a symbol to each head to overwrite the symbol on the tape (in particular, it can give the order to leave it unchanged);

2) it sends one of the commands "MOVE RIGHT", "MOVE LEFT" or "STAY" to each head;

3) it makes a transition into a new state (this may be the same as the old one);

Of course, the heads carry out these commands, which completes one step of the computation. The machine halts when the control unit reaches the "STOP" state.

The time demand of a Turing machine T is a function $timer_T(n)$ defined as the maximum number of steps taken by T over all possible input words of length n. A Turing machine T is called polynomial, if there is a polynomial $f(n)$ such that $timer_T(n) = O(f(n))$. This is equivalent to saying that there is a constant c such that the time demand of T is $O(n^c)$. We can define exponential Turing machines (for which the time demand is $O(2^{n^c})$ for some $c > 0$) in the same fashion. Now we consider a "yes-or-no" problem. This can be formalized as the task of deciding whether the input word x belongs to a fixed language $L \subseteq \Sigma^*$. We say that a language L has a time complexity at most $f(n)$, if it can be decided by a Turing machine with time demand at most $f(n)$. We denote by DTIME($f(n)$) the class of languages whose time complexity is at most $f(n)$. (The letter "D" indicates that we consider here only deterministic algorithms; later, we will also consider algorithms that are non-deterministic). We denote by PTIME, or simply by P, the class of all languages decidable by a polynomial Turing machine. It is standard that the decision problems solvable in polynomial time are also efficiently solvable.

It would be tempting to define the time complexity of a language L as the optimum time of a Turing machine that decides the language. Note that we were more careful above, and only defined when the time complexity is at most $f(n)$. The reason is that there may not be a best algorithm (Turing machine) solving a given problem: some algorithms may work better for smaller instances, others on larger ones, others on even larger ones, etc.

A non-deterministic Turing machine differs from a deterministic one only in that in every position, the state of the control unit and the symbols scanned by the heads allow more than one possible action. We say that a non-deterministic Turing machine T recognizes a language L if L consists exactly of those words accepted by T. If, in addition to this, the machine accepts all words x of size n in L in time $f(n)$, then we say that the machine recognizes L in time $f(n)$. The class of languages recognizable by a non-deterministic Turing machine in time $f(n)$ is denoted by NTIME($f(n)$). We denote by NPTIME, or simply by NP, the class of all languages decidable by a polynomial non-deterministic Turing machine.

We say that a language $L_1 \subset \Sigma_1^*$ can be polynomially reduced to a language $L_2 \subset \Sigma_2^*$ if there is a function $f : \Sigma_1^* \to \Sigma_2^*$ computable in polynomial time by a deterministic Turing machine such that for all words $x \in \Sigma_1^*$ we have the following proposition: any word x belongs to L_1 if and only if the word $f(x)$ belongs to L_2. It is easy to verify from the definition that this relation is transitive. That is, if L_1 can be polynomially reduced to L_2 and L_2 can be polynomially reduced to L_3 then L_1 can be polynomially reduced to L_3. If a language is in P then every language which can be polynomially reduced to it is also in P. If a language is in NP then every language which can be polynomially reduced to it is also in NP. We call a language NP-complete if it belongs to NP and every language in NP can be polynomially reduced to it. The word "completeness" suggests that the solution of the decision problem of a complete language contains, in some sense, the solution to the decision problem of all other NP languages. The NP-complete languages are thus the hardest languages in NP, computationally. The question P =? NP is the most important in computation theory and remains open. However, NP-complete decision problems are intuitively considered hard for computation.

In practice, many problems are of a combinatorial-optimization type; for example, in most cases ecosystem models may be referred to a combinatorial-optimization type: significant control parameters should be chosen (for example, fishing methods) to optimize characteristics of ecosystems (for instance, stabilization or rise in population of some commercial fish species). However, the computational complexity of a combinatorial-optimization problem is strongly dependent on the computational complexity of a decision version ("yes-or-no" version) of the problem.

Combinatorial-optimization problems arise, for example, when using molecular biology methods. Molecular methods offer a fast and sensitive alternative to conventional techniques. In microbiology, for example, traditional cultivation techniques for the enrichment and isolation of microbes yield only a limited fraction of all microorganisms present [Amann et al 1995]. Molecular methods are based on the analysis of single cells, offering an opportunity to analyse the microbial community in its full diversity. The use of molecular biology techniques still provides new and relevant information on the role of microorganisms in oceanic and estuarine environments. Major discoveries in marine microbiology over the past 4 or 5 decades have resulted in major biomass components of marine food webs [Meyers 2000].

One of the most prominent problems in computational molecular biology is multiple sequence alignment. It is used for extracting and representing biologically important commonalities from a set of sequences of polymeric molecules of DNA or protein. Every elementary part of DNA or protein is encoded by a special symbol of the alphabet. We can then define the problem in a formal mathematical language. Let $S_1,\ldots,S_K$, $K \geq 2$ be sequences of length $N_1,\ldots,N_K$ over an alphabet Σ which must not contain the reserved blank character "–" and define a new alphabet $\Sigma' := \Sigma \cup \{-\}$. A multiple alignment of these strings is a $K \times \omega$-dimensional matrix $A = (a_{ij})$ with the following properties:
1. A has exactly K rows,
2. ignoring the blank character, the i-th row is the sequence S_i,
3. there is no column consisting only of blank characters.

We denote with $\omega = \omega(A)$ the number of columns of A and with $A_{i_1,i_2,\ldots,i_k}$ the projection of A to the sequences $S_{i_1}, S_{i_1},\ldots, S_{i_k}$.

The quality of an alignment is often measured with a function over the columns. The cost measurement that is most widely used is the sum of pairs (SOP) cost which is defined as follows: If sub is a fixed symmetrical function $sub: \Sigma' \times \Sigma' \to N = \{1,2,\ldots\}$ with $sub(-,-) = 0$ then we define $c(A_{i,j}) = \sum_{l=1}^{\omega} sub(a_{il}, a_{jl})$ as the cost of the projection of A to the sequences S_i and S_j. Using this definition the sum of pairs (SOP) cost function is defined as

$$c(A) = \sum_{1 \leq i \leq j \leq K} c(A_{i,j}) = \sum_{1 \leq i \leq j \leq K} \left(\sum_{l=1}^{\omega} sub(a_{il}, a_{jl}) \right)$$

The goal is then to compute a minimum cost SOP alignment A. This is a combinatorial-optimization type problem. But we can formulate a "yes-or-no" version of the problem as follows: for a given number n, is there an SOP alignment A such that $c(A) \leq n$? Note that if we can efficiently (polynomially) solve the "yes-or-no" version of the problem for any n then the source optimization problem can be also efficiently solved. However, the "yes-or-no" version of the problem is known to be NP-complete [Carrillo et al 1988].

Thus, the classical theory of computational complexity divides combinatorial-optimization models (tasks) into hard to compute (or NP-hard if a "yes-or-no" version of the optimization problem is NP-complete) and easy to compute (polynomial-time solvable). In this respect a problem arises if the method of assessing computational complexity exists. As shown earlier, if a definitely hard task is polynomially reduced to the task under consideration with undefined complexity, this latter task will also be hard to compute. One cannot always reduce the known definite task to the task with non-defined complexity, while at the same time an efficient algorithm to solve the latter may be not found. The well-known task on isomorphism may serve as an example. In this respect a question arises: is there an algorithm to estimate the computational complexity of a model from its description? If we give a negative answer, it may turn out that our real world is so complicated that we will never be able to determine the degree of its complexity. The arguments favoring this assumption are presented below.

3. Undecidability Results

As preliminarily noted, any model (including an ecosystem model) implies some formalization, i.e. a description of this model in terms of a formal language. Moreover, the model adequately describes a real situation, whether it is an ecosystem or some other complex system only if it meets several parameters (for example, maintenance of the biological diversity of the ecosystem), i.e. from the standpoint of the formal language theory the formalized model belongs to a certain formal language. Thus, according to the formal language theory, models may be formulated with maximum approximation to the real situation, and with the most adequate description of real systems. To prove the algorithmic undecidability of the problem of determining the models' computational complexity, we should refer to the formal language theory.

Let Σ be a fixed alphabet which contains at least two symbols, # a symbol not being in Σ, and L_{NP} a fixed NP-complete language over the alphabet Σ. By G and $L(G)$ we mean the context-free grammar and language, respectively. Let $\propto$ be a metasymbol of polynomial-time Turing reducibility. With every context-free grammar G with the terminal alphabet Σ (see [Ginsburg 1966]), we associate the language $(\Sigma^*\backslash L(G))\#L_{NP}$, i.e., the language $(\Sigma^*\backslash L(G))\#L_{NP}$ is the concatenation of languages $\Sigma^*\backslash L(G)$, $\{\#\}$ and L_{NP} : $(\Sigma^*\backslash L(G))\#L_{NP} = \{x\#y \mid x \in \Sigma^*\backslash L(G), y \in L_{NP}, \#\notin\Sigma\}$. For languages such as $(\Sigma^*\backslash L(G))\#L_{NP}$ the following theorems apply:

Theorem 1. Any language such as $(\Sigma^*\backslash L(G))\#L_{NP}$ belongs to NP.

Proof follows from the fact that any context-free language $L(G)$ belongs to P.

Theorem 2. Provided that NP≠P, the language $(\Sigma^*\backslash L(G))\#L_{NP}$ belongs to P if and only if $L(G) = \Sigma^*$.

Proof. Suppose $L(G) \neq \Sigma^*$. Then $L_{NP} \propto (\Sigma^*\backslash L(G))\#L_{NP}$. Therefore, provided that NP≠P, we have $(\Sigma^*\backslash L(G))\#L_{NP} \notin P$. Suppose $L(G) = \Sigma^*$. Then $(\Sigma^*\backslash L(G))\#L_{NP} = \varnothing$. Consequently, $(\Sigma^*\backslash L(G))\#L_{NP} \in P$ in this case.

Theorem 3. There cannot be any algorithm that determines whether or not the language recognized by a given non-deterministic polynomial Turing machine belongs to P.

Proof. Suppose that such an algorithm is found. Then, for a Turing machine accepting the language such as $(\Sigma^*\backslash L(G))\#L_{NP}$, one can determine whether or not this language belongs to P. Note that such a Turing machine can be effectively constructed from a given context-free grammar G. Therefore, if the algorithm tests whether $(\Sigma^*\backslash L(G))\#L_{NP}$ belongs to P, then using its output one can verify the truth of assertion $L(G) = \Sigma^*$ for a given context-free grammar G (using theorem 2). However, this is impossible because of the algorithmic undecidability of assertion $L(G) =? \Sigma^*$ for an arbitrary context-free grammar G.

Theorem 4. The following problem is algorithmically undecidable: "Is the language NP-complete accepted by a given non-deterministic polynomial-time Turing machine ?"

Proof follows from theorem 3.

Summing up, the problem of determining computational complexity of the models is algorithmically undecidable.

References

[Silvert 1996] Silvert, W: "Complexity"; J. Biol. Systems 4 (1996), 585-591.

[Lovasz et al 1999] Lovasz, L. and Gacs, P.: "Complexity of algorithms"; Lecture Notes (1999), (http://research.microsoft.com/users/lovasz/complex.ps).

[Amann et al 1995] Amann, R.I., Ludwig, W. and Schleifer, K.-H.: "Phylogenic identification and in situ detection of individual microbial cells without cultivation"; Microbiol. Rev. 59 (1995), 143-169.

[Meyers 2000] Meyers, S.P.: "Developments in aquatic microbiology"; Internatl. Microbiol. 3 (2000), 203-211.

[Carrillo et al 1988] Carrillo, H. and Lipman, D.J.: "The multiple sequence alignment problem in biology"; SIAM J. Appl. Math. 48 (1988), 1073-1082.

[Ginsburg 1966] Ginsburg, S.: "The mathematical theory of context-free languages"; McGraw-hill Book Company, Inc./New York (1966).

Chapter 3

Policy/Stakeholder Process in Marine Ecosystem Management

PERFORMANCE METRICS FOR OIL SPILL RESPONSE, RECOVERY, AND RESTORATION: A CRITICAL REVIEW AND AGENDA FOR RESEARCH

T.P. SEAGER
Center for Contaminated Sediments Research
University of New Hampshire, Durham, NH 03824, USA
Tom.Seager@unh.edu

I. LINKOV
Cambridge Environmental Inc, Cambridge, 58 Charles Street
Cambridge, MA 02141, USA
Linkov@CambridgeEnvironmental.com.

C. COOPER
Chevron-Texaco, USA

Abstract

Oil spill response, recovery, and restoration efforts are by necessity an *ad hoc* exercise governed by characteristics of the individual spill and involved or affected parties. However, all spills have certain characteristics in common: they impinge upon shared environmental and ecological resources; they involve multiple stakeholders with differing priorities, concerns, and agendas; they must be managed to mitigate, minimize, or remediate deleterious effects; and they will always involve compromises – there will be no strategy that is viewed as superior by all parties. Despite this complexity, most spill response efforts have been guided by simple metrics developed during the event, such as (in the case of clean-up efforts) the total volume of oil recovered. Such simple metrics chosen during the turmoil of an accident will almost certainly drive nonoptimal decisions. There is an acute need to develop and thoroughly vet reliable, verifiable metrics before a major spill occurs. Of course some of the metrics will need to be region specific to adequately reflect geographical differences in ecology or stakeholder preferences. This paper seeks to lay out a plan for a multidisciplinary research program that examines spill management metrics from both a scientific and a stakeholder perspective and establishes a method that can be used by the responsible government agencies (e.g. Coast Guard and NOAA) to establish scientifically sound, pre-existing, region-specific metrics that are communicable and relevant to area stakeholders.

1. Introduction

Oil spills create both acute and chronic disturbances in coastal and estuarine areas. Even though mechanical countermeasures are still used extensively to remove and recover oil products, experience shows that recovery rarely results in more than 10-20% of the spilled oil being recovered. The low efficiency of mechanical oil removal techniques, coupled with their inability to provide the desired level of environmental protection, has resulted in the development of other response countermeasures and technologies (*in situ* burning, chemical dispersion, etc.) and their regulatory acceptance in many regions. The effective and timely selection of optimal response strategy plays a crucial role in a successful response to a specific oil spill.

Recognizing the need for improved oil spill response, NOAA and other agencies have recently developed protocols and selection guides for the selection of optimal technologies (NOAA 2000, Scientific and Environmental Associates 2000, US Coast Guard, 2001). These guidance documents focus on the technical efficiency of alternative techniques as well as the empirical observations and experiences from past releases to address imminent hazards to public health and environment. However, the approaches described are not instructive on methods of incorporating public participation or selecting technologies that are responsive to or consistent with stakeholder concerns. Partly this is may be due to the fact that human values, like geophysical parameters and ecological habitats, vary regionally. And partly this may be due to the fact that until recently, reliable methods of public participation and stakeholder value elicitation have not been developed or afforded sufficient attention.

This chapter presents a broad framework for interpreting stakeholder concerns in terms of quantitative, verifiable metrics that could be used to plan for, respond to, or gauge recovery from an oil spill. It discusses the attributes that make a good metric and gives examples of good and bad metrics that have been used in the past and how these metrics helped drive decision making. It outlines the advantages of aggregating metrics, and describes general rules for metric aggregation that enhance manageability while minimizing information loss or confusion. Lastly, it proposes an agenda for research that would establish a process for identifying and evaluating effective metrics, and vetting them with critical stakeholder groups. Among the positive outcomes of this research may be an improved framework for technical experts and lay parties to work cooperatively in spill management.

2. The Importance of Quantitative Metrics for Spill Management

Whether catastrophic or chronic, spills can have major environmental and ecological consequences. Spills that garner national (or international) media attention typically also engender strong emotional and economic responses. They invariably impinge upon shared environmental and ecological resources and consequently engage multiple stakeholder groups with differing – even competing or mutually exclusive – objectives for the design of spill response, recovery and restoration efforts. As a result, spill management can be a difficult, contentious, and politically charged task. Inevitably, trade-offs must be made among alternatives that are perceived as redistributing the cost or benefits among different communities. Because not all goals can be satisfied in all instances (and perhaps none can be satisfied completely), no single "optimal" approach that achieves consensus among different groups is likely to be found – especially given the limited resources (including time) available. There is no single *worst case* or *best case* scenario for any given spill. Different groups will describe different worst or best cases that are consistent with their unique set of values.

Then how should the effectiveness of spill response, recovery, and restoration efforts be gauged? Historically, a surfeit of metrics have been employed, including: bird counts, spilled barrels or barrels recovered, slick boundary area, dollars spent, time to ecological recovery, or economic impact. No single metric can be devised that will satisfy every vested party. However, without some systematic method of evaluating the quality of the metrics themselves, it seems that currently the credibility, communicability, verifiability, or relative importance of *any* metric may be questioned on *any* basis by *any* party. It could be of tremendous assistance to spill managers and other decision-makers to understand the goals and values brought to bear in any particular situation and the success measures (metrics) that relate to those goals. With such awareness, the inevitable trade-offs involved may be made consciously, and with respect for the consequences and/or sacrifices required (including perhaps, the wrath of disaffected stakeholder groups).

3. Metrics

The purpose of any metric is to benchmark two or more alternatives, or to assess the changes in a system over time. Without a context for comparison, any metric is meaningless. However, once a metric is implemented, it becomes a tool for prioritization, resource allocation, or intentional structuring of a response towards a set of goals. In short, metrics are a gauge to measure success. They become the focus of management or engineering efforts to shape a system in accordance with organizational objectives. Metrics may be quantitative (such as length), semi-quantitative (such as an ordinal ranking), non-quantitative (such as a favorite color), or qualitative (better or worse). For simple systems, such as board or card games, metrics may be simple to design, easy to enumerate and interpret, and inexpensive to gather data on. However, establishing good metrics for complex environmental and ecological systems presents a significant challenge. Both natural and human systems are multifaceted, and relate to one another in an infinite number of ways. Consequently, any set of metrics for assessing spill management is incomplete and may at best be considered only *representative* of the myriad of decision factors that could be brought to bear. For this reason, environmental and ecological metrics are often referred to as *indicators*, which emphasizes the representational relationship these measures have to the state of complex systems. They are indicative -- but not definitive – gauges, and consequently must be interpreted with their limitations in mind.

Nonetheless, meaningful decision processes involving shared public resources (such as the environment) must inevitably rely upon some credible assessment measures to be accessible or explainable to the public. Even imperfect or incomplete metrics may have currency in a fair and transparent deliberative process -- but what makes a metric "good" or "bad?" An ideal metric would have several characteristics (Graedel and Allenby 2002, Seager and Theis 2002):

- **It would be scientifically verifiable.** Two independent assessments would yield equivalent results.
- **It would be cost-effective.** That is, it would not rely upon technology that was difficult to deploy or prohibitively expensive to obtain. Nor would it require an intensive deployment of labor to track.
- **It would be easy to communicate to a wide audience.** That is, people at large would understand the scale and context, and be able to interpret the metric with little additional explanation.
- **It would relate to something that is important to people.** There is no point assembling a metric no one cares about.
- **It would be relevant to available decision or design variables.** That is, it would exhibit a causal relationship between the state of the system and the variables that are under a decision-maker's control. Metrics that are independent of human action do not inform a management, policy-making, or design process.
- **It would be credible.** That is, it would be perceived as accurately measuring what it is intended to measure.
- **It would be scalable over an appropriate epoch and geographic locus.** That is, it would be indicative of short, medium, and/or long term effects as appropriate. For example, it would not be meaningful to attempt to measure the effects of chronic low-level toxic dosages over a period of weeks or months, just as it would not be appropriate to average local environmental conditions over a widely varying region.
- **It would be ecologically relevant.** That is, it would enhance the ability of spill managers and/or regulators to faithfully execute their stewardship responsibilities.
- **It would be sensitive** enough to capture the minimum meaningful level of change, or make the smallest distinctions that are still significant, and have uncertainty bounds that are easy to communicate.

It may be difficult -- if not impossible – to find metrics that satisfy all of these conditions. Nevertheless, a metric that has several of these characteristics may still prove to be useful. It may be helpful to organize or classify available or historically relevant metrics to help clarify they way we think about them. Virtually all metrics relevant to spill management may be characterized into six broad categories: economic, thermodynamic, environmental, ecological, socio-political, or aggregated. These are summarized below (and in more detail in Seager and Theis 2004):

- **Economic** metrics gauge the state of a particular system in terms of currency so that they may be compared with monetary transactions or industrial accounts. The system in question may be ecological, as in the case of estimating the value of ecosystem services impaired by an oil spill. Or it may be anthropogenic, as in estimating the extent of lost tourism revenues in an economy impacted by a spill. In theory, proper pricing of environmental goods and services could allow market forces to optimally allocate resources between ecological and industrial activities. However, in practice both the calculation methods and the validity of the concept of pricing the environment are recognized as controversial. Because there are no markets for most environmental goods, such as pollution attenuation, external or social costs are highly uncertain, as are the methods and figures reported for the value of ecosystem services. Moreover, monetization may lead to the erroneous assumption that environmental exploitation can be revocable, reversible or reparable in a manner analogous to pecuniary transactions, although in many cases ecological systems are damaged beyond recovery.

- **Thermodynamic** metrics are measures of energy or material resources. Sometimes, they are normalized to *intensive* units such as kg/person or oil equivalents of energy/product, but in the case of oil spills, *extensive* measures such as total barrels lost or recovered may be appropriate. Usually thermodynamic metrics do not indicate the specific environmental impacts associated with resource consumption or loss. On the basis of a thermodynamic measure called *emergy*, which measures energy consumption in terms of the equivalent solar energy required to replace the consumption, Odum (1996) concludes that the extensive clean up efforts that followed the grounding of the *Exxon Valdez* were an unproductive deployment of energy resources. It has also been observed that more diesel fuel was expended on clean up efforts than barrels of oil were lost in the spill.

- **Environmental** metrics estimate the potential for creating chemical changes or hazardous conditions in the environment. Typically, environmental metrics use physical or chemical units such as pH, temperature, or concentration. But not always. A habitat measure such as the number or density of suitable bird nesting sites is one example. However, concentration measures – especially for toxic oil components such as polycyclic aromatic hydrocarbons (PAHs) -- are difficult to put in an appropriate context unless tied to some ecological or human manifestation such as death, cancer, mutation, or even non-health based endpoints such as beach or fisheries closures. Environmental metrics may use physical or chemical units, but they can be distinguished from thermodynamic metrics by the fact that they typically are intended to measure environmental loadings or changes, rather than resource demands. They are generally measures of the residuals created by industrial processes, rather than the raw materials.

- **Ecological** metrics attempt to estimate the effects of human intervention on natural systems in ways that are related to living things and ecosystem functions. The rates of species extinction and loss of biodiversity are good examples, and incorporated in the concept of ecosystem health (Rapport 1999,

Costanza 1998). Oiled bird counts, marine mammal death counts, time to ecological recovery, are all examples of ecological metrics.
- **Socio-political** metrics evaluate whether industrial activities are consistent with political goals like energy independence or eco-justice, or whether collaborative relationships exist that foster social solutions to shared problems. Who can dispute the potentially far-reaching social and political impact of a major oil or chemical spill? And yet, how should these impacts be measured and managed? In property values? Incomes? Non-government organization (NGO) or non-profit donations?
- **Aggregate** metrics are derived by combining other metrics with the purpose of trying to simplify decision-making. Aggregated metrics are described in more detail in the next section.

4. Aggregation

The data potentially available to spill managers is voluminous, multifaceted, and difficult to interpret. The complexity of handling so much information could be so overwhelming – especially in a crisis -- that useful information can be obscured by irrelevant. To make the information more manageable at the level of decision-making, collected data must eventually be *aggregated*. Aggregation is a process of mathematically combining related measures: for example by summing, averaging, or combining by more complex methods such as net present value computation. Data may be aggregated over a geographic area, over time, or other independent variables such as species, habitat type, or demographic profile. Aggregated data are easier to work with, but always contains less information than the original data set from which aggregated measures are compiled. Moreover, the mathematical methods used to aggregate different measures may constrain or confuse the interpretation of those methods. For example, net present value calculations can be extremely sensitive to selection of an appropriate discount rate. (A discount rate is analogus to an interest rate on a savings bond, and is used to estimate changes in value over time). In business (and to a somewhat lesser extent in engineering), this uncertainty is limited by investment time horizons or project design lives that rarely exceed thirty years. However, in ecological systems the time to reestablish climax communities may be twice as long or greater. Consequently, selection of a net present value discounting method that unduly diminishes the influence of long-term consequences on a decison may result in choices that fail to maximize ecological benefits (Ayres and Axtell 1996). Therefore, the method employed for aggregating data over different time periods can have an important effect on how decisions are perceived.

Methodologically unsound approaches to aggregation may render information meaningless or cause managers to reach perverse conclusions. For example, it is not proper to add *intensive* measures (which are expressed as ratios or percent) such as concentration (e.g., mg/l or ppm) or miles per gallon without first converting them to *extensive* measures (e.g., liters, grams, or barrels) with *identical* units. Even then, mistakes may be easy to make – such as in estmating the reduction in concentration due to dilution of two mixtures or average fuel economy of a fleet of vehicles – and calculations carried out with intensive measures must be made carefully. Also, particular attention must be paid to aggregation of data that is expressed in different units. As a general rule, aggregating data that belongs to two different categories, such as economic and thermodynamic, is a dangerous approach. Similarly, aggregating data that has a different qualitative relationship to a decision-makers notion of *merit*, that is progress towards an overall goal, should be avoided. For example, it is not insightful to add a resource-consumption based metric such as tons of disperant deployed or gallons flushing water consumed to an outcomes-based metric such as miles of shoreline cleaned. Consumption-based metrics may be indicative of intensity of activity, but fail to capture effectiveness of resource expenditures towards advancing response or recovery objectives. On the other hand, outcome-based metrics may have only a loose causal relationship to management programs (such as when dispersion of an oil slick occurs spontaneously, independent of human interventions). For these reasons, it is essential that the methods of data or metrics aggregation employed in any spill management plan be completely transparent to other experts and all stakeholder groups.

5. A Dual Heiarchy

Assessment of the utility of any metric exists on two levels. The first relates to the scientific credibility and economy of the metric, as detailed in the section above regarding the characteristics that make a "good" metric, whereas the second relates to the relative importance of any metric in the overall decsion process. About the scientific attributes, there should be little disagreement once the metric is clearly understood. That is, technical experts should be able to agree on the process of collecting data relevant to the metric, on the associated uncertainty bounds, the expense, the sensitivity, etc. However, on the point of relevance there may be considerable disagreement due to legitimate differences in the underlying values held by any individual or stakeholder group. For example, one group may value ecological resources and hold that these metrics hold paramount importance in selecting among available response, restoration, and recovery alternatives. Nevertheless, a second group may value economic resources above ecological, and place greater weighting on pecuniary metrics. To the extent that these two interests overlap (e.g., such as ecological resources that have significant economic benefits), there may be agreement or opportunity for compromise between these groups. However, where these are in conflict, the two groups are much more likely to become estranged. In the latter instance, no reconciliation, consensus, or compromise may be possible. Both groups may have legitimate but mutually exclusive objectives.

6. Research

It is clear that a research program is needed to improve our understanding of the metrics historically available to spill managers, to devise new, improved metrics, and to improve understanding of stakeholder values and the ways in which the public participates in spill management. It is equally clear that such a research program will need to be multidisciplinary and involve experts in the natural, physical, and social sciences. To begin, a review of spill response, recovery, and restoration case studies must be completed to compile a laundry list of metrics currently or previously employed, and comments from scientific experts and spill management practitioners should be solicited discussing the advantages or shortcomings of these. A scientific review panel assessing these metrics with respect to the ideal normative profile detailed above could narrow the list into categories such as preferred, recommended, or not recommended, and significant gaps in any of the five major classification areas could be identified. New or hypothetical metrics could be devised to fill these gaps. Subsequent to the initial screening and classification, the metrics must be vetted with stakeholder groups drawn from different regions of common interest. (It is likely to that the salient metrics – especially ecological or economic -- will vary significantly from one region to the next). The primary purpose of the stakeholder meetings – such as in a workshop or conference – is not to persuade or "educate" but to establish a discourse. As noted, different groups are likely to hold different views. The views expressed by different groups with little in common should not be combined, averaged, or otherwise aggregated. Nor should stakeholders be asked to "vote" on the metrics they favor most dearly. The vetting process must be geared towards eliciting the *values* by which stakeholders judge spill management alternatives, the goals for and constraints they would place upon the process, and discovery of the modes by which these values may be communicated and metrics by which progress towards these goals may be assessed. With these results, researchers must return to expert panels to reassess the relevance and utility of the metrics initially screened. Thus, an iterative process is devised in which stakeholders and experts work in combination to assess metrics on both levels: scientific credibility and stakeholder relevancy. It is likely that this process could be pioneered in one region as a research project to establish a streamlined protocol that can be applied to other regions. The end result would be a greater understanding of the metrics that are "in play" in any given region, the values and objectives held by different groups within the region, the potential for conflict among these groups, and available decision variables that influence progress towards the stated goals.

7. Acknowledgements

This work grew out of a workshop sponsored by the NOOA Coastal Response Research Center at the University of New Hampshire. Holly Bamford (NOAA Office of Oceanic and Atmospheric Research), Don Davis (Louisiana Applied and Educational Oil Spill R&D Program), Mark Fonseca (NOAA Center for Coastal Fisheries and Habitat Research) and Jerry A Galt (Genwest Consulting) contributed to the initial discussions and helped set the agenda for research.

8. References

1. Ayres RU, Axtell R. 1996. Foresight as a survival characteristic: When (if ever) does the long view pay? *Technological Forecasting and Social Change.* 51:209-235.
2. Ayres RU. 1998. The price-value paradox. *Ecological Economics.* 25(1):17-19.
3. Cleveland CJ, Kaufmann RK, Stern DI. 2000. Aggregation and the role of energy in the economy. *Ecological Economics.* (32)2:301-317.
4. Committee on Industrial Environmental Performance Metrics. 1999. *Industrial Environmental Performance Metrics: Opportunities and Challenges.* Washington, DC: National Academy Press.
5. Costanza R, d'Arge R, de Groot R, Farber S, Grasso M, Hannon B, Limburg K, Naeem S, O'Neill RV, Paruelo J, Raskin RG, Sutton P, van den Belt M. 1998. The value of the world's ecosystem services and natural capital. *Ecological Economics.* 25(1):3-15.
6. Global Environmental Management Initiative. 1997. *Measuring Environmental Performance: A Primer and Survey of Metrics in Use.* Washington, DC: Global Environmental Management Initiative.
7. Green L, Myerson J. 1996. Exponential versus hyperbolic discounting of delayed outcomes: Risk and waiting time. *American Zoology.* 36:496-505.
8. Hertwich EG, Pease WS, Koshland CP. 1997. Evaluating the environmental impact of products and production processes: A comparison of six methods. *Science of the Total Environment.* 196(1):13-29.
9. 1Madden GJ, Bickel WK, Jacob EA. 1999. Discounting of delayed rewards in opioid-dependent outpatients: exponential or hyperbolic discounting functions? *Experimental Clinical Psychopharmacology.* 7(3):284-93.
10. NOAA. 2000. Characteristic Coastal Habitats Choosing Spill Response Alternatives. National Ocean Service, Office of Response and Restoration, Hazardous Materials Response Division, November 2000.
11. Olsthoorn X, Tyteca D, Wehrmeyer W, Wagner M. 2001. Environmental indicators for business: a review of the literature and standardization methods. *Journal Cleaner Production.* 9(5):453-463.
12. Odum HT. 1996. *Environmental Accounting.* John Wiley & Sons: New York NY. pp127-132.
13. Rapport D. Gaining respectability: Development of quantitative methods in ecosystem health. 1999. *Ecosystem Health.* 5(1)1-2.
14. Schulze P, ed. 1999. *Measures of Environmental Performance and Ecosystem Condition.* Washington, DC: National Academy Press.
15. Scientific and Environmental Associates. 2000. *Characteristic Coastal Habitats: Choosing Spill Response Alternatives.*
16. Seager TP, Theis TL. 2004. A taxonomy of metrics for testing the industrial ecology hypotheses and application to design of freezer insulation. *Journal of Cleaner Production.* 12(8-10):865-875.
17. Seager TP. Expected 2004. Introducing industrial ecology and the multiple dimensions of sustainability. In *Strategic Environmental Management* edited by B Bellandi. John Wiley & Sons: New York NY. In press.
18. SMART 2001. Special Monitoring Of Applied Response Technologies. U.S. Coast Guard, National Oceanic and Atmospheric Administration, U.S. Environmental Protection Agency, Centers for Disease Control and Prevention, Minerals Management Service.
19. US Coast Guard 2001. Developing Consensus Ecological Risk Assessments: Environmental Protection In Oil Spill Response Planning A Guidebook. by Aurand, D., L. Walko, R. Pond. United States Coast Guard. Washington, D.C. 148p.

THE CHALLENGES TO SAFETY IN THE EAST MEDITERRANEAN: MATHEMATICAL MODELING AND RISK MANAGEMENT OF MARINE ECOSYSTEMS

K. ATOYEV
Glushkov Institute of Cybernetics, Kiev, 03187, UKRAINE

Abstract

The East Mediterranean is one of the bifurcation Zone of World Stability. The location near the arc of geopolitical conflicts is the main basis for challenges to safety in this region. Environmental terrorism provoked by these conflicts is the main threat as the intensity of tanker traffic and the infrastructure of oil and gas pipelines in this region are very high. So any terrorist act may lead to an ecological catastrophe with unpredictable consequences. How may prevention best be accomplished? How can the effects of terrorist acts on marine ecosystems be minimized? How can the marine environment and water resources of the East Mediterranean be protected from great anthropogenic loads connected with the current stage of Globalization? These are the key questions of strategic management. Their solution is impossible without contemporary mathematical methods. This paper illustrates an application of the mathematical modeling, theory of catastrophes and risk analysis to assess and manage the levels of threats. The results of a mathematical investigation of risk dynamics and its dependence on the intensity of fuel flows in the East Mediterranean are presented.

1. The Mediterranean Basin and World Oil Transit

The Mediterranean is a semi-enclosed basin with little flush in. That is why it is one of the seas least able to break down oil, sewerage, and chemical pollutants pumped daily into its waters. The main sources of marine oil pollution are releases from ships, natural slicks and pollution from land. Oil is an extremely toxic substance, containing between 10 and 20 known carcinogens in every 5 tons released into the oceans. The amount of oil spilled from tanker rinsing and "natural" losses in the Mediterranean alone is estimated at 60.0 tons yearly. The Mediterranean Sea is a well-frequented sea route allowing access to Southern Europe, North Africa, The Middle East and The Black Sea. The result of this extensive marine traffic is a high risk of oil pollution. There are two important economical activities that can be affected by oil pollution and which would benefit from any measure taken in the direction of oil spill monitoring and detection: tourism and fishing. Tourism can be considered the most important Mediterranean industry, and the majority of tourism activities are based on coastal resources. The negative effects that oil spills can have on tourism are obvious. In this case, the main concern is focused onto the locations where the pollution reaches the vicinity of the coast. Fishing is a traditional economic activity in the Mediterranean. Although most of the captures occur off the coast, coastal fishing should not be forgotten. Furthermore, the importance of fish farming is steadily increasing. The two latter activities can be seriously affected by the presence of oil pollution.

So, on the one hand oil is vital to the economies of all the world's nations, and for the vast majority, this oil must be transported by sea across great distances before it can be used. On the other hand, oil pollution can cause substantial damage to the environment and traditional economic activities in the Mediterranean. Optimizing this situation is the essential problem of our time. How can we protect the marine environment and water resources of the East Mediterranean from great anthropogenic loads connected with the current stage of Globalization?

According to International Energy Data (IEA) [1] over 35 million barrels per day (bbl/d) pass through the relatively narrow shipping lanes and pipelines discussed below. These routes are known as "chokepoints" due to shutting down risks Disruption of oil flows through any of these export routes could have a significant impact on world oil prices. While oil production is mainly located in the Middle East, the former Soviet Union, West Africa, and South America, a significant volume of oil is traded internationally. This oil is mainly transported by two methods: oil tankers and oil pipelines.

Over three-fifths is transported by sea, and under two-fifths by pipeline. Tankers have made global (intercontinental) transportation of oil possible; they are low cost, efficient, and extremely flexible. Pipelines, on the other hand, are the method of choice for transcontinental oil movements. Pipelines are critical for landlocked crude oil and also complement tankers at certain key locations by relieving bottlenecks or providing shortcuts. Pipelines come into their own in intra-regional trade. They are the primary option for transcontinental transportation, because they are significantly cheaper than any alternative such as rail, barge, or road, and because political vulnerability is a small or non-existent issue within a nation's border or between neighbors such as the United States and Canada.

Oil transported by sea generally follows a fixed set of sailing routes. Along the way, tankers encounter several geographic "chokepoints," or narrow channels, such as the Strait of Hormuz leading out of the Persian Gulf and the Strait of Malacca linking the Indian Ocean (and oil coming from the Middle East) with the Pacific Ocean (and major consuming markets in Asia). Other important sailing "chokepoints" include the Panama Canal connecting the Pacific and Atlantic Oceans, the Suez Canal connecting the Red Sea and the Mediterranean Sea, and the Bab el-Mandab passage from the Arabian Sea to the Red Sea. "Chokepoints" are critically important to world oil trade because so much oil passes through them, yet they are narrow and theoretically could be blocked -- at least temporarily. In addition, "chokepoints" are vulnerable to pirate attacks and shipping accidents in their narrow channels.

One of the most important occurrence was the sinking of the tanker "Prestige" off the coast of Spain. This accident hasized the need for tighter environmental norms, and Europe looked to potential land-based routes to crude oil sources to avoid other environmental disasters. Moreover, exporting the oil by pipeline is safer from terrorist threats. International oil businesses began looking for even more ways to minimize the risks associated with getting oil from the ground to the markets.

Table 1 shows the main characteristics of chokepoints and some pipelines connected with the East Mediterranean.

TABLE 1. Main characteristics of chokepoints and pipelines connected with the East Mediterranean [1].

Chokepoints	Oil Flows (bbl/d)	Destination of Oil Exports	Concerns/Background
Bosporus/Turkish Straits	2.0 million	Western and Southern Europe	Only half a mile wide at its narrowest point, the Turkish Straits are one of the world's busiest and most difficult to navigate waterways
Suez Canal	3.8 million	Predominantly Europe; also United States	Shutting down of the Suez Canal and/or Sumed Pipeline would divert tankers around the southern tip of Africa (the Cape of Good Hope), greatly increasing transportation time and effectively tying up tanker capacity.
Sumed Pipeline	2.5 million		The Sumed pipeline links the Ain Sukhna terminal on the Gulf of Suez with Sidi Kerir on the Mediterranean.
Strait of Hormuz	13 million	Japan, United States, Western Europe	By far the world's most important oil chokepoint, the Strait consists of 2-mile wide channels for inbound and outbound tanker traffic.
Russian Oil Export Pipelines/Ports:		Eastern Europe, Netherlands,	All of the ports and pipelines are operating at or near full capacity, leaving limited
Druzhba	1.2 million	Italy, Germany, France, other Western Europe	alternatives if problems arose at export terminals. With the line reversed to Omisalj, Russian oil exporters will have direct access to the Mediterranean Sea, allowing them to bypass the Black Sea and the increasingly crowded Bosporus Straits.
Tengiz-Novorossiisk Pipeline	564,00		
Baltic Pipeline	240,00		
Druzhba-Adria pipeline	10,00		

Globalization will increase demand for energy consumption, so the potential threat due to increased tanker traffic in the Mediterranean Sea will also increase. Two factors will have a negative effect on this situation.

First, the Caspian Sea basin has attracted considerable attention in recent years, largely due to speculation as to the potential size of the region's natural gas and oil reserves. The development of natural gas and oil there will lead to reduced reliance on Middle Eastern suppliers for both the United States and its European allies. The West has an interest in increasing energy independence. This includes diversifying sources of oil and securing the oil supply. One of the primary points of contention has been the question of how Caspian oil and gas will reach customers. The United States and Turkey have long since supported the construction of the Baku - Tbilisi - Ceyhan (BTC) pipeline, which will bring Azerbajani oil from the Shah Deniz and Guneshli fields to the Eastern Mediterranean for exports to Europe, Israel and the United States. This new oil bridge leads to additional pressure to the environment of the East Mediterranean.

Second, the combined impact of terrorist attacks and ecological catastrophes can have multiple effects on world economic wealth. Terrorism expert E. L. Chalecki [2] agrees that "...at a time when populations all over the world are increasing, the existing resource base is being stretched to provide for more people, and is being consumed at a faster rate". As the value and vulnerability of these resources increase, so does their attractiveness as terrorist targets. The destruction of a natural resource can now cause more deaths, property damage, political chaos, and other adverse effects than it would have in any previous decade». How may prevention best be accomplished?

Federal, state, and local governments can protect environmental resources in situ through more intensive and focused monitoring efforts, in conjunction with increased environmental data gathering, a sort of "early-warning" system to identify future environmental risks. Moreover, the most reliable way for a nation to protect itself against the disruption caused by environmental terrorism is to diversify resource use wherever possible. Having multiple sources of energy means each individual source is less attractive as a target, and even distribution of resources between users contributes to reducing tension over resource scarcity. This may lessen the terrorists' political motivation [3].

So, in order to reduce the damage that can result from environmental terrorism it is necessary to diversify not only oil sources but also its transportation routes. This can be achieved without adding to the East Mediterranean's vulnerability while maintaining oil exports to Europe, Israel and the United States at a sufficient level by by-passing the Mediterranean (across Continental Europe and the Baltic Sea). Now, such a route actually exists. For years, Ukraine has advocated a route through Azerbaijan and Georgia, over the Black Sea, and through Ukraine to Poland. Most of the necessary pipeline already exists. Ukraine's ongoing improvements to its refinery and pipeline infrastructure, given some foreign assistance to speed the process, will make it sufficient to handle the "early" oil, extracted in the next few years. Perhaps most important, the price tag would be relatively small: the cost of a few miles of pipeline is estimated at $40 million, and facility development and improvements are expected to amount to about $60 million.

2. Is it possible to choose the optimal route for Caspian Oil?

The alternative routes for Caspian oil transportation are as follows: 1) through Iran (**Ir**); 2) through Russia, over the Black Sea and through Turkey's narrow Bosporus Strait (**RBS**); 3) through Azerbaijan, Georgia, and Turkey (Baku-Tbilisi-Ceyhan (**BTC**)); 4) through Azerbaijan and Georgia, over the Black Sea, and through Ukraine to Poland (**AUP**); 5) through Russia, over the Black Sea to and through the Balkans (**RBl**). Let us examine the benefits and threats connected with the above routes.

The easiest, most direct route is through Iran, but it is in contradiction with the demand about diversification of oil routes to suppliers formulated above. As O.Oliker noted [4] "it seems ill advised to ship oil billed as an alternative to reliance on the Middle East through one of the Middle East's largest oil producers". Moreover, the

USA have been vehement in their opposition to Teheran's involvement in Caspian development, for political reasons.

A proposal to ship the oil over the Black Sea to and through the Balkans seems equally imprudent, given the instability of that region. Moreover, from the point of view of the Mediterranean's ecological safety, neither examined route reduces the load on the region's environment. Indeed, the first route passes through the Suez Canal and the second one runs through Druzhba and Adria's pipeline system to the Adriatic port of Omisalj in Croatia.

The route through Turkey's Bosporus Straits also has great constrictions. The Turkish Straits are one of the most crowded transportation passages in the world. Increased shipping traffic through the narrow Bosporus Straits has heightened fears of a major accident that could have serious environmental consequences and endanger the health of the 12 million residents of Istanbul that live on either side of the Straits. The Straits--a 19-mile channel with 12 abrupt, angular windings, only 70 meters wide in its narrowest point--have witnessed such an increase in shipping traffic that around 50,000 ships per year (nearly one every 10 minutes) now pass through them. Around one-tenth of these are oil or liquefied natural gas tankers. That is why Ankara is strongly opposed to this oil route.

For economic and political reasons, Russia advocates an expansion of routes through Russia and over the Black Sea and the Bosporus. Russia is opposed to the BTC and AUP routes as they involve Ukraine and Georgia in Caspian development.

The analysis presented in Table 2 shows that both routes have benefits and limitations for the USA and the EC from an economic, ecological and political point of view.

TABLE 2. Benefits and limitations of the Baku-Tbilissi -Ceyhan route

BENEFITS LIMITATIONS

		Benefits	Limitations
		Economic reasons	
B T C		Oil transportation from Shah Deniz and Guneshli fields to the the Eastern Mediterranean for exports to Europe, Israel and the United States.	High cost, upwards of $4 billion, according to some estimates, especially if we take global recession into account .
A U P		Security benefits of increased diversification. If Caspian resources are not as great as hoped, this option remains feasible, because it also offers a short-term solution while the BTC route is in the building.	Ukraine's abysmal investment climate and lack of energy sector reform. Current tax laws penalize rather than invite foreign investors. Ukraine's energy sector is not efficient.
		Ecological reasons	
B T C		Reduces the safety threat to Istanbul and its surrounding area due to decreased tanker traffic through such a narrow bottleneck as the Bosporus.	Increases the load on the region's environment and potential threats due to additional tanker traffic and terrorist activity in the Mediterranean
A U P		Reduces the safety threat to Istanbul and its surrounding area due to decreased tanker traffic through such a narrow bottleneck as the Bosporus.	Increases the load on the region's environment and potential threats due to additional tanker traffic and terrorist activity in the Mediterranean

		Political reasons
B T C	Crosses Turkey, a member of NATO, as well as Georgia and Azerbaijan, key NATO Partnership for Peace states.	Absence of independent routes for oil transportation in Eastern Europe. Instability of Caucasus.
A U P	It will strengthen new members of the EC from a security and economics point of view. Diversifying EC energy imports.	Instability of Caucasus. Abrogation of the BTC route may reduce Turkey's commitment to the West and activate the Muslim factor in this country

The combination of the benefits offered by the two above routes annihilates their negative features. The Ukrainian export route will provide a secure and reliable complement to Baku-Ceyhan for Caspian oil export, one that does not require the United States to abrogate its commitment to Turkey, but which nonetheless serves as an excellent hedge should Baku-Ceyhan fail. Turkey may also decrease the pressure on the Bosporus, as The Bosphorus Straits are reaching their historical limits. Part of the tanker fleet may be unloaded in Odessa without crossing the Turkish Straits. A Bosphorus Bypass is now needed, and Odesa-Brody is available.

It will also strengthen Ukraine from a security standpoint, enabling it to better withstand Russian pressure, thus significantly decreasing the likelihood that it will ask the United States and NATO to defend it from its large neighbor. Furthermore, by diversifying Ukrainian energy imports away from Russia, this political solution creates significant incentives for domestic energy sector reform as well as reform of the overall investment climate, which, in turn, should lead to the development of Ukraine's own oil and gas resources.

The Caspian states also need non-Russian routes for their oil for the same reasons that Ukraine needs non-Russian sources of energy: to break dependence cycles with Russia. Furthermore, Ukraine's involvement with Azerbaijan and Georgia in the GUUAM grouping has paved the way for excellent relations. In fact, as already noted, both Georgia and Azerbaijan have repeatedly spoken favorably of a Ukrainian export route option.

With this option, the USA and the EC would solve some global tasks that will represent main challenges of 21 century: 1) diversifying energy resources and their transportation routes; 2) reducing the threat of environmental terrorism due to decreased tanker traffic through the Mediterranean Sea; 3) strengthening the stability on the EC's borders. The first two items have been examined above. When discussing the third one, it must be underlined that Ukraine is a key transit country using its large network of pipelines to transport a majority of Europe's natural gas supply, and could also provide Europe with an alternative source of oil. Ukraine is also rich in natural resources, possessing a large supply of virtually unexploited energy resources. This explains why Ukraine is clearly of strategic interest to the USA, Russia and the EC as one of the bifurcation zones bridging the European and Eurasian security spheres. It has a direct impact on the eastern and southern security of Europe, especially if we take into account the escalation of the terrorist threat and the existence of zones of instability in the world. Without strengthening Ukraine by integrating it in the world's energy system and helping it develop as a stable, independent, democratic European country within a strong market-economy Europe, the risks of involving Ukraine in the areas listed above will increase.

So, commercialization of the Odesa-Brody pipeline represents a major step along the road toward the stabilization of the European Security Area and European integration. It offers a way to strengthen European ties where energy demand is rising, open up new export opportunities and increase cooperation with Slovakia, the Czech Republic, Germany, Hungary, Poland and beyond.

We have stressed the political, ecological and safety benefits of AUP, but there are also economic benefits. Mrs. Loyola de Palacio, Vice President of the European Commission, officially stated that the Odessa-Brody-Plotsk pipeline project "is of European interest". According to her, a feasibility study has proved the commercial value of the project, which could contribute greatly to the economic development of all the participating countries. The European Commission passed an official document, the Message on the Development of the Energy Policy of the Enlarged European Union, Its Neighbors and Partners. One of the priorities was stated as cooperation with neighbor-countries in ensuring safe transportation of oil by sea, including the extension of the Odessa-Brody pipeline to Poland's Plotsk to be later connected either to the Druzhba or the existing pipeline, which runs to the Polish Baltic port of Gdansk. [5]

The following reasons explain why European experts have issued such a unanimous statement.

1. The changing situation on the European oil supplies and services market. In 2005 the EU plans to introduce new, more rigid standards on the sulfur content in oil products, so European consumers are already looking for lighter sorts of oil with a lower sulfur content. The role of Caspian oil is thus increasing. On the one hand, European businessmen realize that sooner or later they should opt for the more expensive but lighter Caspian oil, because the purification of cheaper sorts would cost them much more, but on the other hand, the existing infrastructure for delivering light Caspian oil to potential markets is insufficient.

2. Authoritative experts maintain that Ukraine has the strategic possibility to make the Eurasian oil transportation corridor a bypass alternative to the Bosporus. Besides, since the cost of tanker shipments has increased two- or threefold, pipelines have become more important to Europe, not only because of lower transportation prices, but also because piping is environmentally safer (the wreck of the Prestige tanker has made an indelible impression on Europe).

3. The Europeans are also worried by shrinking extraction in the North Sea. According to the forecasts made at the Brussels conference, it will have decreased by 87% by 2015, while extraction from Caspian deposits will have increased by 44%.

4. Now that the USA have full access to Iraqi oil, the route via Ceyhan has lost its former importance. Besides, new risks have emerged as the route now lies in an unstable zone. That is why the USA support the AUP route. Moreover, western companies have invested billions in the extraction of oil from the Caspian deposits, and those billions have to be recouped. So they are interested in transporting that oil to high-liquidity and solvent markets via safe and cheap delivery routes. The main market is the EU. The most promising and feasible route that can help solve the Bosporus problem is Odessa-Brody-Plotsk. How can we quantitatively weigh a preference for one of the various routes? Let us use the Theory of Hierarchic Systems developed by T.Saaty [6]. The joint goal of the main actors (the USA, the EC and Russia) is the

security of oil transportation to Western suppliers. We choose economic, ecological and political criteria to weigh various oil export routes.

Weighing the benefits of various routes was accomplished using the analytical hierarchy process (AHP) procedure. Applying multi-criteria such as AHP for ranking various routes of oil transportation involves sorting each of these criteria– economic, ecological and political - into several descending hierarchy levels. The goal of AHP is to select the highest security level in the case where each of the main actors has his own interests and pre-favorable oil traffics (See Fig. 1). The results of weight determination using Saaty's method are presented in Table 3.

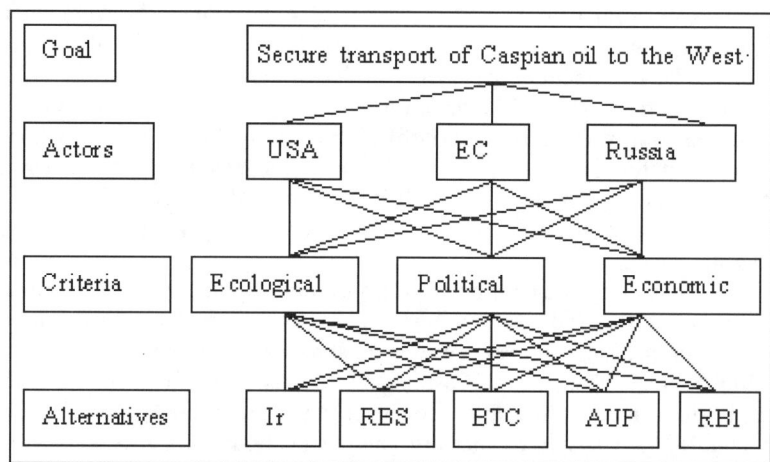

Fig. 1. Structure of analytical hierarchy process used in this research

TABLE 3. Determination of Weight Vector

					Actor's Weight					
		USA			**EC**			**RUS**		
		0.48			0.42			0.1		
					Criteria's Weight					
	Ecol	Pol	Econ	Ecol	Pol	Econ	Ecol	Pol	Econ	
	0.15	0.45	0.40	0.2	0.4	0.4	0.05	0.6	0.35	
					Route's Weight					
Rou-		USA				EC			RUS		
tes	Ecol	Pol	Econ	W_{US}	Ecol	Pol	Econ	W_{ES}	Ecol	Pol	Econ	W_{RUS}	W_{Tot}
Ir	0.2	0.0	0.20	**0.02**	0.20	0.05	0.20	**0.05**	0.20	0.1	0.1	**0.12**	**0.04**
RBS	0.1	0.01	0.10	**0.05**	0.05	0.10	0.15	**0.10**	0.15	0.35	0.35	**0.35**	**0.10**
TBC	0.2	0.35	0.20	**0.30**	0.20	0.15	0.10	**0.20**	0.20	0.10	0.1	**0.08**	**0.24**
EUP	0.15	0.2	0.12	**0.20**	0.3	0.25	0.20	**0.25**	0.15	0.02	0.1	**0.04**	**0.21**
RBl	0.15	0.01	0.10	**0.05**	0.20	0.15	0.10	**0.10**	0.15	0.40	0.25	**0.35**	**0.10**
TBC+ AUP	0.2	0.44	0.28	**0.38**	0.25	0.30	0.25	**0.30**	0.15	0.03	0.1	**0.06**	**0.31**

The following eigenvectors correspond to the final normalization of the above routes from the point of view of the USA, the EC and Russia: W_{US} (0.02, 0.05, 0.30, 0.20, 0.05, 0.38), W_{ES} (0.05, 0.10, 0.20, 0.25, 0.10, 0.30), W_{RUS} (0.12, 0.35, 0.08, 0.04,

0.35, 0.06). The eigenvector W_{Tot} (0.04,0.10, 0.24, 0.21,0.10,0.31) characterizes the rank of routes when the main actors' weights are taken into account.

So we can conclude that a combination of routes over Turkey to Cheyhan and over Ukraine and Poland to Gdansk is the most positive and profitable for all participants. It achieves a strategic goal, it is commercially realistic and it meets the interests of the Enlarged European Union, its Neighbors and Partners as main steps on the road toward the stabilization of the European Security Area and European integration. Moreover the route over Ukraine lowers the intensity of tanker traffic in the Mediterranean sea as well as the risks of ecological disasters in this region.

3. Chaos - a model for the outbreak of instability in the Security Area

Economic "globalization" is a historic process, the result of human innovation and technological progress. It refers to the increasing integration of economies around the world, particularly through trade and financial flows. The term sometimes also refers to the movement of people (labor) and knowledge (technology) across international borders. There are also broader cultural, political and environmental dimensions of globalization.

But there are also some dark sides to globalization and one of them is the paradoxical rise of the vulnerability of rich nations to unanticipated attacks. As noted by T. Homer-Dixon [7] "by relying on intricate networks and concentrating vital assets in small geographic clusters, advanced Western nations only amplify the destructive power of terrorists – and psychological and financial damage they can inflict".

Our open societies have become much too wide-open targets for terrorists because of the growing technological capacity of small groups and individuals to destroy things and people, and because of the increasing vulnerability to carefully targeted attack of postindustrial society's global economic and technological systems. The transformation of the Securities Area is largely driven by the following reasons: more powerful weapons, the dramatic progress in communications and information processing, more abundant opportunities to divert non-weapon technologies to destructive ends. The growing complexity and interconnectedness of our modern societies and the increasing geographic concentration of wealth, human capital, knowledge and communication links may also be added to these reasons [7]. Modern, high-tech societies are filled with supercharged devices packed with energy, combustibles and poisons, giving terrorists ample opportunities to divert such non-weapon technologies to destructive ends. So we can conclude that globalization provides new opportunities for terrorism and creates extraordinarily attractive targets for them. But even as deepening global integration makes nations more vulnerable to exogenous shocks, it only strengthens their resolve to cope with crises.

The decrease in the possibilities opened to national governments in their fight against crime is the inner side of advantages offered by economic globalization today. As a result, 300 –500 billion Dollars (Euros ???) are laundered each year, according to UN experts' data [8]. Part of this money is spent on terrorism support, including environmental terrorism. It allows crime structures to accumulate enormous funds that are commensurable with the cost needed of developing biotechnology based on success in functional genomics, DNA technologies and genetic and protein engineering. That is why the thesis that only well supported, long-term national BW programs would attempt genetic engineering projects for the purpose of weaponizing pathogens is not quite true any longer.

Together with financial support, terrorism has a large moral support on the basis of great migration also partly connected with globalization [9]. According to [10], ethnic human marginalization is caused by the confusion of traditional social and religious norms in human mentality. There is strong correlation between ethno-cultural mental peculiarities and natural-climatic living conditions. The matrixes of Western and Eastern worldviews differ in principle on major ideological directions (truth, perception, god, universe, nature etc.) [11]. As a result of their mutual penetration, great populations of their bearers immersed in alien culture, landscape and reality perception become a powerful source of terrorism. As S.Huntington [12, p.22] has noted "...the fundamental source of conflict in this New World will not be primarily ideological or primarily economic. The great divisions among humankind and the dominating source of conflict will be cultural. The clash of civilizations will dominate global politics. The fault lines between civilizations will be the battle lines of the future" (and the battle lines of the fight against terrorism we can add).

The peculiarities of modern society connected with industrialization, urbanization, technological progress, total population migration, development of mass-media and rise of national and cultural contacts, lead to mental personality disadaptation and form authoritarian personalities inclined to violence [13]. The world's economic globalization also increases the pressure of factors that provoke the rise of violence [9]. Among the main reasons, there are military conflicts, discrimination and violence towards women, children, the disabled and the elderly, corruption (organized and transnational crime, drug and firearm trafficking, money laundering) [8]. All the above concur to increase the number of people who may make up their minds to resort to biological weapons.

On the other hand, terrorism not only limits the pace of globalization, but in various dimensions activates this process. Of course, as stressed in [14], "nations took measures to tighten control of their borders to protect themselves from international terrorist networks". Of course, "the global war against terrorism has provoked fears of a human rights race to the bottom, as nations eager to calm jittery foreign investors by clamping down on terrorist activity might also excessively clamp down on political freedoms". As shown in [14], all these actions reduce the level of economic integration in Western Europe. But beyond the economy, global integration depended on such factors as political enlargement, personal contacts and modern technology. Terrorist activity boosts political integration. The war against terrorism accelerates integration of Eastern Europe and Russia into the global network and leads to significant expansion of International humanitarian and commercial organizations. Moreover, even as people traveled less across borders, they helped push international telephone and Internet traffic to record levels, which significantly increased one of the main factors which characterized globalization – personal contacts.

But terrorism is not the only phenomenon to have a negatively impact on global economics. Princeton historian Harold James cites 3 factors inherent to globalization which he claims could cause it to auto destruct: 1) the instability of capitalism, 2) the backlash among those who did not reap the benefits of global integration, 3) the failure to create institutions that can adequately "handle the psychological and institutional consequences of the interconnected world" [14].

So we can conclude that there are positive and negative feedback loops between levels of globalization and terrorism in modern society and that the dark side of globalization is nothing but the continuation of its lighter side (a modern version of the ancient Chinese theory - Dao).

Figure 2 illustrates these possible interrelations.

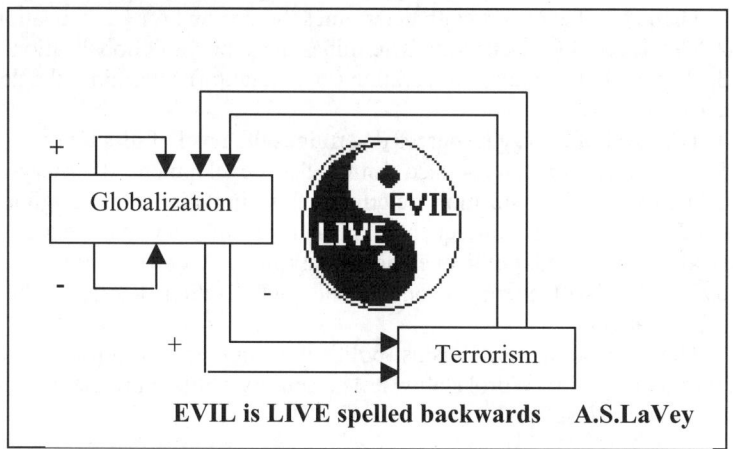

Fig. 2. Interrealtion between the processes of globalisation and terrorism

Globalization leads to more complex and interconnected networks of modern society. The nodes of these networks are acupuncture points of terrorism due to the density of links among the nodes and the speed at which energy, goods and information are circulated along these links. The behavior of such complex systems sometimes becomes unstable. A mistake in one node leads to catastrophic consequences along the whole network, for instance the collapse of electrical systems in the USA and some Western European countries. Terrorists can use these common features of post-industrial networks, especially in energy and communications infrastructures, to amplify their own power.

So, complexity and interdependence create increased possibilities for some fluctuation or minor and routine disturbances to cascade into a disaster with unpredictable ecological, economic and political consequences. Disruption of oil flows along any of the export routes passing through the Mediterranean region due to terrorist acts or disasters on relatively narrow shipping lanes (chokepoints) may also be examined as such disturbance of the system's stable operation. This could have a significant impact on world oil prices (economics), on the quality of marine infrastructure (ecology) and on the security area (politics).

We can conclude that modern society may be examined as a near unstable system in the bifurcation zone, where any fluctuation leads to unpredictable features. As usual, such systems are examined with the help of the modern theories of nonlinear systems, the theory of chaos and bifurcation.

Economic globalization, terrorist activity and intensive oil traffic determine the security area in the Mediterranean region. Our next step will be formalization of interrelations between these processes.

Let X - the level of globalization, determined by Technology, Economic Integration, Personal contacts and Political Enlargement, Y - the level of energy resources and Z - the level of Instability in the Security Area.

The following postulates may be taken as a basis for model development in terms of Demand and Supply.

1. The level of energy wealth determines the demand for Globalization.

2. The level of Globalization determines the supply of Globalization.

3. The level of economic integration (globalization) determines the demand for energy resources.

4. The level of energy resources determines the level of its supply.

5. The level of threats also limits the development of energy sources. Levels of threats and Globalization determine this limitation. A negligible level of threat may have catastrophic consequences, if the levels of economic integration, technological advances and political enlargement (globalization) are high.

6. The levels of energy resources and globalization determine the speed at which the threat arises.

7. There are some mechanisms (political, economic, and military) which limit threat increases and secure World stability. The activity of these protective mechanisms is proportional to the level of threats

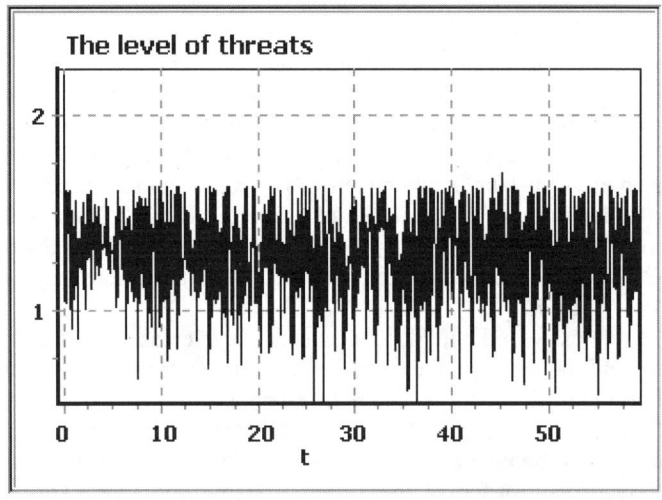

Fig. 3. Model results of chaotic behavior of threat level (variables in conventional

The relationships between Globalization, energy resources and threat of the instability of world order are determined by the following model,

$$dX/dt = a_1(t)[a_2(t)Y - a_3(t)X], \qquad dY/dt = c_1(t)[c_2(t)X - c_3(t)Y] - c_4(t)XZ,$$

$$dZ/dt = d_1(t)XY - d_2(t)Z,$$

where, a_i (i=1,3), c_j (j=1,4), d_k (k=1,2) characterize the rates of processes.

This model may be transformed into **Lorents'** model of Metastable chaos [15]. Under some conditions, sustained chaotic behavior may arise in this model (Figure 3). The transition, from stability to instability, from a state with a high level of Security Area to a state with a low one, could be analogous to the transition from a laminar to a turbulent flow.

4. Risk Assessment of theThreat of Environmental Terrorism.

The problem of minimizing the effects of terrorist acts on the marine ecosystem of the Mediterranean region became very acute after the Irak war. Its location near the arc of geopolitical conflicts leads to instability, even events whose risk value is negligible can indeed occur. In this case the task of early recognition of risks and identification and ranking of critical factors, which determine rare events, played a central part in modern risk analysis. One of the most distinctive features in the study of such events is the difficulty to choose an adequate mathematical apparatus for their investigation. Indeed, the traditional methods of risk assessment were developed on the basis of the theory of probability. However, the theory of probability cannot be correctly utilized for risk assessment in some cases, especially with the absence or incompleteness of data due to the unique character of the event.

Environmental terrorism as a phenomenon is not only connected with ecological threats, but also determined by complex epidemiological, bio-medical, economic, mental and social peculiarities of the actual country and time. Each terrorist act and all the problems associated with its consequences have individual features. That is why the risk of terrorist acts in one country cannot be properly assessed from statistical data obtained under other conditions in various other countries. The unique and single character of such a sophisticated subject as terrorism and its consequences makes it difficult to use the theory of probability for proper risk assessment.

Atoyev [16] developed another approach to risk assessment, which may prove to be more useful here. In this approach the risk assessment is carried out using the theory of smooth functions allowing the determination of critical parameter values which describe the levels of control system intensities and reserve possibilities. The risk is estimated on a scale of the system's parameter approximation of the bifurcation values, which characterize the system's transition from one steady state (norm), to another (catastrophe). This approach allows not only an estimate of the risk of emergency, but also a description of the quantitative characteristic of reserve possibilities of the system and its components. The dynamics of these system parameters is determined with the help of dynamic modeling. The main advantage of this approach is the determination of risk dynamics as the function of dynamic variables of the investigated system. It also allows identifying the weakest link of the system under examination and the areas that need improvement. The developed method may help rank the different regions and routes of oil traffic according to their weakness and vulnerability to environmental terrorism and to an evaluation of their capability to respond effectively to the threat of terrorism and to deal with the consequences of a terrorist attack.

Let us introduce some postulates that we take as a basis for assessment of the threat of environmental terrorism.

1. The state of society is examined as complex interrelations in the "hexagon of security" (HS), which is determined by economic, ecological, epidemiological, medico-biological, mental and social factors. All the threats in the arsenal, including environmental terrorism, have the potential to upset the intricate balance existing within the HS by altering the above mentioned factors. The disturbance of the balance within the HS is the most formidable threat which we must all do our best to prevent, as it leads to transition from one society system, the steady state (norm), to another (crisis or catastrophe).

2. The defense system has some steady states. Using Guastello's [17] idea about the organization safety of complicated systems it is possible to put forward the following suppositions. The first type of states is characterized by the existence of external and internal safety (norm). The second type of state is characterized by external safety alone, as internal safety is breached (intermediate state or middle risk). The third type of state is characterized by the full loss of all safety, external as well as internal (high level of risk).

3. The risk assessment is carried out with the help of the theory of smooth functions, (TSF) allowing to determine a degree of system parameter approximation to their critical values, which characterize system transition from its steady state (norm), to another state with middle or high levels of risk .

4. Safety level X is described by one of the universal deformations of TSF. It is determined with the help of indices, which characterize the interrelations in the "hexagon of security"- economic (a), social (b), epidemiological (c), medico-biological (d), mental (e) and ecological (f) factors

The relationships between safety (X) and the above-mentioned indices are determined for wigwam catastrophe by the following polynomial relation:

$$X^7+aX^5+bX^4+cX^3+dX^2+eX+f=0$$

Economic power (GNP, human development index, unemployment rate etc.) characterizes index *a*. The efficiency of democratic institutes characterizes index *b*. The rates of generation and death of different agents of infectious diseases, their tolerance to drugs or vaccines, the rates of disease spreading characterize index *c*. Total population, annual population growth, the percentage of the population living in urban areas, average annual growth rate, total adult literacy rate, crude birth rate, crude death rate, maternal mortality, life expectancy, total fertility rate and infant mortality characterize index *d*. The levels of suicides, psychical diseases and crime characterize index *e.* The quality of marine water and other environmental monitoring data (Environmental Performance Index) characterize coefficient *f*.

On the basis of works dealing with methods of catastrophe theory, the following risk assessment algorithm can be suggested:

1. Information characterizing the above-mentioned indices is inputted from modern databases.

2. The indices characterizing the appropriate group of parameters are estimated by means of developed mathematical models with the help of inputted data.

3. The bifurcation values of the parameters at which the number of system states is changing are calculated.

4. Restoration possibilities of each of the considered systems are estimated by remoteness of parameter characterizing the appropriate index from its bifurcation value.

So, the risk is estimated on a degree of system parameter approximation to their bifurcation values, which characterize system transition from one of its steady states (norm), to another (crisis or catastrophe).

With the help of the developed computer method the preliminary ranking of a number of countries on a degree of environmental terrorism threat escalation was carried out. On fig. 4 projections of surfaces describing the catastrophe on surfaces *a* (economical index) and *b* (social index) for the Netherlands and the USA are presented.

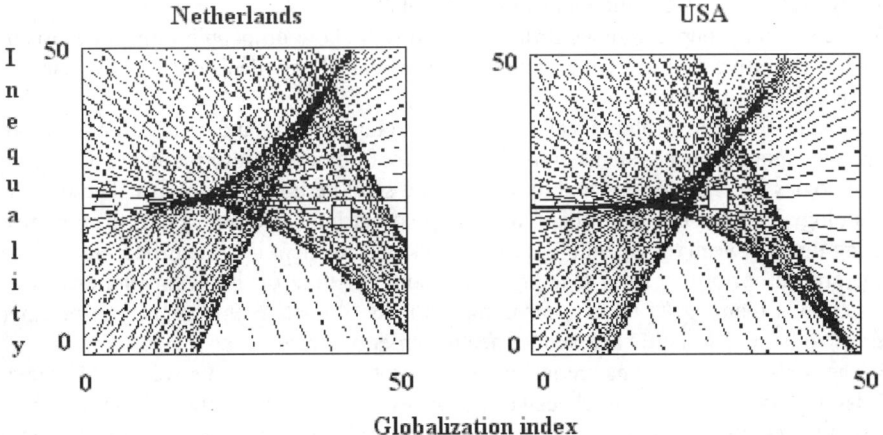

Fig 4. Example of ranking of threat escalation by proposed methods for the Netherlands and the USA

The data from [14] were utilized. The Netherlands has more high value of globalization index and more low level of inequality. The distance from the current state (white square) to the curve of bifurcation that separates the areas of small and high levels of risk is larger for the Netherlands than for the USA. So the Netherlands has more safety reserves.

5. Where is the Powerful Source of Terrorism?

Pooling of enormous finances, advanced technologies and authoritarian personalities gives a dangerous explosive mixture. So, we can conclude that the threats of terrorism have a universal character. They include the following types of threat: 1) local and regional threats, 2) transnational threats, 3) the spreading of dangerous technologies, 4) violation of governmental structure work.

These threats may be classified according to their origins as follows. First, advanced technologies applied to terrorism. Second, the increase in corruption and money laundering. Third, the strengthening of authoritarianism and spreading of fundamental ideology of various genesis due to problems of world economic globalization (migration, unemployment, vast disparities between rich and poor, collision of cultures). Fourth, the contradictions between society's need for increased liberalization while switching from the industrial age to the computer network age, when more effective horizontal structures push out vertical integration and the fight against corruption becomes more difficult. Finally, antiterrorist and anticrime legislation is not effective and efforts of different poles of terrorism fight are not sufficiently coordinated. The analysis of the main difficulties encountered by the fight against terrorism allows to emphasize two groups of problems.

The first group combines the following questions: 1) what are the main threats, connected with advanced technologies applied to terrorism; 2) what targets of environmental terrorism are the weakest links, which lead to the maximum waste for society; 3) how can we fight against these threats?

The second group combines further questions: 1) how do people turn to terrorism; 2) what cognitive and affective factors have provoked this decision; 3) how can we preserve them from this choice?

Solving the first problem calls for the development of and innovation in more sensitive monitoring systems on the basis of modern information and space technologies that may be assisted by ground and satellite tracking systems. These high-technology systems can be utilized to solve two cardinal question. First, to minimize the lag time between the beginning of the attack and moment it is identified. Second, to transform the priorities from reacting to the consequences of the event (the tactics of "yesterday's planning" [9]) to controlling the risks of such events occurring, in other words "to being constantly adapted to meeting emerging challenges" [18].

The first-group problems are technological problems that may be solved if sufficient funds are available. Comprehensive threat evaluations integrating information on economic, ecological, epidemiological, medico-biological, mental and social factors allow to determine their impact on a country's vulnerability to possible environmental terrorism. They also allow assessing the capability to respond effectively to the threat of environmental terrorism, which is an essential baseline tool for developing a global strategy on terrorism, which the world currently lacks. Risk analyses and ranking of various factors will improve our understanding of the threat, and should help identify the weakest link under various scenarios of environmental terrorism, and make it possible to redistribute help to the national services responsible for the prevention of environmental terrorism and the elimination of their consequences.

The second-group problems are caused by major trends of the third millennium: economic globalization, social disruptions, corruption, mental disadaptation, disparities between the speed at which political institutions develop and the rate at which economics and social systems undergo transformations in accordance with Huntington's conception [12] and finally the cultural clash between civilizations. Solving these problems requires more efforts and is connected with the bases of how humankind operates. With the end of the Cold War, many countries have switched from autocratic forms of government to democracy. But even then, when the political and economic elite of these countries wish their country to join Association of Democratic Nations, there are many historical, cultural, and traditional indigenous features that push them behind. Some openings in such countries with weak democracy may be utilized by terrorist movements, which become one of the beneficiaries of these changes. So every country has its own way of solving the problem of maximizing the development of democracy and liberty and of minimizing the possibilities for crime to benefit from this freedom. In short, it is necessary to find the optimal trajectory for society's development when democracy is on the rise without strengthening destructive terrorist movements. This will decrease the number of reasons pushing people to resort to terrorism, and weaken the cognitive and affective factors leading people to decide to use violence as a means to their ends.

6. Conclusions

There are two main challenges to safety in the East Mediterranean: environmental terrorism and marine oil pollution. The first threat originates from common reasons connected with the strengthening of authoritarianism and the spreading of fundamental ideology of various genesis due to the problems of world economic globalization (migration, unemployment, vast disparities between the rich and the poor, collision of cultures). Moreover, geopolitical conflicts located near the area are also major threats. The second threat originates from of global economics demand for fuel, so transportation of this fuel across great distances by sea remains a vital necessity. Oil pollution can cause substantial damage to the environment and to traditional Mediterranean economic activities: tourism and fishing. Additional pressure on the environment of the East Mediterranean may affect oil traffic from Caspian Sea basin. So, economic globalization, terrorist activity and intensive oil traffic determine the Security Area in the Mediterranean region.

How may the Eastern Mediterranean region be protected without adding to its vulnerability while maintaining oil exports to Europe, Israel and the United States at sufficient levels?

The first answer to the above challenges is connected with diversifying oil traffic. In order to reduce the damage that can result from environmental terrorism and potential disasters due to intensive tanker traffic in the Mediterranean Sea, there is only one option – the oil route around the Mediterranean (across Continental Europe/Baltic Sea).

The benefit of various oil traffic routes was estimated in this work from an economic, ecological and political point of view, using the procedure of analytical hierarchy process. The analyses that were carried out show that a combination of routes over Turkey to Cheyhan and over Ukraine and Poland to Gdansk is the most positive and profitable option for all participants. It increases the energy independence of the West because it diversifies oil sources and traffic and secures the oil supply.

The second answer is connected with lowering the threat of terrorism. Terrorists utilize such common features of modern society as intricate networks and the concentration of vital assets in small geographic clusters, especially in the energy and communications infrastructures, to amplify their own power. There are positive and negative feedback loops between levels of globalization and terrorism in modern society that allow to determine modern society as a system nearing unstable steady state in bifurcation zones, where any fluctuation leads to unpredictable features.

The mathematical model required for investigating the dynamics of the above loops was developed. The principle of how this model could be transformed into **Lorents** model of Metastable chaos was demonstrated. Under some conditions, sustained chaotic behavior may arise in this model. The complexity and interdependence of the above feedback loops create increased possibilities that some fluctuation or minor and routine disturbance can cascade into a disaster with unpredictable ecological, economic and political consequences. So, events whose risk value is negligible can indeed occur.

That is why the task of early recognition of risks and identification and ranking of critical factors, which determine rare events, now plays a central part in modern risk analysis.

In this work the approach to risk assessment was carried out. The risk is estimated on a degree of the system parameter approximation of the bifurcation values, which characterize the system's transition from one steady state (norm), to another (catastrophe). With the help of the method developed, different regions and oil traffic routes may be ranked according to their weaknesses and vulnerability to environmental terrorism.

The peoples of the West, in Jean Baudrillard's phrase, base themselves on positive values and lose their immunity to such viruses as authoritarianism and fundamental ideology, and their extreme form – terrorism. So we need a tool to measure this threat.

This paper illustrates how mathematical modeling, the theory of catastrophes and risk analysis may be applied to assessing and managing the levels of such threats.

Further development of the suggested approaches would allow developing such a mathematical tool. It is necessary to develop such a system as soon as possible, because in Ralph Waldo Emerson's words, "in skating over thin ice our safety is in our speed."

7. References

1. Feld L. (2002) World oil transit "chokepoints", Energy Information Administration, URL http://www.eia.doe.gov/emeu/cabs/choke.html
2. Chalecki E.L. (2001) A new vigilance: identifying and reducing the risks of environmental terrorism, A report of the pacific institute for studies in development, environment, and security http://www.pacinst.org/environmental_terrorism_final.pdf
3. Toman M.A. (2002) International oil security: problems and policies, *Brookings Review 2*, 20-23
4. Oliker O. (2002) Ukraine and the Caspian: An opportunity for the United States, *RAND Issue Paper 198*, www.imi-online.de/download/IMI-Analyse-02-52-akIrak.pdf
5. Silina T. (2003) Whose hnterests are being piped up?, *Mirror-weekly 20*, 1-2 http://www.mirror-weekly.com/nn/show/445/38695/

6. Saaty T.L. (1998) Multicriteria decision making–analytic hierarchy process, University of Pittsburg, Pittsburg.
7. Homer-Dixon T. (2002) The rise of complex terrorism, *Foreign Policy 1*, 52-62.
8. Report on the World Social Situation , United Nations, New York, 2001..
9. Probst P. (2001) Antiterrorism – some aspects of organization, National Defense 9-10, 21-25.
10. Stonequist E.V. (1961) The marginal man, Russel & Russel, New York.
11. Gilgen A., Cho J.(1979) Questionnaire to measure Eastern and Western thought, *Psychol. Reports 44*, 835 –841.
12. Huntington S.P. (1993) The clash of civilizations, *Foreign Affairs 3*, 22-49.
13. Ducat J. (1989) Authoritarianism and group identification: a new view of an old constrict, *Political Psychology 10*, 63-84.
14. Kearney A.T., (2003) Measuring globalization: who's up, who's down, *Foreign Policy 1*, 60-72.
15. Lorenz E. N. (1963) Deterministic nonperiodic flow, *Journal of Atoms.Sci..20*, 130 –141.
16. Atoyev K.L. (2001) Risk assessment in Ukraine: new approaches and strategy of development, in I.Linkov and J.Palma-Oliviera (eds.), Assessment and management of environmental risks: methods and applications in Eastern European and developing countries, Kluwer Academic Publishers, pp. 195-202.
17. Guastello, S.J. (1988) The organizational security subsystem: some potentially catastrophic events, *Behavioral Science 33*, 48-58
18. Berger S.R. A foreign policy for the global age, *Foreign Affairs* 6,.22-39.

STRATEGIC MANAGEMENT OF MARINE ECOSYSTEMS USING WHOLE-ECOSYSTEM SIMULATION MODELLING: THE 'BACK TO THE FUTURE' POLICY APPROACH

Tony J. PITCHER, Cameron H. AINSWORTH, Eny A. BUCHARY,
Wai lung CHEUNG, Robyn FORREST, Nigel HAGGAN,
Hector LOZANO, Telmo MORATO and Lyne MORISSETTE
*Fisheries Centre, University of British Columbia, Vancouver
Canada V6T 1Z4*

Abstract

'Back-to-the-Future' (BTF) attempts to solve the 'fisheries crisis' by using past ecosystems as policy goals for the future. BTF provides an integrative approach to the strategic management of marine ecosystems with policies based on restoration ecology, and an understanding of marine ecosystem processes in the light of findings from terrestrial ecology. BTF employs recent developments in whole ecosystem simulation modelling that allow the analysis of uncertainty, tuning to past biomass estimates, and responses to climate changes. It includes new methods for describing past ecosystems, for designing fisheries that meet criteria for sustainability and responsibility, and for evaluating the costs and benefits of fisheries in restored ecosystems. Comparison of ecosystems before and after major perturbations, including investigation of uncertain ecological issues, may set constraints as to what may or may not be restored. Understanding how climate and ocean changes influence marine ecosystems may allow policies to be made robust against such factors. A new technique of intergenerational discounting is applied to economic analyses, allowing policies favouring conservation, as the same time as addressing economic standard discounting of future benefits. Automated searches maximise values of a range of alternative objective functions, and the methodology includes ways to account for uncertainty in model parameters. The evaluation of alternative policy choices, involving trade-offs between conservation and economic values, employs a range of economic, social and ecological measures. BTF policy also utilizes insight into the human dimension of fisheries management. Participatory workshops attempt to maximise compliance by fostering a sense of ownership among all stakeholders: ideally, collaboration by scientists, the maritime community, managers and policy-makers may build intellectual capital in the model, and social capital in terms of increased trust. BTF may help to reverse the shifting baseline syndrome by broadening the cognitive maps of resource users. Some challenges that have still be met include improving methods for quantitatively describing the past, reducing uncertainty in ecosystem simulation techniques and in making policy choices robust against climate change. Critical issues include whether past ecosystems make viable policy goals, and whether desirable goals may be reached from today's ecosystem. Examples are presented from case studies in British Columbia, Newfoundland and the Gulf of St Lawrence in Canada; the Gulf of California, Mexico; the Bali Strait and Komodo National Park in Indonesia; and the South China Sea.

1. Introduction: Rationale for 'Back-to-the-Future'

Commercial fish populations world-wide are in disastrous state (Hilborn et al. 2003, Pauly et al. 2002), recent evidence suggesting that depletions and collapses may be even worse than had been thought (e.g., fish biomass: Christensen et al. 2003; whales: Roman and Palumbi 2003; large fish: Myers and Worm 2003; sharks: Schindler et al. 2003, Baum et al. 2002; turtles: Hays et al. 2003). The changes in marine ecosystems wrought by these depletions prejudice the sustainability of human fisheries in a profound way. Fisheries are prosecuted today with high-tech gear and have developed huge overcapacity: fishery management's endeavors to hold this fishing power in check have been too late, too little and ineffective. These events have occurred just as risks from climate changes are thought to be very large. Any attempt at the strategic management of marine ecosystems must address these problems. This paper summarizes recent work on an idea that may provide an answer; 'Back-to-the-Future' (BTF) is an integrated policy for strategic management of ocean ecosystems (Pitcher 2004a). The policy attempts to harness understanding of the history of ecosystems and insight of ecosystem processes to developments in ecosystem modelling and an appreciation of the human dimension of fisheries management to try to solve this 'fisheries crisis' with a strategic management approach.

The fundamental ecological issues here are only partially understood. Although we have been well aware for a long time how changes in biomass affect the population dynamics of a single species (indeed, the discipline of fisheries stock assessment is built upon the quantitative expression of that understanding), it has only recently been realised that all fisheries change the ecosystem in which they are embedded (Walters and Martell 2004, Pitcher 2001). It is now becoming evident how large these changes might be, when they may be reversible, and what their consequences are for the sustainable extraction of benefits from aquatic ecosystems.

A historical and archeological review of the evidence (Pitcher 2001) has characterized the processes at work. Three ratchet-like processes have eroded aquatic biodiversity and hence compromised fisheries. First, Odum's ratchet (named after E.P. Odum's concerns with extinctions caused by humans; Gibbons and Odum 1993), shows how harvesting acts as a selective force within ecosystems by removing long-lived, slow-growing life-histories in favour of those endowed with higher turnover; the ratchet operates both within and among species. Only a few marine fish have become globally extinct during the recent fisheries crisis, but many have suffered local extinctions as a result of overexploitation. When species become locally extinct (= 'extirpation'), the past becomes hard to restore, like a ratchet. Rapid local extinctions appear to occur in

ANCIENT PAST   PAST   PRESENT   ALTERNATIVE FUTURES

Figure 1. Diagram illustrating the 'Back-to-the-Future' concept covering the restoration of past ecosystems. Triangles at left represent a series of ecosystem models, constructed at appropriate past times, where vertex angle is inversely related and height directly related to biodiversity and internal connectance. Time lines of some representative species in the models are indicated, where size of the boxes represent relative abundance and solid circles represent local extinctions. Sources of information for constructing and tuning the ecosystem models are illustrated by symbols for historical documents (paper sheet symbol), data archives (tall data table symbol), archaeological data (trowel), the traditional environmental knowledge of indigenous Peoples (open balloons) and local environmental knowledge (solid balloons). Alternative future ecosystems, restored 'Lost Valleys', taken as alternative policy goals, are drawn to the right. (Modified from Pitcher et al. 1999, Pitcher 2001, Pitcher 2004a, Pitcher et al. 2004).

the early stages of exploitation of marine ecosystems (Christensen and Pauly 1997), probably because of narrow niches and k-selected life history parameters. Dulvy et al. (2003) found that most marine global extinctions were overexploited mammals or birds, or a result of habitat loss for sessile invertebrates and small fish. Since they are the result of a series of local extinctions, the risk of global extinctions may be significant where fishing is widespread, or for large, slow-growing organisms, irrespective of high fecundity that may have evolved to buffer habitat volatility (e.g., Chinese bahaba, *Bahaba taipingensis*, Sciaenidae; Sadovy and Cheung 2003).

Secondly, Ludwig's ratchet (Ludwig et al. 1993) describes the generation of overcapacity in fishing power through pressure from repayment of financial loans that require continued catches that, on account of stock and ecosystem depletion, can be generated only by ratchet-like further investment in fishing technology. It is hard to go back to using yesterday's fishing gear. Thirdly, Pauly's ratchet refers to the psychological tendency for scientists, and others, to relate changes in the system to what things were like at the time of their own professional debut, regarding earlier accounts of great abundance as anecdotal and methodologically naïve ('the shifting baseline syndrome'; Pauly 1995). This syndrome affects the cognitive map of resource users in a more insidious fashion than is commonly realized (e.g., Jackson et al. 2001), resulting in over-optimistic behaviour by resource users and with serious implications for how attempts at strategic management may be regarded.

The combined effect of these three ratchets, acting over historical time, has been not only to bring about the collapse of many major commercial fisheries, but also to shift the structure of ecosystems towards lower trophic levels (Pauly et al. 1998), favour simpler organisms and energy pathways (Parsons 1996), and compromise biodiversity (Sadovy and Cheung 2003, Dulvy et al. 2003) in ways that might be hard to reverse (Jackson 2001). Many of these events have caught fisheries science, managers, and the public by surprise (Haggan 2000).

2. The 'Back-To-The-Future' Policy and Research Agenda

Back-to-the-Future (BTF) is a science-based restoration ecology that uses past ecosystem states as candidate policy goals for the future (Pitcher et al. 1998a, Figure 1); it aims to create sustainable food and wealth from capture fisheries in aquatic ecosystems (Pitcher et al. 1999). BTF's strategic management is to design future fisheries (a), to be responsible according to set criteria, and (b), to be sustainable according to simulations that take account of risk and uncertainty. BTF fisheries are embedded in aquatic ecosystems that, by quantitative analysis and with the consent of stakeholders, trade-off wealth and food with a specified degree of retention of biodiversity, trophic structure and resilience against change.

In practice of course, BTF policy goals are subject to a number of practical constraints from species, habitat, climate changes and the human dimension of management. The logical choice of policy goal is the one that maximises benefits set against the costs of restoration and management. Other considerations, however, complicate this task. For example, the way in which expected future benefits are calculated greatly affects the outcome, an issue that is discussed in more detail below.

The BTF process commences with the construction of descriptive models of past ecosystems. This in itself is a major task of historical ecology. The second step is to devise sustainable and responsible fisheries that can operate within each of these ecosystems if they were to be restored, goes on to compare forecast benefits among these systems, and then selects candidate policy goals by setting them against the likely costs of restoration using suitable instruments of restoration. Finally, BTF attempts to achieve consensus on agreed restoration goals through fostering a sense of ownership of the process amongst all principal stakeholders. Once in place, adaptive management procedures are set up using the quantitative BTF procedures, in order to insure against unexpected changes in resulting from poorly-understood ecology, from climate and

TABLE 1. Summary of six stages in the 'Back to the Future' process. Workshops are in *italics*. (Modified from tables in Pitcher 1998a, Pitcher *et al.* 2004, and Pitcher 2004a.)

Stage	Goal	Steps
1	Construction of models of past and present aquatic ecosystems	Assemble present-day ecosystem simulation model Assemble preliminary past models using compatible structure and parameters Search data archives, historical documents, archaeological information *Workshop with scientists knowledgeable about system* Interviews for traditional environmental knowledge, and for fisher's opinions and behaviour Assemble and standardize historical and interview database *Workshop with scientists and managers to compare and standardise ecosystem models* Test and validate suite of ecosystem simulation models
2	Design future fisheries	Devise fisheries according to criteria of responsibility Search for sustainable fisheries with which to exploit reconstructed ecosystems ('Opening the Lost Valley') Challenge model scenarios with uncertainty and likely climate changes *Workshops to evaluate fishery policies with fishing communities* Adjust 'Lost Valley' fisheries scenarios after full evaluation
3	Choice of fisheries and ecosystem that maximises benefits to society	Identify trade-offs among economic, ecological and social criteria Ecological and economic evaluations, including analysis of risks *Workshop with communities, managers, scientists, NGOs, and government* Participatory policy choice
4	Design of instruments to achieve policy goal	Determine Optimum Restorable Biomass (ORB) for 'Lost Valley' scenarios Quantify risks to ORB policies Exploration of management instruments such as MPAs, effort controls, quotas, times and places for fishing, etc. Evaluation of costs of the desired management measures
5	Participatory choice of instruments	Community, stakeholders, managers and scientists choose instruments to achieve policy goals *Workshops with communities, managers, scientists, NGOs, and government* Participatory policy choice
6	Adaptive management: implementation and monitoring	On-going monitoring, validation and improvement of forecasts using adaptive management procedures On-going participatory guidance on instruments and policy goals

other factors. These six steps in the development of a BTF policy are summarized in Table 1 (Pitcher 2004a).

2.1 WHAT CAN WE LEARN FROM TERRESTRIAL RESTORATION ECOLOGY?

Many changes seen in the oceans parallel those long studied in terrestrial ecology, such as destruction and fragmentation of habitat critical to small and juvenile animals; local extinctions of species, and some global extinctions; huge depletions of large grazing animals with loss of the fertilising effect of their waste products; and reductions in top predators and consequent increases of prey species, leading to trophic cascades. In terrestrial ecology, changes to wild habitats have been so great that they are scarcely ever noted as such: refuges from predators for small animals are reduced, breeding areas shifted and food supplies for grazing herbivores destroyed. In the past 8000 years, human agriculture and the use of wood for building and war has brought about immense changes in habitat (Diamond 1997). For example, at the time of expansion of the Roman state in the first century AD, it is said that, without touching the ground, a squirrel could travel in continuous woodland from north of the Alps to the Baltic. Today most terrestrial landscapes throughout the world have been greatly altered by humans as a consequence of agriculture. Freshwaters have experienced similar large changes. Over the past 200 years, wetland habitats in North America have been so drastically altered as to devastate natural populations of salmon and almost eliminate several of the most widespread wetland herbivores (beaver, moose, wood buffalo; Lichatowich 2001, Callenbach 1995). And so, in terrestrial and freshwater environments, management tries to conserve what is left of ancient wild ecosystems by creating parks and protected corridors, severely limiting hunting, restoring habitats, and re-introducing locally extinct species (e.g., wolves). Accompanying these practical efforts, a whole new field of restoration ecology has arisen (e.g., Morrison 2002) that attempts to systematize how these things are done.

In marine environments, however, a restoration ecology perspective has been neglected until recently, and as a consequence, the study of restoration ecology in terrestrial environments seems further advanced than its aquatic equivalent (Dobson et al. 1997) and has developed a powerful set of analytical tools to aid recovery of degraded systems. Terrestrial reserves provide baselines against which to judge human impacts (Arcese and Sinclair 1997), and restoration may be viewed as a necessary hedge against loss from natural causes (Sinclair et al. 1995). Habitat is the essential template upon which species conservation is founded: "habitats can only be preserved if they are treated as a renewable resource; otherwise all habitat will decay to zero" (Sinclair et al. 1995: page 585). It is interesting that, in terrestrial ecology, there is no disagreement that hunting of wild animals has to be strictly regulated or else extirpations, and ultimately global extinctions, will surely occur (e.g., Ward 1997; and see Magna Carta 1215). Moreover, for wild areas, the economic benefits of conservation and restoration have recently been estimated as outweighing further depletion (Balmford et al. 2002).

Most marine organisms lack the tangible habitat made of plant architecture that we are familiar with in terrestrial animals. Some major exceptions are coral reefs (McClanahan 2002), rocky shores and oyster reefs (Lenihan and Peterson 1998), and kelp forests (Steneck et al. 2002). But the structural habitat concept itself needs extending only a little to encompass oceanographic structures in ecosystems; for example, the great marine populations of fish are bounded by tangible ocean structures (Bakun 1996), and there is a clear association between habitat complexity, biodiversity and fisheries 'hotspots' (Worm et al. 2003, Ardron 2002). Today, the majority of terrestrial ecosystems are highly modified agricultural habitats in which food organisms are grown, and so restoration or conservation of wild areas usually represents some form of loss to food production. In contrast, with the exception of aquaculture, in most marine ecosystem habitats provide homes for the wild food that is hunted there. Hence, in marine systems, the interests of both exploitation and conservation may be met by efforts to restore and preserve habitat productivity and resilience. Most marine structured habitats are suffering considerable losses: changes to open ocean habitats are more subtle but are probably no less damaging.

3. New and Adapted Methodology in 'Back-To-The-Future'

This section introduces the new and adapted methods that have been developed to deal with the new concepts and procedures used in our BTF research. They are divided into five groups: methods required to describe and model past ecosystems; ecosystem-based methods to determine sustainable fisheries; methods that set out a rational basis for choosing appropriate ecosystem restoration goals; methods that set out how these goals may be achieved; and finally, realistic techniques that attempt to secure compliance and consent through participation (Pitcher 2004a).

3.1. MODELLING ECOSYSTEMS OF THE PRESENT

The present-day ecosystem at the start of the BTF process is represented by mass-balance and dynamic simulation modelling (at present using *Ecopath* with *Ecosim*; Walters et al. 2000, 1997), using techniques that have received approval by marine ecologists (e.g., Hollowed et al. 2000, Whipple et al. 2000). Christensen and Walters (2004) discuss the methods, capabilities and limitations of the approach. In BTF, the present-day model attempts to capture the structure and form of the ecosystem as it exists today, usually after fishing impacts have occurred. It is likely to be the most 'data-rich' of the set of BTF models, and so models of the past are often based upon it. Construction of ecosystem models, whether they represent the past or the present, requires prior consideration of several important factors including the area of the model, its complexity, data-availability, and important spatial and physical processes.

3.1.1 Area of the Model. Determining the spatial extent of an ecosystem model is a non–trivial exercise because ecological and physical processes operate from scales of millimetres to scales of hundreds of kilometres. Many marine species, especially top predators such as sharks, marine mammals and seabirds have large spatial ranges that may extend far beyond the immediate area of interest to management. Also, of particular relevance to the BTF approach, some species likely occupied large ranges in the past but may have undergone a range collapse (e.g., Pitcher 1997) or extirpation due to fishing activities or other factors. Most species of fish and many invertebrates undergo ontogenetic migrations at one or more stages of their life-history (e.g., Zeller and Pauly 2000, 2001), and it is usually desirable to attempt to include all stages in an ecosystem model. Nevertheless, for some species whose range or migration route is very large (e.g., some species of Pacific salmon, eels, many sharks and whales) it will be necessary to leave some part of the life-history outside of the model dynamics and assume that variables beyond the boundaries of the model are constant (Walters and

Martell 2004), an assumption that may not always be appropriate. Alternatively, it may be possible to model the dynamics in a separate, adjacent model (Martell 2004), although this can introduce more complexity and uncertainty than is desirable for the problem at hand.

At the opposite end of the spectrum, details of fine-scale ecological interactions, such as those operating in small patches of habitat like seagrass beds and rocky reefs may be lost if the extent of the model is too large. In very large models, such as those incorporating whole EEZs, near-shore coastal and estuarine processes in general may be poorly represented if the majority of the model covers the continental shelf, slope and deep pelagic waters. Very often, inshore zones are the site of important subsistence and small-scale fisheries that support local communities (and these are often those with a vested interest in the research), so it is undesirable to oversimplify that part of the model in order to extend the boundaries to include life-history stages of a few species.

Furthermore, fisheries may span several ecological or political boundaries, or reporting zones may not overlay one another. For example, foreign fleets operating inside the EEZ of another country may extend their fishing activities beyond its jurisdiction. In south-eastern Australia, recreational fisheries, state-operated commercial fisheries and federally-operated (Commonwealth) commercial fisheries have reporting zones with very different northern and southern boundaries, even though the same stocks of many species are caught in all three types of fisheries.

Modellers must consider the trade-off where, on the one hand, the full spatial range of some species or fisheries is not included in the model and, on the other, finer-scale detail is sacrificed in favour of a larger model. The modeller must apply 'expert-judgment' to determine the scale at which most of the dynamics of interest are adequately represented. It may sometimes be useful to compare models at different spatial scales: Ainsworth (2004 in prep.) compares the utility of a spatial model of the whole of northern British Columbia, with a smaller-scale model of one reserve area (Salomon et al. 20002).

<u>3.1.2. Aggregation of Species for Modelling.</u> Over 1000 species of fish are listed for Hong Kong in the global database, Fishbase (www.fishbase.org). Indeed, marine ecosystems contain many hundreds or even thousands of species, and so some degree of species-aggregation within the model is usually required. Although there is no theoretical limit, most *Ecopath* models can be built with 50-90 functional groups. The appropriate degree of aggregation for an ecosystem model depends in part on the particular research questions at hand. For example, in a model of the southern plateau of New Zealand where the researchers were interested in modelling the sources of primary production and subsequent microbial cycling, 'fish' were aggregated into just three groups – juveniles, adults and mesopelagics (Bradford-Grieve et al. 2003).

In most cases, however, researchers will be interested in simulating effects of policy on particular fished species, especially those species or groups of species that are known to have undergone dramatic changes in abundance. While these species should be modelled explicitly, decisions must still be made about the level of aggregation of other species with which the primary species directly and indirectly interact. Guidelines that have emerged from a number of studies (reviewed by Fulton et al. 2003) suggest that predators and prey should not be grouped, and neither should species or age-classes with rate constants that differ more than 2- to 3-fold. Violation of these guidelines can lead to greatly reduced model performance. *Ecopath* with *Ecosim* allows users to split groups into multiple life-history stages to represent more effectively species with strong ontogenetic shifts in trophodynamics (Christensen and Walters 2004), although models can become very large and complex very quickly if this is applied to all species.

There is likely a dome-shaped relationship between model-complexity (i.e. level of species-aggregation, *sensu* Fulton et al. 2003) and model-performance. Over-aggregated models can fail to capture important dynamics of the system, whilst in overly complex models, errors and uncertainty can become compounded, especially if input data are noisy (Fulton 2001, Fulton et al. 2003). While several quantitative routines have been developed to evaluate optimum model-complexity (see review in Fulton et al. 2003), Walters and Martell (2004) suggest that the measure of an ecosystem model's performance should be in its ability to provide useful policy insights rather than the degree of certainty in its outcomes (i.e., if certain policies consistently outperform others over a range of parameter-estimates or levels of species-aggregation). This highlights the point made earlier that the structure of a model will depend upon what the researchers and collaborators wish to do with it. Iterative refinement of the model and sensitivity analysis will be an important part of the process of evaluating the 'right' level of model-complexity

3.1.3. Data Availability. When first designing an ecosystem model, it is often tempting to include as many species as possible as separate entities. Bearing in mind the effects of model complexity on performance as discussed above, consideration should also be given to the availability of data for different species in the system. This is not to say that models should focus only on species for which there are the most data, but rather that the model should be compatible with available data. For example, catch records often include aggregated suites of species (e.g., 'rockfishes', 'flounders', 'crabs'), usually reflecting the level of taxonomic resolution appearing on fishers' logbook forms. If there is no evidence to suggest that within-group dynamics are important to the policy questions of interest, it may be better not to disaggregate such groups. It is important to note that 'evidence' here refers not just to 'hard' data, such as diet studies or trends in estimated biomass, but also to other forms of knowledge, such as local and traditional ecological knowledge. These other forms of knowledge become especially important when reconstructing ecosystems of the past (e.g., Haggan et al. 1998, Salas et al. 1998).

3.1.4. Uncertainty in Models. Most whole-ecosystem models have many parameters: however, in *Ecopath*, values are constrained by the need to achieve mass-balance. Nevertheless, modelling needs to take account of the considerable uncertainly remaining in many of the parameter values. As an example, Figure 2 shows an

	Biomass	P/B	Q/B	Diet	Catch
Prim. prod	3	3			1
Zooplankton	3	6	6	6	
Carn. Zp.	3	2	2	6	
Dep. feeders	5	1	6	6	
Sus. feeders	5	7	7	3	
Shrimps	8	2	7	8	3
Whelk	5	1	7	3	3
Echinoderms	5	4	2	3	3
Bivalves	5	1	6	6	3
Scallops	5	3	6	6	3
Crab	5	4	6	8	3
Comm. crab	3	2	6	8	3
Lobster	5	1	6	5	3
Sm. dem.	8	5	5	1	3
Sm. gads	8	5	5	1	3
Mullet	8	5	5	3	3
Sole	3	1	5	3	3
Plaice	3	1	5	3	3
Dab	5	1	5	3	3
O. flatfish	3	1	5	3	3
Gurnards	5	1	5	3	3
Whiting	3	1	5	3	3
Cod	3	1	5	3	3
Hake	3	1	5	3	3
Rays/dogs	3	6	5	3	3
Pollack	3	1	5	3	3
Lg. bottom	3	1	5	3	3
Seabream	3	1	5	3	3
John Dory	5	5	5	3	3
Sandeels	8	5	5	5	
Herring	5	1	3	3	3
Sprat	5	6	3	3	3
Pilchard	5	2	3	5	3
Mackerel	3	5	5	3	3
Overw. mac.	3	5	5	3	3
Scad	3	5	5	1	3
Bass	3	2	5	3	3
Sharks	5	5	5	5	3
Basking Shks	5	5	5	5	3
Cephalopods	3	5	6	5	
Seabirds	3	6	5	5	
Toothed cet.	5	6	5	5	7
Seals	3	6	5	3	
Juv bass	5	7	5	3	5
Juv sole	5	7	5	3	5
Juv plaice	5	7	5	3	5
Juv cod	5	7	5	3	5
Juv whiting	5	7	5	3	

Figure 2. An example of a model pedigree for an Ecopath-with-Ecosim model of the English Channel (from Stanford and Pitcher 2004). Functional groups in the model are listed in the left hand column. Main parameters are shown in columns. The different shading and numbers refer to confidence limits (+/-%) as follows: 1=10, 2=20, 3=30, 4=40, 5=50, 6=60, 7=70, 8=80. Hence scores of (6,7,8) indicate data that are less trustworthy. Blank rectangles refer to groups where there were no data.

uncertainty table for a model of the English Channel (Stanford and Pitcher 2004). Specified levels of uncertainty are entered for each parameter of the ecosystem model, and the probability distributions are then used to drive Monte Carlo selection of the main parameters used in mass-balance ('Ecoranger': Christensen and Walters 2004). This can be used to explore the likelihood of alternative mass-balanced ecosystems (see Morissette 2004).

Additionally, in dynamic simulations using *Ecosim*, repeated simulations may be performed using Monte Carlo sampling of the main parameters. Not only does this allow the robustness of various policy scenarios to parameter uncertainty to be tested, but it also provides a way of performing a whole-ecosystem based population viability analysis.

3.1.5. Improving Models: Spatial Aspects and Physical Processes. Ecosystem models of the present day can increase in complexity when taking into account spatial aspects, or physical processes in the system. In contexts where fisheries assessment methods and traditional regulatory methods have failed, there is a strong demand for policies that are more conservative and somehow will prevent past mistakes from being repeated. One key suggestion that has been offered in response to such demands is spatial management (Walters et al. 1999, Giske et al. 2001, Meester et al. 2001, Pelletier et al. 2001). Traditional single-species stock assessment tools are incapable of asking questions about the ecosystem consequences of spatial management options and so spatially explicit, multispecies assessment models have to be developed (Walters et al. 1999). However, while the need for spatial models of fisheries dynamics is widely acknowledged, few such models have been devised. One of the recently developed tools, *Ecospace* (Walters et al. 1999) is a spatially explicit ecosystem-based model for policy evaluation that relies on *Ecosim* dynamic modelling for most of its parameterization. For example, it enables the evaluation of the impact of MPAs in an ecosystem context.

Over the past 100 years, fisheries oceanography has attempted to describe and understand relationships between the physical environment and the spatio-temporal distribution and abundance of marine organisms. Although we still have much to learn about the interactions between biology and physics in the sea, a number of important themes have emerged (see Dower et al. 2000). Among these are: (a) recruitment of organisms during the early life-history stages is controlled by a suite of environmental factors; (b) the distribution patterns and abundance of fishes are determined by a mix of environmental factors and fisheries impacts; (c) growth and migration are driven by the local abundance of suitable food organisms and predators, which are themselves subject to environmental drivers; and (d) biomasses in ecosystems oscillate on a variety of time scales due to both natural and anthropogenic causes. It may be hard to disentangle the effects of fishing from those of environmental variability. However, the scope of the problems that may be addressed is wide: Ecosystem modelling endeavours to capture the complex interactions between biology and physics in the sea occurring over very different time and space scales. The ocean environment is highly variable across a range of spatio-temporal scales, and the physical processes relevant to fishes range from those that affect interactions between individual larvae and their prey and occur over scales of seconds and centimetres, to those that affect basin scale levels of productivity, and which may affect 1000s of kilometres over decades. Ecosystem modellers have to take into account some of these spatial aspects and physical processes.

3.1.6. Improving Models: Climate Factors. We know that changing environments can change ecosystems and consequently, one of the major challenges facing fisheries and scientists is to understand how the organization and dynamics of large marine ecosystems are affected, including key interactions and feedbacks between the ocean and the atmosphere. These changes can be manifested by shifts in productivity, abundance and distribution of many species (Polovina et al. 1994, Francis and Hare 1994, Hayward 1997). A notable example of such shifts occurred during the largest recorded negative anomaly (4oC) observed in the Gulf of California, Mexico, during La Niña of 1988-89 (Soto-Mardones et al. 1999, Lavin et al. 2003). An alternation or replacement between primary consumers was observed in this ecosystem; the Pacific sardine, Sardinops sagax caeruleus, the most important fishery in the gulf, declined dramatically (Cisneros-Mata and Hammann 1995), while the population of northern anchovy, *Engraulis mordax*, a newcomer to the Gulf, increased (Hammann and Cisneros-Mata 1989). This pattern of replacement has also occurred in the past (at least in the last 200 years), where according to the sedimentary record, high abundances of fossil anchovy scales of were associated with the cool periods of the nineteenth century, while major sardine peaks occurred during the warming events of the last century (Holmgren-Urba and Baumgartner 1993). The sediment record demonstrates that populations of these small pelagics were fluctuating long before fisheries commenced (Holmgren-Urba and Baumgartner 1993), showing the indisputable role of climate in the organization of large marine ecosystems such as the Gulf of California. When ecosystem models are driven credibly by such factors, the additional effects of fishing may be investigated. For example, the collapse of the California sardine in the 1950s was due to both factors operating together (MacCall 1990).

One of the ultimate goals of ecosystem modelling is to make chosen policies more robust by detecting changes in structure and functioning of large marine ecosystems, and identifying their underlying oceanic and atmospheric causes. Biological responses to climate variability are complex and not always well understood, although some recent progress has been made (McPhaden and Zhang 2002, Pitcher and Forrest 2004). To achieve this task, we need knowledge gained from non-biological fields, such as climatology, in order to look for early warning indicators of change in the oceans. Also, it is useful to try to use ecosystem modelling to evaluate worst-case climatic scenarios and assess risks in a quantitative fashion, attempting to forecast likely responses not only for commercial fish populations, including their food and predators, but also for charismatic or protected bird, mammal and fish species.

Two critical strategies that allow us include the influence of climate into the policies for the strategic management of fisheries are to devise spatial management schemes able to respond rapidly to early warning, and to ensure stakeholders and coastal communities are aware of risks and agree with contingency plans. Moreover, many fish are affected by climate, but there is no doubt that commercial fisheries have the immediate capability of driving slow-growing 'table' fish to extinction, a process that climate can merely slow or accelerate. The response of marine ecosystems to changing ocean and climate factors is not only relevant for local fisheries and food supply, but also for the functioning of basin-scale biogeochemical cycles (Miller and Schneider 2000).

3.2. MODELLING ECOSYSTEMS OF THE PAST

Reconstruction of ecosystems that resemble the past is a fundamental part of the BTF approach. Although estimation of model input parameters for past ecosystems is both highly uncertain and difficult, the mass-balance assumption of the *Ecopath* modelling approach provides a useful constraint with which to screen out unrealistic parameters. The resulting alternative ecosystems are thermo-dynamically realistic hypotheses about past states.

A model of the current ecosystem is a good starting point for past ecosystem reconstruction. Availability of present day data is often relatively better than that for the past, which enables the modelling of present ecosystems to be more confident. Although current ecosystem structure may have changed from its past state due to either anthropogenic (e.g., fishing, coastal development) or natural factors (e.g., climate change, large scale change in ocean productivity), we can obtain a qualitative picture of the past ecosystem based on the present. Hypotheses about factors leading to changes can be developed from both quantitative and qualitative evidence obtained from literature, archives, or descriptions from local communities. Moreover, comparable model structure between the past and present day ecosystem facilitates ecosystem comparisons in the BTF process (Pitcher 2004), although biomasses and fluxes can be vastly different over time. Global extinctions of species, such as the great auk in the North Atlantic (over half a million birds before 1830; Montevecchi and Kirk 1996) or Steller's sea cow in the North Pacific (Anderson 1995), mean that we have to eliminate these species that are included in models of the past from future restoration goals.

Data for modelling a past ecosystem may be assembled from published or unpublished scientific literature, government reports, archaeological data, historical information, and local and traditional ecological knowledge (Christensen 2002, Pitcher et al. 2002). In the absence of relevant local publications, interviews, conducted under suitable partnership agreements, have been employed to gather Traditional and Local Ecological Knowledge for use in the modelling (e.g., Ainsworth and Pitcher 2004a, Salas et al. 1998, Haggan et al. 1998). Qualitative description from historical archives or local communities can be particularly useful when formal scientific data are rare. Example of such descriptions can be:

> "*Sharks are found everywhere; they are especially common off Tai O, Cheung Chau Island, Lamma Island, Junk Island and Tunglung Island all the year. Groupers of 2 to 10 or more pounds in weight, snappers, chicken-grunt, spotted grunt, sea-breams, etc., are usually caught in the areas (Hong Kong waters)....*" Descriptions of Hong Kong marine ecosystem in the 1940s by Lin (1949).

Multiple sources of data help to fill in gaps and can provide cross-validation between uncertain data. For instance, Cheung (2001) reconstructed Hong Kong's marine ecosystem as it was in the 1950s based on scientific literature, government survey data and knowledge from local fishers and other marine resources users. Heymans (2003a) combined archaeological and anthropological data on human diets and populations from the ancient past to estimate aboriginal fishery catches in reconstructing an ancient marine ecosystem of Newfoundland, Canada. An integrated example for a species in the heavily-depleted Strait of Georgia system in BC, Canada is presented by Martell and Wallace (1998). A history of inshore habitat changes can be valuable (e.g., McLoughlin 2002).

Information about the history of local fisheries, with analyses and surveys, and about local aquatic fauna and flora is usually easily found, and may be encouraged as an output of 'science workshops' comprised of research partners and local scientists with expert knowledge of the area and the taxonomic groups (see Table 1). One of the principal problems here is data that has been gathered on either a very small or a very large scale compared to the area of focus. Another issue often requiring a lot of work is the concordance of measurement units, since specialists on different taxa often work in very different fields.

Scientists who generously make the relevant information available, often from a lifetime's work on a group of organisms, are encouraged to publish a paper in one of the project reports, so that they retain a recognised ownership of material that otherwise could easily vanish into model parameters. For example, in four BTF projects in Canada, the output from this process of consultation with the scientific community has been presented in a series of reports (Heymans 2003a, Ainsworth et al. 2002, Pitcher et al. 2002a and 2002b, Haggan and Beattie 1999, Pauly et al. 1998), where information essential to the modelling process, such as biomass, relative fishing mortalities and diets are assembled together with bibliographies. The breath of knowledge required to model an ecosystem is considerable, particularly for the past state. Therefore, contribution from experts on different fields relating to the marine ecosystem can greatly increase our understanding on the past ecosystem state. An expert workshop facilitated by ecosystem modellers, may be useful for the experts to share their knowledge and help estimate input parameters for the model. These experts can be from both the scientific and local (e.g., fishers or fishing-related practitioner) communities; the latter can contribute their local/traditional historical and ecological knowledge.

Assembling the collated information in relational database, together with evaluations of its scope and quality, facilitate the retrieval of relevant information for the models (e.g., Erfan 2004). Even so, a significant task is systematising the way in which information is collated for use in the models. The reason is that, once documented, all this information has to be expressed in a form that can be used in determining ecosystem model structure, in setting parameters, or in shaping dynamic responses to changes. Although presence and absence of a species is easily dealt with, the models require us to know actual biomasses, size and growth parameters, and items in the diet.

Similar ecosystems with available data that may resemble the past state of the targeted ecosystem provide useful references for past ecosystem reconstruction. In particular, an ecosystem that is protected from fishing or other human impacts may be similar to past under-exploited states of the modelled ecosystem. For instance, the Brunei marine ecosystem, historically protected from fishing, can provide valuable information when reconstructing the past ecosystem of the South China Sea continental shelf (Cheung 2001).

Comparison between simulated and observed time-series data facilitates testing of hypothesis put forward on the model parameters. Changes in ecosystem from the past to the present can be simulated with changes in fishing mortalities and climate forcing using *Ecosim* (Walters et al. 1997). Results from the simulation model can be compared with observed time-series data on abundance of ecosystem groups (e.g., Gulf of Thailand, Christensen 1998). However, this often requires a great deal of data on past fisheries and climate. Some practical solutions to the issues discussed here may be found, for a Canadian example see Heymans and Pitcher (2004).

When species have gone locally extinct ('extirpation'), one has the choice of eliminating them from models of the future ecosystem altogether, or allowing the possibility of them returning, either through natural migration or though active reintroduction (Pitcher 2004e). Examples of the former are the more than 200 Humpback whales resident until the 1920s in the Strait of Georgia (Winship 1998, Merliees 1985), or almost quarter of a million walrus resident in Newfoundland before 1800 (Mercer 1967). An example of the latter is sea otters reintroduced to Vancouver Island in British Columbia (Watson 2000, Watson et al. 1997). To accommodate dynamic ecosystem modelling, biomasses of groups have to be set at very low levels rather than at zero, and this creates some technical problems as they may undergo unexpected resurgence. Simulating changes to habitat can be tricky, moreover, when they are keystone species that cause large changes in habitat structure, such as the sea otter (Pitcher 1998b, Simenstad et al. 1978). Sea otters alter the type of kelp cover available to a suite of juvenile fishes and invertebrates by foraging on kelp-eating invertebrates that themselves graze selectively (Riedman and Estes 1990).

Another frequent problem is that reconstructions of the ancient past may suggest the presence of top predators in such large numbers that they are not able to be supported by what are thought to be realistic levels of forage organisms (e.g., reconstruction of the North Sea as it was in 1880: Mackinson 2001). We can likely rule out the answer that the apparent high abundance of top predators was a false impression, but perhaps such a high abundance of prey is actually acceptable. For example, the diet of abundant old predators would have been broader in the ancient past because of intense competition. Moreover, it is also likely that there were more forage fish species filling a diversity of niches in the ancient past (Pitcher 2004d). To answer these questions, we might look for evidence of ancient diets using archaeology and stable isotope analysis, and look for archaeological evidence of more forage fish species.

Representing changes in ecosystem structure over long periods of time represents a major challenge. Clearly, the effect of climate change has to be accommodated in the forecasts as much as possible. Primary production, and other parameters of the ecosystem model, such as stock-recruitment relationships, may be driven by a variety of forcing functions. In the major oceans, inter-annual variation may seek to accommodate El Niño/La Niña alternation, or proxies such as a local upwelling index, decadal oscillations or temperature anomalies (e.g., in the English Channel, Stanford and Pitcher 2004). Longer term climate cycles may be included in the forcing function, like the 62-year 'La Vieja/El Viejo' [old man/old woman] alternation between warm/cold eastern boundary current sardine/anchovy regimes (Chavez et al. 2003). In some cases, silica deposits may accurately reveal the past annual abundance of diatoms (e.g., Johnson et al. 2001). Although precise forecasts of inter-annual climate changes are not possible, randomized selections of such data may be used to drive forecasts on the basis of likely scenarios.

To emulate changes in species composition, the modelling system could perhaps be modified to use a 'cast of players', members of which might be brought on- and off-stage when conditions are appropriate (Pitcher and Forrest 2004). For example, in the Cueva de Nerja, Andalusia, middens reveal the fish that early Mediterranean people were eating over a 9000-year sequence (Morales et al. 1994, Rosello-Izquierdo and Morales-Muniz 2001). Early in the sequence, from about 14000 BP, fish in the human diet consisted of a sparid fauna similar to the present, but during a pluvial period between 11000 BP and 9000 BP humans were eating large cod and haddock, fish typical of Norway today. By 8000 BP, a typical Mediterranean fauna had returned.

Early periods of depletion by human exploitation also had significant impacts on ecosystem structure and function (e.g., Wallace 1998). Recent reconstruction work on North American inshore ecosystems by Jackson et al. (2001) shows what may be possible in this respect.

Ideally, the timing of the series of snapshot ecosystem models for BTF may depend on the locality, the dawn of quantitative documentary evidence, and major shifts in resource and ecosystem history such as the introduction of new fishing gears, damming of rivers and collapses of fish stocks. But because of the large amount of work involved in drawing up each ecosystem model, the gaps in time between a series of BTF models may end up being quite large. So an ideal choice of the time snapshots to use is generally constrained by the resources available for the research. This raises a significant methodological problem in that failure to cover important changes that occurred within these time gaps can prejudice the choice of appropriate policy goals at the end of the BTF process. In the event, the choice of the time periods to model in a BTF analysis is something of a compromise (Heymans and Pitcher 2004), and, in case studies to date, has entailed 3-4 ecosystem models spread over a hundred years or more.

In many cases, additional informative models might be drawn up for pre-modern humans in the late Pleistocene post-glacial era (Neolithic). Although such ancient ecosystems would be unlikely ever to become practical policy goals, they have the advantage of providing a 'pristine' baseline against which all more recent changes might be assessed. In fact, for some areas of the world only recently colonised by Europeans, such as Australia, New Zealand and the Pacific coast of America (Diamond 1997), models of 'pre-contact' ecosystems may serve this 'baseline' purpose well.

In models of the distant past, the estimation of the size and impacts of ancient fisheries presents many problems. Although the history of fishing technology is quite well known from archaeology and from Traditional Knowledge, its likely fishing power may be estimated, and ancient diets may be calculated, nevertheless, the size of the human populations that engaged in fishing is often hard to assess. Estimates of ancient human population sizes are often the subject of considerable controversy among archaeologists and anthropologists. Heymans (2003b) and Wallace (1998) present examples from Newfoundland and British Columbia respectively. Aboriginal fisheries in these ecosystems are described in order to provide an accurate picture of an ancient ecosystem. The same fisheries would not necessarily be chosen for a future BTF restoration policy; although aboriginal fisheries in some form may well be appropriate members of a future sustainable fisheries portfolio, as outlined below.

Finally, many of these problems may be eased if we are able to run a model of the past forward to simulate its change into a more recent ecosystem. Performing this using *Ecosim* is difficult and requires a great deal of data on past biomass changes and climate, but it has been possible for some ecosystems that have undergone rapid change, such as the Gulf of Thailand (Christensen 1998) and, as described in the next section, the Northwest Atlantic.

3.2.1. Exploring Ecosystem Change Through Time: A Northwest Atlantic Case Study.
The BTF approach explores changes in marine ecosystems over long periods of time in order to try to find systems that might serve as suitable policy goals. However, when a clear event happens and is known to represent a drastic perturbation to an ecosystem, a related approach that can be used, generally focussed on trying to understand changes in ecology, is to set out to compare the states of the system before and after the perturbation. This is the aim of the Comparative Dynamics of Exploited Ecosystems in the Northwest Atlantic (CDEENA) program for the east coast of Canada.

The CDEENA program was established in 1999 by the Department of Fisheries and Oceans of Canada in order to do a comparative analysis of changes in the structure and function of different Northwest Atlantic shelf ecosystems, and to determine how these may have affected the productivity of living resources. The objective was to develop individual ecosystem models for the northern Gulf of St. Lawrence (NAFO statistical zones 4R and 4S), southern Gulf of St. Lawrence (NAFO zone 4T), the Newfoundland and Labrador coast and offshore region (NAFO zones 2J3KLNO), and the Scotian Shelf region off the coast of Nova Scotia (NAFO zones 4VsWX). For each region, two models were built: one just prior to the cod collapse of the early 1990s, and one after the collapse in the mid-1990s. One of the desired outcomes of the project has been to determine why the collapsed demersal stocks have failed to recover, and why this failure seems to be more pronounced in some areas than others. To this end, CDEENA brought together the expertise of field scientists and modellers to: (1) describe the changes in time and space; (2) identify and fill critical data gaps in the knowledge base; and (3), develop models to investigate ecosystem-level hypotheses (i.e., environmental variation, predation, fishing effects) concerning changes in reproduction, mortality, growth, and feeding of cod and other species.

Morissette et al. (2004) developed models of the northern and southern Gulf of St. Lawrence (NAFO zones 4RS and 4T) ecosystems during the 1980s and the 1990s using *Ecopath*-with-*Ecosim* (Christensen et al. 2000). The models constructed for the Gulf of St. Lawrence ecosystems are well documented and their uncertainties fully explored (Morissette 2001, Morissette et al. 2003, Savenkoff et al. 2004 a,b), and so they may be used with some confidence to attempt to explain the causes of important

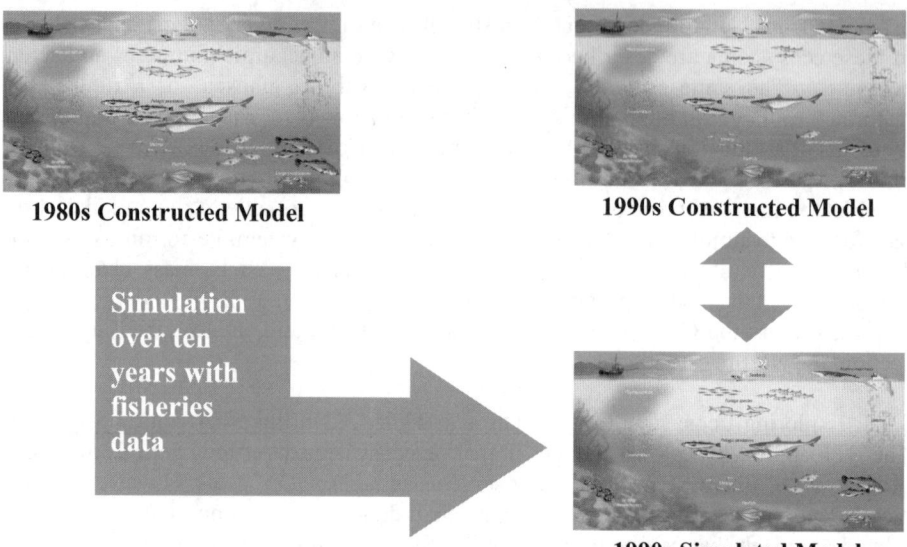

Figure 3. Conceptual framework of the ecosystem modelling approach in the Gulf of St. Lawrence, Canada, with the CDEENA program.

changes in the ecosystems between the 1980s and the 1990s.

To explore this issue, *Ecosim* models of the northern and southern Gulf of St Lawrence in the 1980s were run forward over a period of 10 years, driven by temporal forcing functions for fishing mortality and environmental effects. Once this was done for both ecosystems, the simulated 1990s version, driven by the same set of factors in *Ecosim*, was compared with the models constructed using 1990s data. It was then possible to evaluate the hypothesis that forcing functions of estimated fisheries mortality and environmental changes acting on specified biomass pools could explain the changes in the ecosystems (Figure 3).

Cod was formerly one of the most important commercial species in the Gulf of St. Lawrence. The cod stocks suffered a huge collapse in the late 1980s: a phenomenon that has been well documented (M. Harris 1998) but poorly explained. In the literature, there is no consensus for the primary cause of the collapse of cod fishery in the Gulf of St. Lawrence. Some authors suggest that it was mainly the result of a combination of environmental changes which led to a reduction in recruitment and an increase in natural mortality from predation, in combination with over-exploitation (Dutil et al. 1999). Others believe that the cod collapse can be attributed solely to overexploitation (Hutchings and Myers 1994, Walters and Maguire 1996). In nearby areas, it has been suggested that, whatever the reasons for collapse, predators such as seals may prevent recovery of the collapsed stock (Fu et al. 2001). The ecosystem models reveal that, in the Gulf of St Lawrence, a large part of the cod mortality in the models remains unexplained: Savenkoff et al. (2004c) suggested that much of this resulted from under-reporting of catches. This modelling suggests that overfishing, caused by undetected illegal catches, was sufficient to explain the collapse of this cod stock; historically the second largest in the northwest Atlantic.

This modelling research now aims to explore these different hypotheses as forcing functions in *Ecosim* in order to determine their relative importance in explaining the collapse of groundfish species in Gulf of St. Lawrence ecosystems. Factors causing important perturbations to the ecosystem are adjusted to improve the fit to the data between the two models. This work is still in progress, and the results are not yet published. However, the technique represents an important step in using ecosystem simulations as way of testing alternate hypotheses about what has caused large changes.

3.2.2. Exploring Ecosystem Change Through Time: A Gulf of California Case Study.
The Gulf of California in northwest Mexico is recognized as one of the most productive marine ecosystems in the world; it generates 40% of Mexico's total fish catch, providing jobs to nearly 30,000 fishers (INEGI, 2001). A critical and unique area of this large marine ecosystem is its northernmost area, the upper Gulf of California (UGC), which itself provides 15% of the national landings and has supported important fisheries since the 1920s. The historical richness of this area is explained by the nutrients and sediments provided by the Colorado River, and by extreme tidal and upwelling processes (Carraquiry and Sánchez 1999). However, since 1908 the U.S.A. initiated a series of dams and irrigation projects (1935: the completion of the huge Hoover dam; 1963: completion of Glen Canyon dam) that divert water flow from the Colorado River (Figure 4), and result in a dramatic decrease of nutrients and sediments (from 195 x 106 ton/year in 1930's to 86 x 106 ton/year during 1950's; Van Andel 1964). This has resulted in a cascade of ecological impacts, starting with a complete loss of the estuarine environment, a key element for spawning and nursery grounds of hundreds of species, the reduction of refuge habitats for more than 100,000 migratory

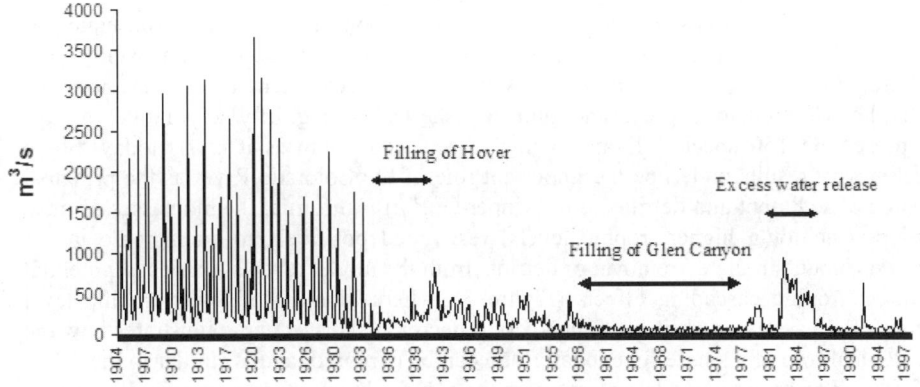

Figure 4. Total annual river discharge of the Colorado River, Mexico in the Gulf of California. The arrows represent the years when the two main dams in the Colorado River were constructed, and the excess water released during strong precipitation events associated with El Niño conditions 1982-83.

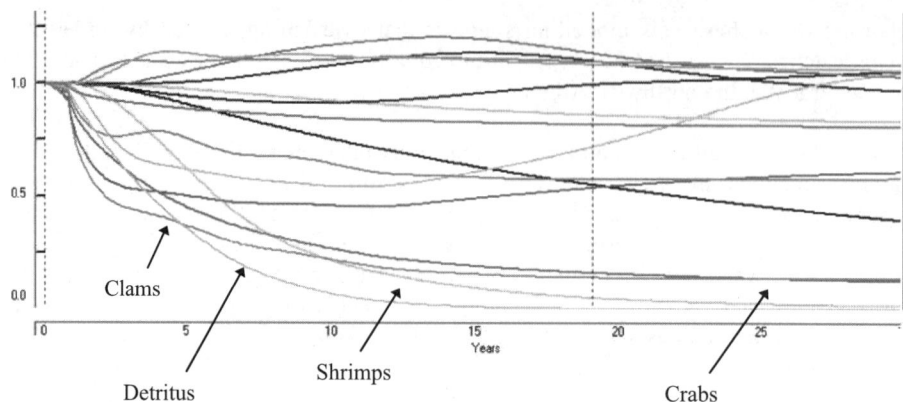

Figure 5. Simulated removal of detritus in the Upper Gulf of California ecosystem model; revealing not only potential cascading effects, where biomass of lower trophic level groups such as clams, shrimps, squids, crabs decrease, but also the relative importance of historical detritus accumulation from the Colorado River as a critical attribute in this ecosystem. Vertical axis indicates biomass relative to the present. Horizontal axis shows simulation run over 30 years. Lines indicate ecosystem model groups.

birds visiting the area each winter, and the loss of a traditional way of life for the indigenous Cocopah people. Large changes in physical properties of the UGC (Lavin 1999, Caraquiry and Sánchez 1999) have brought about a 94% reduction of benthic macrobenthos productivity; according to paleontological information, this ecosystem supported at least 50 clams/m2 during the last 1,000 years, in contrast to the present value of 3 clams/m2 (Glenn et al. 2001). Moreover, uncontrolled fisheries have brought several endemic species such as Giant Gulf croaker (*Totoaba macdonaldi*) and the vaquita porpoise (*Phocoena sinus*) near to extinction (Román-Rodríguez and Hamman 1997, Jaramillo-Legorreta 1999, D'Agrosa et al. 2000). The totoaba has the unenviable distinction of being the first marine fish listed under CITES and ESA. The economic and ecological crisis in the area during the late 1980's led to the UGC being declared a Biosphere Reserve in 1993 (Gómez-Pompa and Dirzo 1995).

In order to answer the critical questions about environmental consequences deriving from the presence of the huge dams along the Colorado River, a Back-to-the-Future project was initiated in 2002. A present-day ecosystem model describes the interplay of predators, prey, and human fisheries among fifty key model groups (representing 150 species) living in this marine ecosystem as it exists today. Some preliminary results underline the important role of the Colorado River as the principal source of sediment and detritus to the Upper Gulf of California; the biomasses of most groups (including higher trophic levels) responded positively to an increase in the detritus input; simulated removal of detritus from the model reveals a series of potential indirect trophic cascading effects (Figure 5), where biomass of lower trophic level groups such as clams, shrimps, squids, crabs decrease. This example illustrates how the ecological role of a single functional group can be explored: and is similar to the effect of adding or removing salmon carcasses from a freshwater/forest system (Watkinson 2001). Dynamic simulations may also emulate changes in fishing effort and other disturbances. In the next phase, the project will reconstruct past stages of the UGC ecosystem attempting to capture changes in its structure and form, and begin to quantify ecological impacts attributable to diversion of the Colorado River.

3.2.3. Exploring Ecosystem Change through Time and Space: A South China Sea Case Study. Although the BTF approach puts a strong focus on the reconstruction to remediate temporal ecosystem changes, exploration of spatial dynamics are useful particularly when ecosystem changes have been strongly influenced by spatial factors, such as habitat destruction or ocean current changes and to understand the implementation of spatial management tactics (e.g., Marine Protected Areas or MPAs). Here, we present a case study that uses spatially explicit ecosystem modelling tool (*Ecospace*) to generate alternative hypotheses about the effect of various spatial fishery management tactics in the northern South China Sea (NSCS) ecosystem and its implications for conservation of marine biodiversity (Cheung and Pitcher, submitted).

We defined the NSCS as the continental shelf (below 200 m depth) ranging from 106o53'-119o48' E to 17o10'-25o52' N (Figure 6). The area falls within the Exclusive Economic Zone of the People's Republic of China and, therefore, fishery resources have mainly been exploited by Chinese fishing fleets. Over the past five decades, dramatic expansion of these fleets, accompanied by mechanization and other technological advances, has resulted in over-exploitation of the near-shore, and later, the offshore fisheries resources (Shindo 1973; Cheung 2001). For example, catch rates of Chinese trawlers in NSCS dropped by more than 70% from 1986 to 1998 (Lu and Ye 2001).

Dynamics of the NSCS ecosystem were modelled using *Ecopath*, *Ecosim* and *Ecospace*. The model includes 41 functional groups ranging from phytoplankton, zooplankton, benthic invertebrates, demersal and pelagic fish groups, to marine mammals and birds, which are fished by 7 different fishing fleets (for earlier versions of this model see Buchary et al. 2003; Cheung et al. 2002, Pitcher et al. 2000). We defined the model area on a 72 x 22 grid (Figure 6a), each cell representing a 25 km2 area Four habitat types were included: (1) natural reef (coral and rocky reef), (2) estuary area, (3) shelf less than 50m depth, (4) shelf between 50 m and 200 m depth. In the spatial simulation model, each functional group had different affinities for particular habitats based on their ecology, movement of the functional groups among habitats was determined by a habitat gradient function (Walters et al. 1999).

To represent spatial differences in primary productivity, we used a map of primary production of the NSCS[1] on our grid map (Figure 6b), dividing the data into four bands representing 1, 0.75, 0.5 and 0.25 times the baseline value (used as *Ecosim* input). Stochastic spatial and temporal changes in productivity were, however, not included in model.

---

[1] Using data from the Inland and Marine Waters Unit (IMW), Institute for Environment & Sustainability, EU Joint Research Center (JRC), Ispra, Italy as processed by S. Lai, Sea Around Us Project, Fisheries Centre, University of British Columbia, Vancouver, Canada.

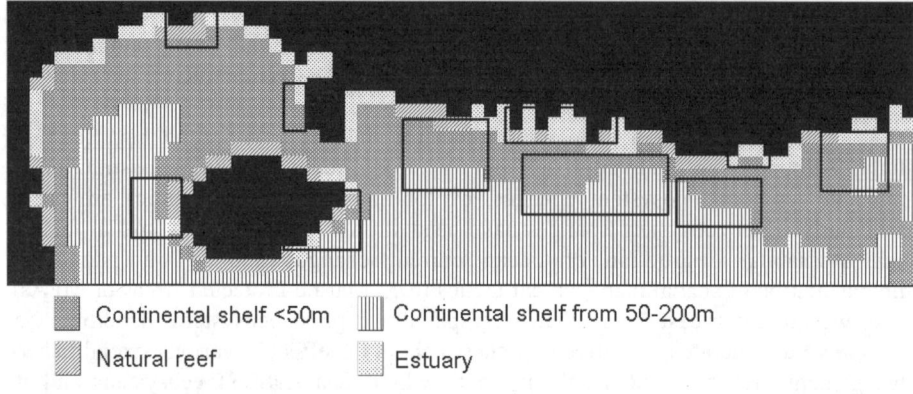

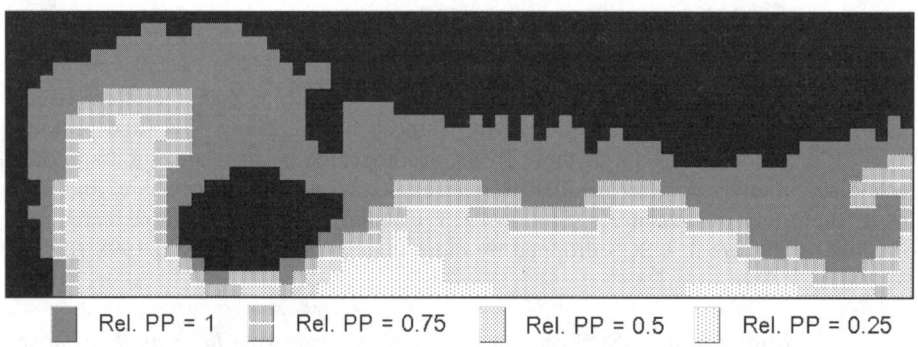

Figure 6. Base map of the northern South China Sea used for the spatial dynamic simulations in Ecospace. Figure 6a (upper) Map of habitat types. The rectangles represent the approximate locations where MPAs were designated in the simulation. Figure 6b (lower) Map of relative primary productivity (Rel. PP) ranging from 1 to 0.25 times to the base level

We explored the possible effects of a range of fisheries management scenarios with different spatial management tactics on the ecosystem. The scenarios included: (1) a two-month seasonal trawl moratorium as currently implemented by the Chinese Authorities; (2) small MPAs representing 5% of total area; (3) medium MPAs representing 10% of the area; and (4) large MPAs representing 20%. Various indexes (biodiversity, Q90; Ainsworth and Pitcher 2004b; local extinction index; Cheung and Pitcher 2004) were used to evaluate the changes in biodiversity and the risk of local extinction associated with the marine fishes.

The modelling results suggest that large marine protected areas and a considerable reduction of the current level of fishing effort would be needed to restore the biomass of most functional groups back to the levels obtaining in 1990. It is interesting that, in the model, the 2-month seasonal trawl ban does not help reduce biodiversity loss or extinction risk. Also, fish groups with different habitat associations showed differential responses to the policy scenarios. Increased benefits in terms of biomass recovery with larger MPAs are more apparent for reef-associated fishes.

This case study suggests that a spatially-explicit ecosystem model can be used to explore strategic management options for the NSCS marine ecosystem, in attempting to restore a degraded ecosystem back to a more desirable state.

4. Ecosystem-Based Methods For Devising Sustainable Fisheries

A marine ecosystem restored to some semblance of its past state might be thought of as a 'Lost Valley' Pitcher et al. 2004), an ecosystem, like Arthur Conan Doyle's Lost World (Doyle 1912), discovered complete with all of its former diversity and abundance of creatures. The BTF process aims to describe a series of such 'Lost Valleys' as a set of potential restoration goals.

The 'Lost Valley' must be fished sustainably if there is to be any point to the restoration; if today's fishing fleet were to fish a restored ecosystem, massive depletion would soon re-emerge (e.g., Hong Kong; Pitcher et al. 2004). Nor is it realistic to expect the fishing gear and methods of former times, including those of aboriginal fisheries, to be re-employed. Nevertheless, some of today's fisheries may target highly desirable species, have low by-catch, low operating costs, or ease of construction and use, and so it is evident that rational criteria for the selection and operation of sustainable fisheries need to be operated.

4.1. CRITERIA FOR SUSTAINABLE AND RESPONSIBLE FISHERIES

Rational criteria for opening a portfolio of responsible and sustainable fisheries have been devised as part of the BTF process (see tables in Pitcher 2004, Pitcher et al. 2004), based on the principles embodied in the FAO Code of Conduct for Responsible Fisheries (FAO 1995) and in WWF's guidelines for Ecosystem-Based Management of Marne Ecosystem (Ward et al. 2002). The following list of nine criteria for opening 'Lost Valley' fisheries is meant to be operated in a hierarchical fashion.

1. Fisheries should produce minimal by-catch discards. Over the past decade, trawl, trap and purse seine fisheries have attained impressive reductions in by-catch through the use of separators, lifters, gates and excluders (review: Kennelly and Broadhurst 2002), and by altering fishing practices (e.g., dolphins released in tuna purse seine fisheries, Hall 1988; long-line sets modified to reduce hook mortality in seabirds, Brothers et al. 1999). These technological advances may be successfully employed to reduce greatly the by-catch of non-target species of fish, marine mammals, reptiles, birds and invertebrates. Moreover, in some jurisdictions, discards have become illegal (e.g., Norway).

2. Fisheries should minimise damage to habitat. Unmodified bottom trawls and dredges cause great harm to sessile benthic invertebrates (e.g., sponges, gorgonids, corals), whose architecture acts as refuge habitat for small fish and invertebrates important in the food chain, and the juveniles of many commercial fish species (Kaiser et al. 2001, Hall 1999). To meet this criterion, technological improvements to the fishery will have to be employed to minimise damage, for example by permitting only trawls that fish above the bottom. Where some collateral damage to benthos is inevitable, such as in prawn trawls, large and progressive reductions in damage, say ten-fold, might be mandated, perhaps with some permanent closure of areas to preserve bottom structure and biodiversity.

3. Fisheries should include aboriginal, indigenous or customary fisheries. Many indigenous or aboriginal peoples operated fisheries that were sustained over thousands of years (e.g., eulachon, salmon, and halibut in the Pacific Northwest, Richardson 1992). For reasons of equity, these fisheries should be included in the candidate portfolio, provided the catch is sustainable, and in jurisdictions where customary rights are recognised. These fishers often have an intimate knowledge of coastal marine ecosystems (e.g., Johannes et al. 2000), and their support for BTF policy can enhance compliance with regulations.

4. Fisheries should include traditional target species. Provided Criteria 1, 2 and 3 above are satisfied, this criterion is included because of demand for traditional desirable fish species in local fishing communities. For example, the historic Atlantic halibut fishery has not proven sustainable, but halibut would for sure be in demand in a restored ecosystem.

5. Fisheries should minimise risk to charismatic species. Many 'charismatic' species, such as seabirds, whales, seals and sirenians, are very sensitive to exploitation by humans (e.g., Dulyy et al. 2003, Roman and Palumbi 2003, Mowat 1984). This criterion can easily be in conflict with Criteria 3 and 4 above, since indigenous peoples traditionally exploited seals, sea lions, whales, porpoises, dugongs, turtles, ducks, gulls, petrels, and auks (e.g., in British Columbia: Brown et al. 1997, in Australia: Williams and Baines 1993). Where customary rights are recognised, an aboriginal take of these species in 'Lost Valley' fisheries would be allowed under Criterion 3, where there is appropriate consent under Criterion 7 below. On the other hand, many marine mammal, bird, and even shark species have only recently become 'charismatic' to the conservation movement, and in some places legal bans on their capture reflect public revulsion at their use for human food. But such views may be quite volatile and local, so in the last resort, the choice of whether to exploit these types of animals should be locally or nationally determined. From a fisheries perspective, the only rational criterion is avoidance of excessive depletion, and minimizing the risk of extirpation.

6. Fisheries should not catch juvenile groups. Generally, heavy fishing on juveniles increases the risk of recruitment failure, so such fisheries would not normally be allowed in the 'Lost Valley' portfolio. Some traditional fisheries allowed under Criterion 4 may include eggs, fry and juveniles of highly fecund species such as herring, anchovy, sardines, milkfish or hake. These fisheries would be permissible where impacts can be proven to be minimal. Likewise, fisheries might be permitted on restricted numbers of juveniles where adults live and spawn in refuges from fishing, as in traditional Mediterranean fisheries (Caddy 2000).

7. Fisheries should be vetted and approved by all participating stakeholders. The local fishing community must vet and approve the candidate portfolio of 'Lost Valley' fisheries, notwithstanding Criteria 3 and 4. In addition, the management agency must be convinced with science-based evidence that chosen gears are appropriate (Criteria 1 and 2), that management and monitoring (Criterion 9) are feasible for the chosen fisheries, and that the scientific basis of the forecasting (Criterion 8) represents best practice. Such wide ownership of the fishery process will make for better compliance with the fishery regulations that seek to control the 'Lost Valley' fisheries.

8. Fisheries should be demonstrated as sustainable in an ecosystem context using appropriate simulations. Assessments must show that under the impact of the chosen fishery portfolio, the biomass of the main ecosystem groups, biodiversity, risk of local extinctions, and catches in the 'Lost Valley' will not fall below agreed reference levels over a 100-year period. These assessments must evaluate changes in the entire food web, preferably using indices based on whole-ecosystem simulations tuned to field estimates of biomass of as many species groups as possible. Forecasts should be made robust against anticipated climate changes, and define uncertainty to a specified level of risk. The crucial importance of ecosystem-based analysis is evident here, since, on their own, single-species stock assessments cannot show risks to charismatic or non-target organisms, or to sessile organisms that provide important structural cover. Criterion 8 describes a critical part of the process: examining trade-offs of ecological with social and economic objectives using as a wide a range of indices as possible.

9. Fisheries should have an adaptive management plan in place. Environmental changes such as climate, pollution, and our ignorance of fundamental ecological processes, often lead to the unexpected in natural ecosystems. It is therefore prudent for the restored 'Lost Valley' and its fisheries to be subject to regular monitoring of the indices from criterion 8. This would allow adaptive shifts in fishing, much like the way that catch quotas and fishing locations are regulated today, but driven by an ecosystem approach.

A candidate portfolio of fisheries designed with Criteria 1-9 can be evaluated by assessing its conformity with the FAO Code of Conduct for Responsible Fisheries (FAO 1995) using a rapid appraisal technique (Pitcher 1999). Although a set of fisheries may be designed and evaluated in this way for the 'Lost Valley' modelling process, in practice it is likely that continual adjustment will take place once a BTF policy is implemented.

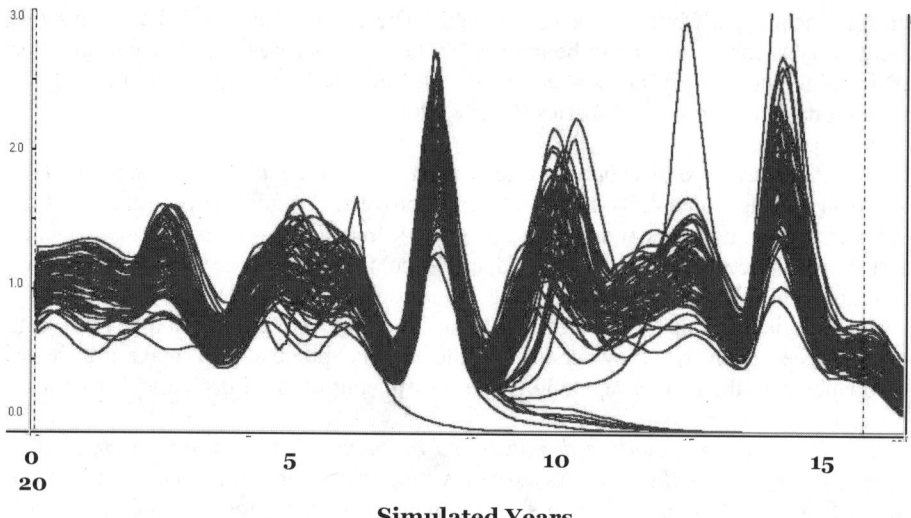

Figure 7. Example of ecosystem-based Population Viability Analysis. Lines shows forecast biomass of a small pelagic zooplanktivorous fish (Usipa: Engraulicypris sardella), one functional component of a 36-group ecosystem model of Lake Malawi driven for 20 years by a random selection of annual primary production values taken from historical series of biogenic silica in lake sediments (Pitcher and Nsiku, in prep: silica data from Johnson et al. 1998). Biomass is plotted relative to starting biomass of 1. Each line represents a simulation based on one selection of parameters, taken from prior probability distributions of parameter values that achieved a mass-balance criterion: 100 such lines are superimposed. Six out of the hundred runs lead to extinction of this species. The starting values for this analysis came from a 'Lost Valley' optimisation for ecological values.

4.2. OPENING THE 'LOST VALLEY'

After an 'ideal' set of fisheries have been selected according to the Criteria 1-7 discussed above, simulations are used to forecast fishing and its effect over a long time period, typically 50 or 100 years (Criterion 8). Relative fishing mortalities over the set of fisheries are adjusted until catches are sustainable, and impacts on the ecosystem meet specified criteria: this process has been termed 'opening the Lost Valley' (Pitcher 2004c). The adjustment is carried out iteratively using an automated search routine.

Using a multi-dimensional Davidon-Fletcher-Powell search algorithm, a 'policy' search routine in *Ecosim* seeks to maximize a specified objective (Christensen and Walters 2004, Walters et al. 2002). The search iteratively varies the fishing mortality per gear type to maximize the specified objective function over the simulated time horizon, usually 50 or 100 years. Alternative fishery objectives may be selected, including economic value, numbers of jobs, the biomass of long lived species, or a conservative portfolio utility function. These may be used as single objectives, or attempts made to mix then in a desired ratio: for example, it may be thought that an equal balance of jobs, profit and ecosystem values would be a desirable objective (Cochrane 2002).

When a policy search is started in the BTF 'Lost Valley' process, the species landed by each of the chosen fisheries are related to the model groups. Initial small catches for each fishery are entered (e.g., 2.5% of starting biomass), along with any discarded by-catch, ex-vessel prices by species and gear, and relative operating costs by gear. Before the search is started, the basic parameters of the underlying *Ecosim* model may have to be readjusted slightly to achieve mass-balance: for replicability, this may be performed with an automated procedure (Kavanagh et al. 2004). Often, searches have to be repeated many times fro different starting points to reduce the chances of finding a local optimum, or to distinguish among multiple optima in some cases (see Ainsworth et al. 2004).

The results of the search maximise the chosen objective and provide forecast fishery catches, biomass, economic values, numbers of jobs, and biomass changes in all other groups in the fished 'Lost Valley' ecosystem. Results are examined by running the ecosystem simulation for 50 or 100 years. Scenarios that cause extirpation, or severe depletion of species, are eliminated from consideration as viable policy goals. In fact, the searches may be re-run with the biomass of designated species protected from large changes as part of the policy search objective function (Cochrane 2002). Adjustments to the weightings in the objective function enable (after some iteration) policies that attempt to balance economic with ecological or social values. The search procedure is repeated for a wide range of policy objectives and for each candidate restored ecosystem, producing a number of forecast scenarios that may be compared.

In addition, we may seek to challenge these results with climate changes that might realistically be expected for the locality in question, and taking account at the same time of the principal uncertainty in the simulation modelling. These can be achieved by driving the simulations with various types of climate forcing functions, and with semi-Bayesian Monte Carlo simulations. When this is done, in effect we have a population viability analysis on each of the main groups in the ecosystem (Figure 7). Fished 'Lost Valley' scenarios that aim to minimise extinction risks can therefore be chosen.

5. Choosing Ecosystem Restoration Goals

Using the techniques described above, the BTF process provides us with a snapshot of what a set of alternative restored ecosystems, complete with sustainable fisheries, might look like. The next problem is to choose the best restoration goal from among them. We use an objective method to compare the benefits that will accrue to society from each alternative future represented by a fished 'Lost Valley' ecosystem (Figure 8). The present day ecosystem, with a new portfolio of fisheries designed to be sustainable, is included in the comparisons in order to show the full range of options that may be considered. It might also be useful to include ecosystems even further depleted than that of the present day, especially if we wish to evaluate the advantages and risks associated

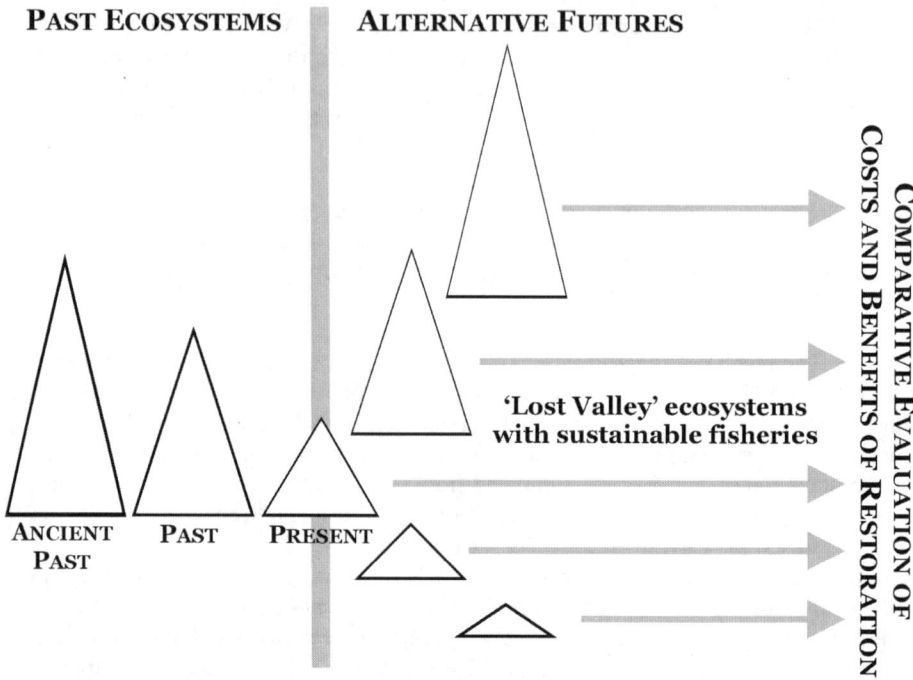

Figure 8. Diagram illustrating the concept of sustainable fishing of restored 'Lost Valley' ecosystems, and the comparative evaluation of their costs and benefits. Triangles represent ecosystem models of the past and possible futures, as in Figure 1. Arrows represent sustainable fishing by responsibly designed fisheries in restored ecosystems.

with increased food fishing for large amounts of lower trophic level organisms (e.g., krill).

A fundamental way to evaluate the benefits of alternative restored ecosystems is to compare the economic value of their fisheries, information that is readily estimated from the *Ecosim* simulations mentioned above. But how we determine the economic "success" of restoration will depend greatly on how we value the expected stream of benefits.

In cost-benefit analysis (CBA), economists use conventional discounting to summarize the expected stream of benefits from an investment into a single term, the net present value (NPV), which can then be compared among alternative investments.[2] In this calculation, the value that future benefits carry in the present is discounted exponentially through time; this reflects the investor's preference for immediate consumption and delayed payment. The practice of discounting takes account of lost opportunity costs, uncertainty, impatience, and other factors that make a unit of money today worth more than the same unit of money tomorrow. In this fashion, discounting

---

[2] It is unrelated to the inflation rate.

acts a model of human behaviour; it is an analytical tool to help make our value-based decisions.

The discount rate applied in the cost-benefit analysis has a critical influence in determining whether an investment is seen as worthwhile. However, in valuing a long-term conservation plan like an ecosystem restoration project, or a sustainable fishery, immediate costs tend to outweigh far-off benefits, so that only 'myopic' policies can be viable at any practicable level of discounting (Sumaila 2001). As detailed below, it has been suggested that over-exploitation of Canadian Atlantic cod may be blamed, at least partly, on high discount rates applied by fishing consortia, and not just to other commonly stated problems like open access, bad management etc. (Clark 1973).

For this reason, Sumaila and Walters (2003, 2004) have devised a new form of discounting to evaluate long-term environmental initiatives. It is set apart from other intergenerational discounting techniques (e.g., Fearnside 2002, Weitzman 2001, Chichilnisky 1996) because it allows us to divide explicitly the stream of benefits between ourselves and our children. The method incorporates into each year of the analysis an annual influx of investors, who bring a renewed perspective on future earnings; this partially resets the 'discounting clock'.

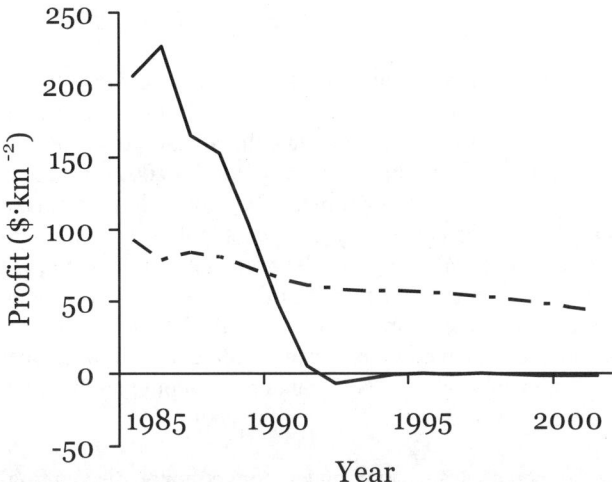

Figure 9. Profit profile from the actual Grand Banks cod harvest (solid line) versus estimates from a conservative optimal policy (broken line). The conservative policy is optimized for NPV at $\delta = 9.3\%$ (the rate of return for an alternative investment - see text). Data from DFO statistical areas 2J3KLNO off the Grand Banks, Canada..

If this system were adopted, not only does it meet the Aboriginal criterion of planning for the '7th generation' (Clarkson et al. 1992, Haggan et al. 2004), but it

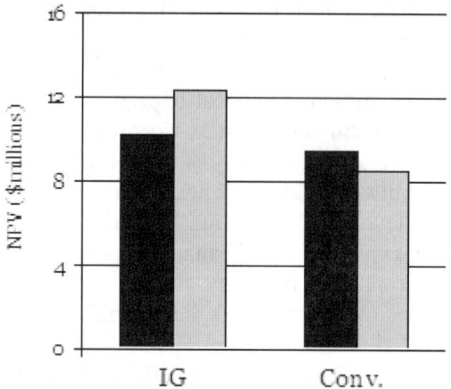

Figure 10. Conventional (Conv) and intergenerational (IG) net present value (NPV) of actual Newfoundland cod harvest profile (black) versus a simulated optimal sustainable harvest profile (grey). Under conventional valuation, depletion of the resource is better than conservation, but under intergenerational valuation, sustainability becomes worthwhile.

would satisfy mandated national and international requirements to manage for the benefit of future as well as present generations.

5.1 ANALYSIS OF THE NEWFOUNDLAND COD COLLAPSE

A cost-benefit analysis of the Atlantic cod fishery prior to the 1992 collapse provides an example of the ability of the intergenerational method of Sumaila and Walters (2003) to preserve resources. Figure 9 shows the actual stream of profits taken from Canadian government records (solid line) from 1985 to 2001, and the predicted profits from a sustainable harvest strategy optimized for economic performance (dotted line). The optimal profile was calculated by *Ecosim*'s policy search routine (see above; Walters and Christensen 2004) using the Newfoundland ecosystem model of Heymans (2003a). It represents a theoretical maximum: the greatest total value that could be removed from the system (see Ainsworth and Sumaila 2003)[3]. Although the optimal plan generates more catch overall, the front-loading of revenues seen in the actual harvest profile results in a higher NPV under conventional discounting due to the immediacy of benefits seen from the 1985 vantage point.

Figure 10 demonstrates that under conventional discounting it made more sense for fishers to deplete the stock than to conserve it (at a standard discount rate (δ) of 9.3%, meant to represent the potential return from an alternative investment[4]). That

---

[3] The optimal policy search delivers a single fishing mortality, which is continuously applied to the harvest simulation throughout all years. This is why the optimal policy is sustainable, and does not include sudden profit taking towards the final years of the simulation.

[4] The value used to represent the rate of return for an alternate investment ($\delta = 9.3\%$) corresponds to the average annual rate of return for Bank of Canada long-term (10+ years) marketable bonds between 1981 and 2001 (GOC 2002).

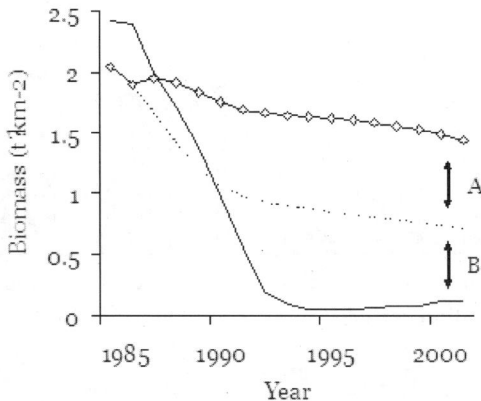

Figure 11. Biomass profiles of actual cod fishery (solid line), intergenerational discounting optimum (diamonds), and conventional discounting optimum (dotted line) at $\delta=9.3\%$. B= Depletion that may be blamed on ineffective management; A = depletion that may be blamed on application of conventional discounting.

is, the actual harvest profile (black) is worth more than the sustainable harvest profile (shaded) under conventional valuation. Under intergenerational discounting, however, future benefits carry more weight in the NPV term, and so the sustainable policy becomes the more valuable option.

Figure 11 compares the actual cod biomass profile (solid line) with biomass resulting from two optimal solutions: one that maximizes conventional NPV (dotted line) and one that maximizes intergenerational NPV (diamonds). If we can assume that industry operates at a discount rate of 9.3%, then the difference in end-state biomass between the actual profile and the conventional optimum (B) represents depletion that may be blamed on ineffective management. The difference between the intergenerational and conventional results (A) may be blamed on the application of conventional discounting.

Conventional discounting is said to serve as a model of human behaviour, but Ainsworth and Sumaila (2003) use a cost-benefit analysis of education to demonstrate that it does not adequately account for value that people place on their children's welfare. Using intergenerational discounting to value future benefits may enable environmental policies that favour conservation and rebuilding of natural resources to be more easily selected, and hence provide for the needs of future generations. In BTF analysis, both conventional and intergenerational discounting methods are used and the results compared, to try to take account of the 'intergenerational externalities' identified by Padilla (2002).

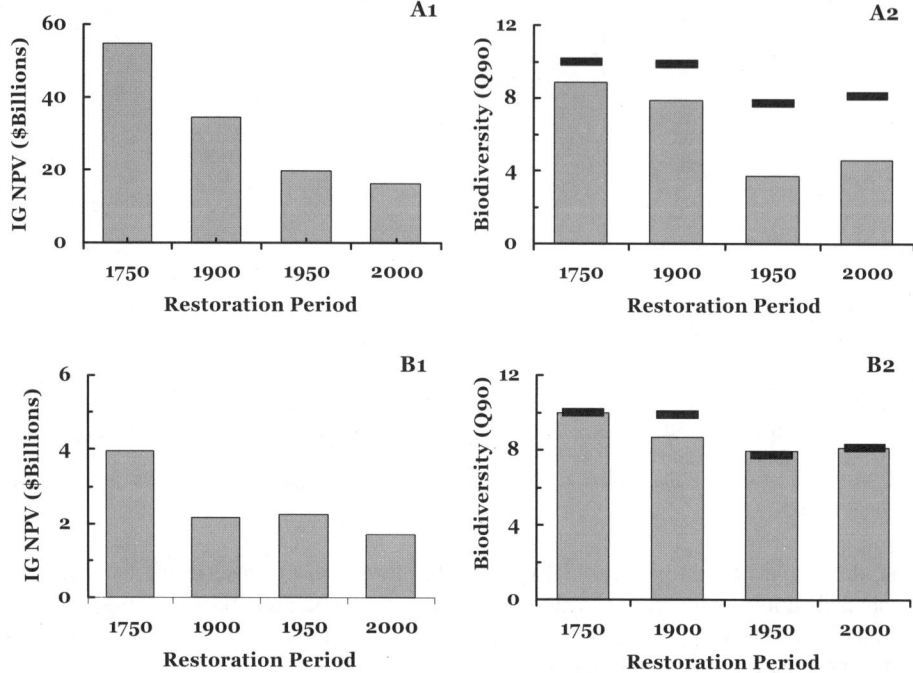

Figure 12. Economic and ecological potential of each BTF restoration target in Northern British Columbia, representing restored ecosystems of 1750, 1900, 1950 and 2000 fished with responsible and sustainable fleets. A1 and A2: Harvest policy optimized for economic performance (economic harvest objective). B1 and B2: Harvest policy optimized for ecological maintenance (ecological harvest objective). A1 and B1: Maximum sustainable fisheries benefits over 100 years (intergenerational NPV). A2 and B2: Equilibrium biodiversity after fishing harvest, cross bars show initial system biodiversity (Q90).

5.2 EXAMINING TRADE-OFFS

For policy aimed at the strategic management of marine ecosystems, purely economic considerations, however, are not considered sufficient. The simulations in *Ecosim* also provide an estimate of the number of jobs that might be supported directly by in the fishery, at least those directly involved: fishery processing and marketing sectors are not included in the simulations, but might, for a particular fishery, be estimated from the total catch, and they could then be added in.

The simulation technique covers all biological components of an ecosystem; it is also possible to examine the effect of the 'Lost Valley' fisheries on a number of measures of system integrity and biodiversity. For example, system resilience may be estimated by a whole-ecosystem index (Heymans 2004, Ulanowicz 1999). A biodiversity index modified for use with this type of modelling (a modified Kempton

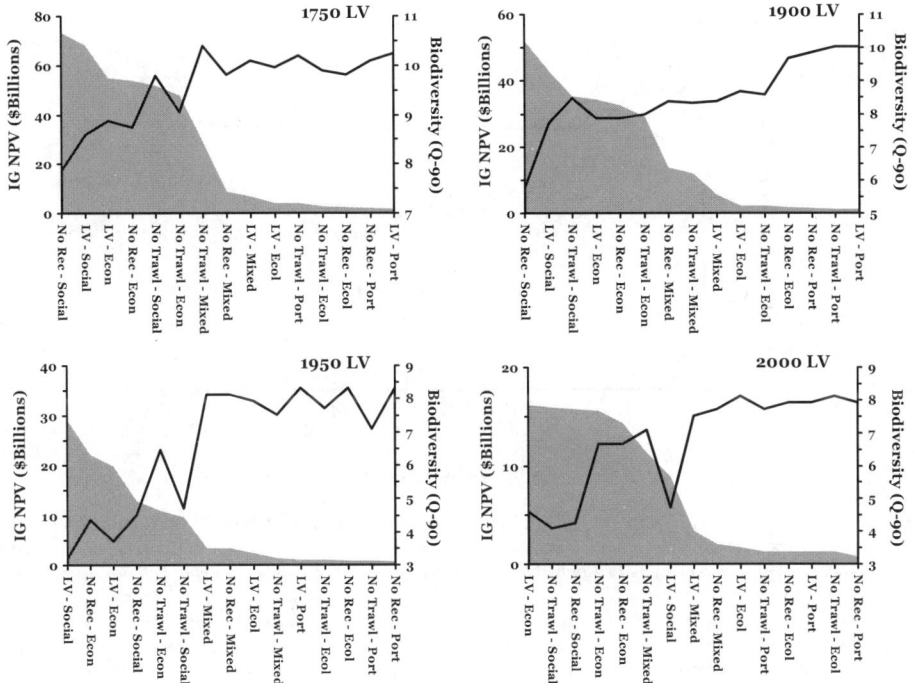

Figure 13. The trade-offs between economic value (intergenerational net present value, IG NPV, left Y-axis, shaded area), and biodiversity (Q90, right Y-axis, solid line) for fifteen fished 'Lost Valley' scenarios based on the northern British Columbia marine ecosystem restored to its state in 1750, 1900, 1950 or 2000, and fished with fleets that meet stated BTF criteria for responsibility. Each scenario represents the result of a single optimal policy search that maximises an objective function (see text). Scenarios are sorted in order by total monetary value. IG NPV is in US$ billions over 100 years of fishing ($\square$ = 4%; $\square IG$ = 10%); end-state biodiversity is represented by the modified Kempton Q-90 statistic. In the scenario labels, 'social' maximised jobs, 'econ' maximised IG NPV, 'ecol' maximized B/P ratios, 'mixed' represents roughly-balanced three-way objectives, and 'port' was a portfolio utility optimisation. Scenarios that generate large revenues tend to sacrifice biodiversity and vice versa.

index, Q-90, Ainsworth and Pitcher 2004b) may be used for comparison among scenarios. Another index, expressing the risk of local extinction, can be calculated from simulation results based on life history and fishery parameters (Cheung and Pitcher 2004).

By expressing the 'success' of each restoration and fishing plan in diverse economic, social and ecological terms, we can examine inherent trade-offs in the scenario. The final fishery plan we choose, and the 'Lost Valley' historical period to which our conservation measures should aspire, will depend on what relative weightings that policy makers choose to assign to these competing interests. The

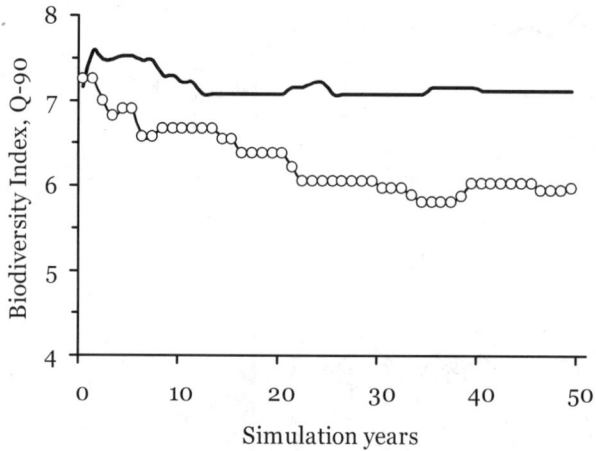

Figure 14. Example of biodiversity (modified Kempton index, Q-90) tracked over 50-year optimal harvest scenarios in a Northern British Columbia marine ecosystem restored to a candidate 'Lost Valley' and then fished (from year 0) with a designed responsible fishing fleet. This 'Lost Valley' example is the ecosystem in its estimated state in 1750 before contact of native Peoples with Europeans. Objective functions have been maximised by searches. Open circles show results for a mixed ecological/economic objective, solid line shows results for an ecological objective. Note that fishing when an economic objective is included progressively depletes biodiversity – see also Figure 13 (Modified from Ainsworth et al. 2004).

quantitative BTF methodology described makes those trade-offs explicit. The participation of all interests in developing the models and future scenarios increases likelihood that restoration targets will be met and management regimes observed.

5.3 'BACK-TO-THE-FUTURE' CASE STUDY FROM NORTHERN BRITISH COLUMBIA, CANADA

For the ecosystem of northern British Columbia, we have developed four historic models that represent significant periods in the development of the marine resource: prior to European contact (c. 1750), before the introduction of steam trawlers (c. 1900), during the heyday of the Pacific salmon fishery (1950), and at the present day (2000). Having designed an idealized fleet based on responsible criteria as described above ('Lost Valley' fleet; Pitcher et al. 2004), we use the optimization routine in *Ecosim* to determine the combination of fishing mortalities per gear type that will maximize a stated fishing objective. We apply those optimal mortalities to a 100 year simulation (50 years dynamic, 50 years equilibrium), and evaluate the resulting harvest profile and ecosystem integrity to gauge the merits of each restoration goal.

Ainsworth et al. (2004) test five harvest objectives that, together, span the spectrum of human use versus conservation. A social objective maximizes employment, an economic objective maximizes profit, an ecological objective increases

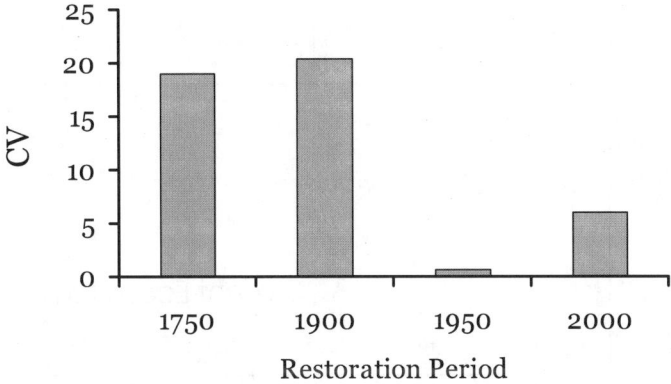

Figure 15. Variability in biomass (CV) for four restoration goals for northern British Columbia, averaged across 54 functional groups at the end state of 50-year ecosystem simulations. The model was subjected to Monte-Carlo sampling from even distributions of all biomass parameters, based on quality of the data used in each a case, and filtered by the need to achieve mass-balance (From Ainsworth et al. 2004).

the abundance of long-lived species in the system, a mixed objective combines priorities, and a conservative portfolio function, which also combines priorities, includes risk aversion (Walters et al. 2002; Walters and Christensen 2001). For a complete description of BTF methodology applied to northern British Columbia see Ainsworth et al. (2003).

Figure 12 shows the economic and ecological potential of each restoration target. The monetary value of the optimal harvest profile is measured using intergenerational net present value (Sumaila and Walters 2003, Ainsworth and Sumaila 2004); end-state equilibrium biodiversity is measured using the Q90 statistic (Ainsworth and Pitcher 2004b) and describes the condition of the ecosystem after 100 years of optimal fishing. Additional indices are examined in Ainsworth et al. (2004), who determine the robustness of policy recommendations through examination of the optimization response surface.

Overall, the restored-and-fished 1750 ecosystem emerges as the most attractive restoration goal. In monetary terms, the sustainable take from the system is more than ten times the current real-world profit enjoyed from the fisheries of northern British Columbia. Most of that benefit can be attributed to restored piscivore and shellfish biomass.

The 1750 'Lost Valley' ecosystem is the least similar to the present, however, and so it represents the most ambitious restoration project. Restored 1900, 1950 and 2000 ecosystems follow in economic potential. Under the harvest strategy designed for economic performance, the 1750 system sees the least relative and absolute decline in biodiversity upon harvest. Under the harvest strategy designed to minimize ecological impact, the economic value of the fisheries (even from the almost-pristine 1750 system)

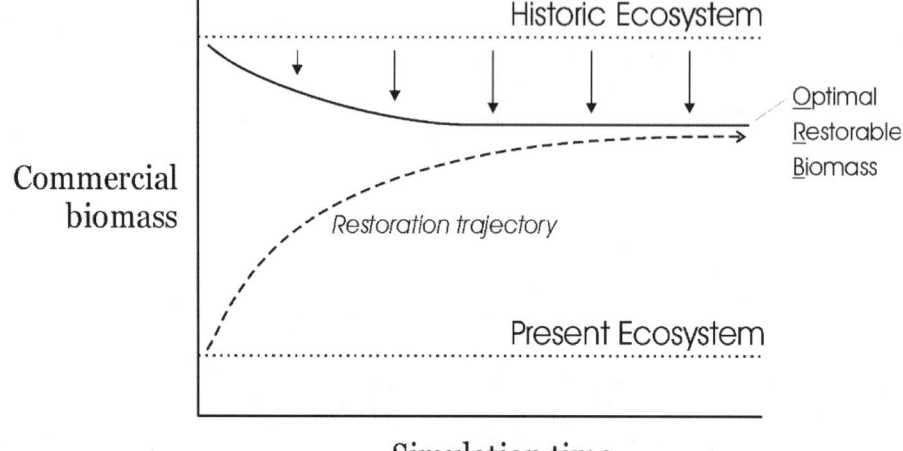

Figure 16. Conceptual diagram showing development of an optimal restorable biomass (ORB) restoration target, and a possible restoration trajectory ORB is the theoretical biomass equilibrium that would result after long-term optimal harvesting of the historic ecosystem (down arrows). A restoration plan (dotted line) would see the present ecosystem changed to resemble the ORB state. Note that simultaneity is not implied between ORB determination and actual restoration. (From Ainsworth and Pitcher 2004).

is only a fraction of today's real-world profits, indicating that present-day levels of extraction cannot be sustainable over a long time-frame.

When the optimal fishing solution is designed to maximize economic performance from the restored system (economic objective), harvest value is an order of magnitude greater than under the more conservative plan (ecological objective), but biodiversity is sacrificed. When the policy is designed to preserve the environment, the restored systems lose little biodiversity over the course of harvest, long-live species increase in abundance and food chains lengthen improving resilience, but less catch is permitted.

Figure 13 describes this fundamental trade-off between exploitation and conservation. The x-axis indicates optimized fishing simulations conducted, varying fleet structure and harvest objective. Three fleets were tested: the idealized 'Lost Valley' fleet, including only responsible gear types, and two abbreviated fishing fleets, one with no recreational sector and one with no trawlers (Ainsworth et al. 2003).

On the left of each graph, we see scenarios that generate wealth, but neglect the environment: these are typically the social and economic runs. Exploitation rates are high and the fishery tends to be concentrated in a few profitable gear types. On the right are scenarios that preserve the environment through conservative harvests, spread evenly across gear types: these are ecological and portfolio utility runs. Mixed runs are intermediate between these extremes.

Figure 14 illustrates the economic versus ecological trade-off throughout the course of a dynamic simulation. The 'Lost Valley' in this case is the marine system of northern British Columbia in about 1750, prior to European contact. Two fishing scenarios are applied for comparison: one gives priority to conservation (ecological/economic mixed objective - solid line) and one gives priority to exploitation (economic objective – open circles). Note that the economic objective progressively reduces biodiversity while the other preserves the biodiversity of the ecosystem more or less in its restored state.

Simulation results may be challenged by uncertainty in the principal model parameters using Monte-Carlo sampling. Figure 15 shows the coefficients of variation for four restoration goals for northern British Columbia, averaged across 54 functional groups at the end state of 50-year ecosystem simulations (from Ainsworth and Pitcher 2003). The value for the 1950 restoration goal is anomalously low because of deficiencies in the model and the data upon which it is based. It would be unwise therefore to include this goal in any realistic discussion of policy until this ecosystem model is improved.

Having determined the benefits available from each Lost Valley we are able to rate past each of the past ecosystems we have modelled as possible restoration goals for the future. Having described the specific trade-offs between competing socioeconomic and ecological interests inherent in potential fishing solutions, we may maximize the utility of the restoration according to the specific needs of stakeholders. The next phase in our BTF work will examine the practicalities of finally achieving restoration to 'Lost Valley' conditions, and determine how far into the past we must reach to maximize cost-effectiveness of reparative measures.

6. Implementing Back-to-the-Future Policy: How Do We Get There?

The intention of the BTF process is to provide a clear cognitive map of a future ecosystem that, as far as possible, resembles one from the past, and its likely sustainable benefits. For a full evaluation, the costs of restoration should be considered alongside these benefits, because the policy goal cannot be determined until a full cost-benefit analysis is performed. But there are two problems with this apparently logical approach.

First, there is a fundamental problem in that the costs of restoration may depend on precisely what techniques are adopted, and the actual instruments of restoration may themselves generate conflict. For example, marine protected areas set up adjacent to a traditional fishing community will often trigger protests from inshore and local fishers, who may not be easily convinced of long-term advantages. Moreover, reducing quotas for some sectors as fisheries are modified to become more sustainable is also likely to be controversial.

Secondly, there are psychological advantages in emphasising the possibility of considerable restored biomass in long-lived target species like halibut or cod. Hence, it may be easier to achieve agreement by putting forward an attractive long-term goal rather than by diverting attention to the means by which one might get there. Agreement on one element of a policy generally eases subsequent steps, although final agreement may always be difficult. However, once management aims to make progress towards a specific goal, the use of passive adaptive management (e.g., Walters 1986), will help to reduce difficulties caused by changes in the environment and imperfectly-understood ecology.

Surprisingly, implementation of BTF may engender some conflict between managers and the conservation community as the system moves towards a 'fished Lost Valley' objective. Higher biodiversity may entail reductions in some species while others increase. For example, over the last 100 years in the North Sea, seabird populations have increased at least two-fold as a result of human activities, including fishing (Furness 2002, Mackinson 2001). So attempts to restore older systems might involve active or passive management to reduce some species. In this manner, it is interesting to contemplate some of the trade–offs that may have to be faced in a BTF restoration process. For example, restoration of habitat and wild populations in terrestrial systems usually has to be strictly managed. The common taboo on killing mammals and birds in marine ecosystems does not extend to terrestrial systems, and we see many examples of elephants, kangaroos, elk, wolves and crocodiles whose populations are controlled by active culling at the same time as they are protected from hunting. So it is important to realise that, once marine mammal, bird, fish or carnivorous turtle populations have recovered to sustainable levels within the meaning of the target 'Lost Valley' ecosystem, they may have to be controlled within the management boundaries. Similarly, Walters and Martell (2004) show that active culling may lead to higher fishery values when predator and prey are linked in a depensatory relationship. Monitoring programmes should be set up to ensure that all changes, including charismatic fauna, are within the expected bounds of the transitional path towards new management objectives.

When BTF is implemented, the 'Lost Valley' ecosystems are reduced from their original state by the sustainable fisheries designed to exploit them. So, rather than allow expensive restoration to go all the way to the 'Lost Valley' state only to be subsequently depleted, it would clearly be expedient to try to restore the system directly to one of these optimally-fished ecosystem states. Consequently, we may regard the 'fished Lost Valley' as comprised of a suite of organisms at their Optimum Restorable Biomass (ORB, Ainsworth et al. 2004). The quantitative basis for this new concept in the strategic management of marine ecosystems is discussed below.

6.1 OPTIMAL RESTORABLE BIOMASS (ORB) AS A NOVEL STRATEGIC MANAGEMENT GOAL

The most productive state of a stock is not in its unfished condition, but when older, less productive individuals have been removed from the population; by manoeuvring

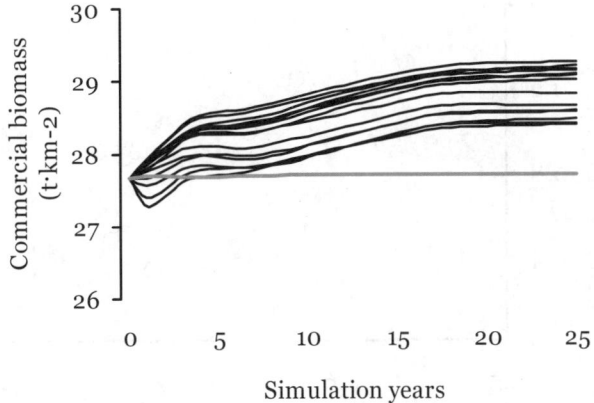

Figure 17. Biomass trajectories of various restoration plans. Black lines show various restoration trajectories, slow to fast; grey line shows model baseline. The fastest plan achieves restoration quickly but produces negligible income; the slowest plan continues to provide for resource users during restoration.

stock abundance to an optimal size, surplus production can be maximized (Pitcher and Hart 1982). This is pertinent when considering historic ecosystems as rebuilding goals; we would not wish to undertake a costly restoration to achieve a historic system just to fish it down to a more productive state. Instead, we should restore that optimally productive state directly - the biomass equilibrium that theoretically results after long-term, ecosystem-based optimal harvests of the historic system (Figure 16). The biomass of each ecosystem component in that new equilibrium is referred to here as the optimal restorable biomass (ORB); an ecosystem containing ORB attuned components may be called an ORB ecosystem. Unfortunately there is no unique solution: the specific ORB configuration we prefer will depend on what benefits we want out of the historic system, monetary or otherwise.

Where stocks interact through predation or competition, it may be impossible simultaneously to achieve BMSY for multiple stocks (Walters et al. 2004). From a whole-ecosystem perspective, it becomes necessary to choose between stocks, holding the biomass of some close to their optimal levels while sacrificing the productivity of others. According to our choice, total catch from the ecosystem can be maximized - or profit, or biodiversity, or any other measure of socioeconomic or ecological utility (e.g., ORB goals in Figure 13). If our management goal is simply to maximize catch, for example, BMSY should be sought only for the most productive and massive stocks, while maximum productivity of low-volume fisheries may need to be sacrificed. An alternative goal may maximize system biodiversity. In most cases, a practical management policy will contain a balance between socioeconomic and ecological priorities. ORB calculation based on historic systems (Ainsworth and Pitcher 2003) therefore satisfies two requirements: it increases production of commercial groups by reducing their biomass, and optimally balances harvests between stocks to provide maximum benefit.

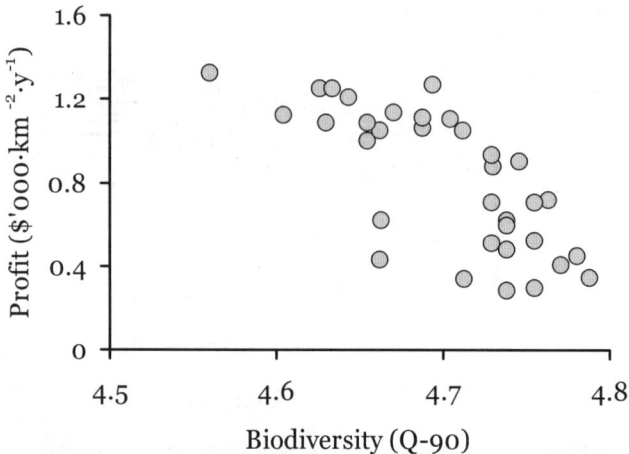

Figure 18. Restoration trade-offs. Equilibrium tradeoffs resulting after 25 years of restoration. Not shown: baseline system provides profit 0.7 and biodiversity 3.9; goal of ORB system provides profit 1.2 and biodiversity 8.4.

In a first attempt to develop ORB restoration strategies, Ainsworth and Pitcher (2004c) have modified an existing routine in *Ecosim*, 'mandated rebuilding', to accept a vector of ORB biomasses as targeted rebuilding goals. The policy search routine (Walters et al. 2002) will try to satisfy those goals through selective fishing, cultivating and sculpting the ecosystem into the desired configuration. The authors selected for their demonstration a rebuilding goal based on the 1900 historic model – the ORB ecosystem which had been maximized for a mix of economic and ecological benefits. They generated a variety of restoration plans to rebuild commercial biomass (Figure 17). All plans recover biomass, but "fast" plans neglect the needs of resource users in favour of rebuilding, while "slow" plans continue to generate significant annual income. A preliminary cost-benefit analysis of restoration suggested that, under ideal circumstances, ecosystem restoration can offer a financial return comparable to bank interest. In absolute terms, 25 years of restoration achieves only a fraction of the 1900 ORB potential, but the convex relationship of trade-offs between end-state profit and biodiversity (Figure 18) suggests an exciting possibility. There may be an optimal rate of recovery where human needs and ecological needs can both be met. Future BTF research will further investigate this optimal rate of restoration, taking into account costs associated with fleet restructuring, and pit candidate rebuilding strategies against environmental and management uncertainties.

7. Implementing Back-to-the-Future Policy: Participation

The serial depletion of large, high trophic level fish and compromising of ecosystem structure and function parallels the extinction, extirpation and depletion of terrestrial animal that were good to eat, dangerous or merely incompatible with resource extraction, mining, agriculture, industry and the sheer collateral damage associated with sharing a planet with six billion humans. The difference is that anthropogenic terrestrial extinctions commenced tens of 1000s of years, while our technical ability to catch the last fish in the sea has developed only over the last 100 years. Reduction of 'table' fish in the Atlantic to 10% of their abundance in 1900 (Christensen et al. 2003), can be ascribed to discounting the future (Clark 1973), and to the fact that, outside of the very small world of fisheries science and some NGOs, there is almost no awareness of the extent of the damage. Even within the fishing industry, there is little collective awareness of the cumulative and long-term effects of current catch rates, a situation exacerbated by many management agencies distracted from a 'longer view' (Beverton 1998) by day-to-day issues and quotas. Hence we feel that the key to the strategic management of marine ecosystems has to be a change in the way that the problem is perceived and a broad awareness and willingness to be participants in reconstruction.

The BTF process involves the collaborative construction of whole ecosystem models by scientists, the maritime community, managers and policy-makers. The process of model construction builds intellectual capital in the model, as it goes from rudimentary to robust and social capital in terms of increased trust among the collaborators. Reconstruction of past abundance refocuses attention from allocation disputes to evaluation of what it would be worth to us today, if we could restore some level of past abundance. The end goal is to build support for the re-investment in natural capital necessary to restore systems. The short term goal is to build stakeholder trust in the model to the point where they can collectively evaluate the cumulative effect of their individual catch share aspirations. This is a very different approach from the bilateral consultation that characterizes national and international allocation processes today, but is congruent with ecosystem-based management, precautionary management and widespread requirements for broad based consultation.

Implementing a policy goal that has been chosen using any science-based process, including BTF, is, of course a difficult matter. When fishing communities and other essential stakeholders actively participate in the policy agenda, compliance and consent may be high (Hart and Pitcher 1998, Harris 1998). Sometimes even voluntary agreements with a strong local base may operate surprisingly well (e.g., English Channel area agreements by gear sectors: Blyth et al. 2003).

Unfortunately, much recent fisheries management operates at very large scales. Degnbol (2o03) considers that this scale is often too large to reflect local needs of fishing communities (Newell and Ommer 1999). Pauly has called for "putting fisheries back in local places" (Pauly 1999). Since ecosystem simulations generally work on a per-km-square basis, and wide-ranging important species that are seasonal visitors can be emulated with a range of techniques that deal with migration, these methods may be adapted to quite small local areas (e.g., Hong Kong, <2000 km2; Pitcher et al. 1998). This means that a BTF analysis can hope to reflect local needs and issues reasonably well. An example of pilot work in a local community in British Columbia is presented in Pitcher et al. (2002c).

TABLE 2. Summary of participatory elements from fishing and scientific communities in the BTF process. TEK = traditional ecological knowledge, LEK = local ecological knowledge. All stages are intended to work in concert with science-based decision making.

	Maritime community	Scientific community
Members	fishers, community members, managers, policy makers	ecologists; marine biologists, archaeologists, oceanographers, anthropologists
Model development phase	TEK: in model construction	Published and unpublished archival and archaeological data for model groups and their parameters
	LEK: in model construction	Biomass time series data for model tuning
	TEK/LEK/Community: model credibility and validation	Ocean climate change and forecasts for region
Policy development phase	Community choices: how to rebuild	Present policy, stock assessment, and quotas
	Community choices: choice of best benefits to cost ratio for policy goal	Fishery by-catch and discards Fishing gear research
	Community choices: choice of acceptable and sustainable fisheries	Economic, social and ecological evaluations a wide range of indices
Operational phase	Consent and compliance	Stock assessment and quotas
	Monitoring	Monitoring and adaptive management

Resource users are often deeply suspicious of 'models', as something that they, a) do not understand, and b) because of bitter past experience of the consequences of models applied to their fisheries.

Moreover, in near-shore areas in tropical developing countries where restoration efforts have been put in place (e.g., through MPAs), a top-down approach has adversely impacted many small scale fisheries, mostly women and children who glean for fish and invertebrates for their subsistence. BTF differs from the classic 'black box' view of models insofar as workshops explain what the model does, i.e. simulate ecosystem connections that they understand or can readily grasp, and critique from their vastly greater time and experience on the water. Haggan (2000) identifies four elements as critical to successful participation:

- Recognition of the scope of the problem;
- Acknowledgement of collective responsibility of fishers, scientists, managers and policy makers;
- Respect for and agreement to share different systems of knowledge essential to harnessing the collective talents of the collaborators; and,
- Commitment to evaluate and share the benefits of restoration.

The first element is achieved, in part, by the audit function of BTF in comparing past and present systems. This tries to reverses the 'cognitive ratchet' of the shifting baseline (Pauly 1995) by assessing what has been lost (see case study described below). Understanding the scope of the problem also means coming to terms with the need to build consensus in a climate of scarcity. It is not the number of police that keeps us alive in our cars, on foot, bicycles or (God help us) skateboards, but the consensus amongst road users to abide by a set of rules (Haggan 1998).

The second element has proved to be helpful when scientists approach stakeholders as learners as well as teachers, able to admit their mistakes (Power et al. 2004). Fishers often openly acknowledge unease about their former involvement in industrial fisheries (e.g., the Canadian east coast fishery for cod and west coast fishery for herring; Coward et al. 2000).

In the third point, respect for different systems of knowledge is critical for building trust. Time must be set aside in initial workshops and the early stages of collaboration to understand worldviews and technical terms other than one's own. The desired result is to be able to share knowledge in the interest of conservation, and to express it in terms that are accessible to the public.

The last element involves calculation of the ecological and social as well as economic values of both past and present systems. The methodology for doing this is discussed elsewhere in the paper, as are the real short-term costs associated with moving from a depleted state to a restored system.

In BTF the aim is to encourage a greater chance of success because a sense of ownership of the process is fostered and developed from the earliest stages of the work. The BTF process includes the broadest possible base of collaboration in building models of the past, in the choice of sustainable fisheries and in the evaluation of the costs and benefits of alternative restoration goals. Moreover, the mental maps shaped by awareness of past abundance and diversity develop in BTF process may serve to assist consent and compliance with a restoration agenda (Pitcher and Haggan 2003). Participatory groups and elements that are integral to three phases of the BTF process are summarised in Table 2. A case study of a pioneering attempt to use the BTF approach with resource users is described next.

7.1 TOWARD REVERSING 'SHIFTING BASELINES' IN FISHERIES: A CASE STUDY LEARNING FROM STAKEHOLDERS IN THE BALI STRAIT AND KOMODO NATIONAL PARK, INDONESIA

In Indonesia, the 'fisheries crisis' includes overoptimistic behaviour by fisheries policy makers (Pet and Mous 2002) that is leading to unsustainable levels of resource exploitation (Venema 1997). Moreover, stock assessments are irregular and employ single-species approaches (Widodo et al. 1998), while fisheries statistics are very unreliable (Venema 1997). Compliance with fisheries regulations is poor, as reflected in widespread destructive fishing practices (e.g., Erdman and Pet 1999), large illegal, unreported and unregulated fishing (e.g., Erdmann 2000), and poor implementation of marine protected areas (MPAs) (e.g., Fauzi and Buchary 2002). A recent report to Indonesia's Ministry of Marine Affairs and Fisheries suggested that the Indonesian government should:

> *Create, build and arouse awareness to change the perception and mindset of the people to stop romanticizing that the country's seas have over-abundant or*

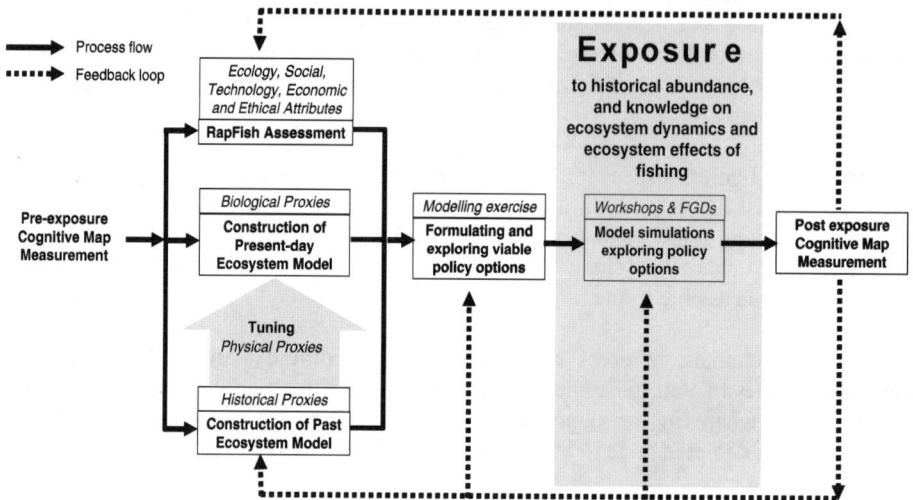

Figure 19. A schematic representation of the methodological framework used in an attempt to reverse the shifting baseline syndrome by assessing resource user's cognitive maps before and after exposure to Back-to-the-Future material about their local fisheries and ecosystem. This work with coastal communities is based at two sites in Indonesia.

overflowing resources, in particular fisheries resources (Pacific Consultants International 2001, cited in Pet and Mous 2002).

Nevertheless, fisheries policy direction seems to lean in the opposite direction, such as advocated in a recent speech delivered by the Minister of Marine Affairs and Fisheries at the 'International Seminar on Sustainable Development in the EEZ and the EEZ as an Institution for Cooperation or Conflict' (Dahuri 2002, cited in Pet and Mous 2002).

Strategic management of marine ecosystems is important in Indonesia because this region is endowed with the richest marine biodiversity in the world, with for example, more than 450 species of scleractinian corals (Tomascik et al. 1997a), 15% of the world total (Daws and Fujita 1999), and a third of all fish species in the world (Daws and Fujita 1999, Froese et al. 1996). Hence, losses to marine biodiversity of Indonesia also mean significant losses to global marine biodiversity. In addition, about 60% of the total Indonesian population (estimated mid-year total 2002 is 211.7 million; World Bank 2003) is coastal people (Burbridge et al. 1988, cited in Tomascik et al. 1997b), and almost 80% of these coastal people engage in marine resource-dependent activities such as fishing and mariculture (World Wide Fund for Nature 1994). Without addressing the underlying root causes of the problems, the gross depletion, and ultimately the demise, of many fisheries resources in Indonesia is likely unavoidable. This will be the dawn of a bleak episode for these coastal people.

In light of these issues, an empirical study in Indonesia is testing whether the cognitive maps of resource users can be readjusted by exposure to historical accounts of great abundance (Pitcher and Haggan 2003). The work focusses on two case studies: the Bali Strait between Bali and Java, and, further to the west, the area around and in Komodo National Park. In the Bali Strait, productivity from an ocean upwelling supports a large medium-scale industrial fishery for a small pelagic clupeid fish called the lemuru, which is mainly reduced to fishmeal. In Komodo artisanal fishers exploit reef fish in and around the park, which includes designated no-take zones, and have a sometimes tense relationship with the Park authorities.

A BTF approach endeavors to describe present-day and past marine ecosystems comprises both quantitative - ecosystem dynamics using ecosystem modelling techniques - and qualitative approaches, aiming to describe perceived interactions between stakeholders and the ecosystem. The qualitative methods rely on tools such as semi-structured interviews to draw on local and traditional ecological knowledge, focus group discussions, workshops, pile-sorting techniques, questionnaires and content analyses of various archival records (e.g., historical expedition monographs).

The central feature of the work is the empirical assessment of the cognitive maps[5] of the stakeholders before and after exposure to historical accounts of great abundance and imparted knowledge about the inter-relationships of organisms, ecosystem dynamics and the ecosystem effects of fishing (Figure 19).[6]

The cognitive map assessment employs a set of explicit questions that asks the respondents their views on historical abundance, ecosystem dynamics and ecosystem effects of fishing. Similar sets of questions are asked of the same selected respondents before and after the exposure sessions.[7] Results will be used to feedback to the BTF process for future development when needed (Figure 19). In this case, it is hoped that

---

[5] The term 'cognitive map' was first defined by Tolman (1948) to denote a mental map of spatial relationships in the environment. Cognitive map is here more broadly defined as "the interpretive framework of the world which exists in the human mind and affects actions, decision making and knowledge structures". This definition is taken from (www.webref.org/anthropology/c/cognitive_map.htm).

[6] The exposure sessions are presented in a form of workshops and focus group discussions to various stakeholder groups. During these exposure sessions, models of the past and present-day states from the BTF process (notably step 1 through 4; see Table 1) are presented. Some simulation results that depict ecosystem dynamics and ecosystem effects of fishing will also be communicated in an attractive manner and non-intimidating way for lay people (viz., small scale fishermen and fisherwomen who may be illiterate and know nothing about modelling). In addition, audio visual technique using screening of relevant video clips (subtitled with lingua franca of the study areas) such as those documented by "The Blue Planet" series of BBC will also be used to enhance the points to be made.

[7] Respondents are stratified based on geographic, demographic, and fishing sector characteristics to ensure that all stakeholders are adequately represented. Parameters for geographic characteristics include, among others, location of villages where respondents reside, and location of fishing grounds or fishing-related activities. Meanwhile, demographic characteristics include occupation types (viz. fishers, non-fishers, and experts), gender and ethnic origin. Fishing sector characteristics include types of fishing fleet (inshore, near-shore or offshore), whether the fishers are migratory or permanent, whether the fishers are original fishers or fisher-converts from other livelihood sectors, etc. A snowball sampling method is used to canvas the respondents.

the BTF approach can be used as a tool to help gradually readjust the cognitive maps of stakeholders involved in the fisheries, ultimately aiming to reverse 'Pauly's ratchet'.

8. Problems and prospects for back-to-the-future policy in the strategic management of marine ecosystems

At this juncture, our research on Back-to-the Future is work in progress, and so it is not surprising that there are a number of methodological challenges to the work, particularly in relation to the ecosystem modelling.

The quantitative ecosystem modelling we have employed relies almost exclusively on *Ecopath* and *Ecosim* techniques. Yet many of the assumptions in this modelling system, while plausible, remain unvalidated. Of especial concern are the *Ecosim* 'vulnerability' parameters, to which specific results often appear very sensitive. Moreover, these parameters not only shape predator-prey interactions (which they do in a credible fashion for evolutionary ecologists), but also pre-determine the scope for further biomass growth in relation to current levels. For any series of 'snapshot' BTF ecosystem models, this problem creates a conflict between the need to compare the outcomes of various fisheries options while other parameters remain fixed, and setting parameters correctly for biomasses that were closer to unexploited levels in the past. The series of ecosystem modelling case studies in this paper shows some of the ways that these modelling problems have been approached, but it remains true that many issues have yet to be resolved.

Our models of past times are not built by 'winding back the present', but by using specific historical information on the presence and absence of species, or trends in past biomasses from stock assessments or surveys. We have used information from archives, historical documents (Erfan 2004), archaeological investigations (e.g., Orchard and Mackie 2004, Heymans 2002), interviews that collate traditional or local environmental knowledge (e.g., Simeone 2004, Haggan et al. 1998, Salas et al. 1998), language and vocabulary (Danko 1998) and even ancient art work (Williams 1998). The many sources of uncertainty in this material need to be addressed and the rigour with which historical information is turned into inputs for the ecosystem modelling needs considerable improvement. As pointed out by Heymans and Pitcher (2004), our past ecosystem models may resemble the actual past as a painting by Picasso resembles reality. A critical question is whether our comparative restoration policy scenarios can be made robust against such distortions. A deeper insight of the dynamics of ecosystems under change will be required before we can answer this question.

The considerable logistics and costs of mounting a quantitative, robust and credible BTF analysis are hard to find. An inter-disciplinary team needs to be assembled to gather, validate and analyse the historical, archaeological and ecological information. The scope of BTF work appears to be beyond the capacity of one graduate student thesis, and therefore hard to find resources for.

One common criticism of BTF is that 'ecosystems do not rewind'; even if the past would form a desirable goal, it is not possible to get there starting with today's state, because the environment and the ecosystem have changed too much. The ecosystem will fail to 'rewind', thereby fatally compromising the BTF policy goal. This is unlikely, however, because the experience with fisheries management is that the majority of exploited species do indeed 'rewind' if they are not too depleted (e.g., Hilborn et al. 2003, Hall 1999 page 202; Hilborn 1996). It is also worth reflecting that the assumptions of conventional single species stock assessment would never be valid if the "cannot rewind" argument were often true.

Nevertheless, there are some instances where a 'rewind' does not happen as expected. This may be because of habitat destruction or pollution, loss of keystone species or predator-prey cascades in the trophic web (e.g., Newfoundland after the cod collapse, Fu et al. 2001), and these factors will vary among the different life histories of species in the ecosystem. The populations of small pelagic fish are likely to governed by ocean cycles (e.g., Chavez 2003) as well as fishing (e.g., Oliviera et al. 1998). Alternate stable states may exist for some ecosystems (e.g., a switch between cod and clupeids in the Baltic; Rudstam et al. 1994), and the spread of exotic organisms may alter ecosystem structure irreversibly (e.g., Black Sea, Zaitsev 1992). We would hope that the possible operation of such irreversible changes or switches could be anticipated by those knowledgeable about a particular ecosystem, so that constraints on any restoration may take what is feasible into account. Monitoring what happens as restoration begins is crucial.

So the answer to this criticism is that we can easily allow a possible BTF policy to be compromised by unexpected one-way ecological changes. In these circumstances we can hope to have early warning and continue to try to monitor progress, and model what may be different using insight of local experts. The rest of the BTF procedures, in devising sustainable fisheries and evaluating the most beneficial reconstruction that may be achieved, remain valid.

A variant of this criticism is that, since the past was very different, models of the past, even if accurate, could not represent viable restoration goals. The past was different in terms of species composition and climate, primary production was likely lower because of recent eutrophication, and pollution was lower from smaller populations of humans. While all these things are true, differences between the past and the present can be dealt with in a similar way as for the first criticism. Ecological processes in the past must very likely have obeyed the same sets of rules as we see to day, even if climate and other factors had different values. Past BTF 'snapshot' models often deliberately span a number of years to minimise some of the shorter-term changes. Although it would be satisfying to have models of the past that could be derived directly by 'winding back' models of the present using past time series of fisheries and climate, beyond a certain date, this process becomes increasingly uncertain. As shown above, a continuously modified ecosystem model like this is not necessary for BTF; a series of snapshots models through time will suffice. Moreover, if the influence of climate on elements of the ecosystem is reasonably well understood, past models may be adjusted appropriately when prepared as candidates for 'Lost Valley' protocols.

Back-To-the-Future provides for the strategic management of marine ecosystems by setting up an explicit, quantitative long-term policy goal and the restoration trajectory needed to achieve it. Short-term variation in ecology, climate and human influences need not deflect the ultimate policy goal, which may only slowly be approached and never reached.

The process involves the collaborative construction of whole ecosystem models by ecologists and other environmental scientists, the maritime community, managers, policy-makers and members of the public. The collaboration builds intellectual capital in the model as it transforms from a rudimentary to a more robust form, while social capital is created in terms of increased trust among the collaborators. Moreover, for the public, model reconstruction of past abundance refocuses attention from today's allocation disputes to an evaluation of what that past state could be worth to us today if we could restore some level of it. Hence, in societal terms, the BTF process builds support for the re-investment in natural capital necessary to restore systems. A broad participation in BTF is critical for the success of BTF, and might perhaps be assisted by an explicitly ecosystem-based policy that could be seen to employ inputs of information from all sectors of society (Pitcher 2000). So far, we have barely scratched the surface of the deep issues raised by the need for this level of participation.

Furthermore, we are not yet sure how to convey both the utility and the uncertainty of our modelling work, which to many may seem arcane. Jeremy Prince has envisaged a cadre of 'barefoot ecologists', the equivalent of rural development generalists for fisheries (Prince 2003), who might be able to help. The next step in this work is to analyse in more detail the trade-offs represented by BTF policies: early signs are encouraging that restoration may sometimes allow for win-win solutions in allowing long-term policy goals that are less painful for all.

9. Acknowledgements

Back-to-Future research has been supported by small grants from the Peter Wall Institute for Advanced Studies and from the Fisheries Centre, both at the University of British Columbia; the World Wildlife Fund, Canada; and the Department of Fisheries and Oceans, Canada. A major interdisciplinary grant from National Science and Environmental Research Council and the Social Science and Humanities Research Council of the Government of Canada to Coasts under Stress (CUS) supported the research that made possible most of the quantitative work, interviews and workshops outlined in this paper. We are also grateful for enthusiasm and commitment from BTF team members Dr Sheila (J.J.) Heymans, Dr Marcelo Vasconcellos, Dr Rashid Sumaila, Dr Melanie Power and Aftab Erfan. Dr Villy Christensen and Professor Carl Walters provided invaluable and long-suffering support for the modelling.

Cameron Ainsworth is funded by Coasts Under Stress project and the Natural Sciences and Engineering Research Council of Canada. Eny Buchary is partially funded by a Doctoral Research Award of the International Development Research Centre (IDRC) of the Government of Canada, a University Graduate Fellowship (UGF) from the University of British Columbia, Canada, and by a Natural Science and Engineering Research Council of Canada (NSERC) Research Fellowship awarded through Prof. T.J. Pitcher and her research is in partnership with The Research Centre for Capture Fisheries (PURISPT) of the Indonesian Ministry of Marine Affairs and Fisheries, and the Southeast Asia Center for Marine Protected Areas of The Nature Conservancy Indonesia Program. Wai Lung Cheung is partially supported by the Sir Robert Black Trust Fund Scholarship for Overseas Study, Hong Kong. Robyn Forrest is partially supported by the Charles Gilbert Heydon Travelling Fellowship in Biological Sciences, awarded by the University of Sydney, Australia; and by a grant awarded to the UBC Fisheries Centre under a Memorandum of Agreement between the Fisheries Centre and New South Wales Fisheries, Australia. Nigel Haggan is partially funded by Coasts Under Stress. Hector Lozano is supported by a Memorandum of Agreement for Human Resources Development between National Council for Science and Technology (CONACYT) of Mexico and The University of British Columbia (UBC) of Canada, according to the CONACYT/UBC graduate Scholarship programme. Telmo Morato acknowledges support from the "Fundação para a Ciência e Tecnologia" (Portugal), the European Social Fund through the Third Framework Programme, and from Professor Ricardo S. Santos, Portugal. Lyne Morissette acknowledges financial support from the Department of Fisheries and Oceans, Canada (Science Strategic Fund) and carried out this study as a contribution to the Canadian CDEENA (Comparative Dynamics of Exploited Ecosystems in the Northwest Atlantic) programme, from which many colleagues made data available for the models.

10. References

1. Ainsworth, C.H. and Sumaila, U.R. (2003) Intergenerational Discounting and the Conservation of Fisheries Resources: A case study in Canadian Atlantic cod. Pages 26-32 in Sumaila, U.R. (ed.) Three Essays on the Economics of Fishing. Fisheries Centre Research Reports 11(3): 33pp.
2. Ainsworth, C.H. and Pitcher, T.J. (2003) Evaluating goals for restoration in the marine system of northern British Columbia. Submitted to 1st Wakefield Fisheries Symposium, Assessment and Management of New and Developed Fisheries in Data-limited Situations, Oct 22-25, 2003. Alaska Sea Grant, Anchorage, Alaska, USA.
3. Ainsworth, C.H. and Pitcher, T.J. (2004a) Using Local Ecological Knowledge as a Data Supplement for Ecosystem Models. Proceedings of the 21st Lowell Wakefield Fisheries Symposium, Oct 22-25, 2003. Alaska Sea Grant, Anchorage, Alaska, USA. (in press).
4. Ainsworth, C. and Pitcher, T.J. (2004b) Modifying Kempton's Biodiversity Index for Use with Dynamic Ecosystem Simulation Models. Pages 91–93 in Pitcher, T.J. (ed.) Back to the Future: Advances in Methodology for Modelling and Evaluating Past Ecosystems as Future Policy Goals. Fisheries Centre Research Reports 12(1): 158pp.
5. Ainsworth, C.H. and Pitcher, T.J. (2004c) Back to the Future: Restoring Historic Ecosystems in Northern British Columbia. Submitted to Proceedings of the Fourth World Fisheries Congress. Vancouver, Canada. May 2-6, 2004. American Fisheries Society, Bethesda, USA.
6. Ainsworth, C. and Sumaila, U.R. (2004) Economic Valuation Techniques for Back-To-The-Future Optimal Policy Searches. Pages 104-107 in Pitcher, T.J. (ed.) Back to the Future: Advances in Methodology for Modelling and Evaluating Past Ecosystems as Future Policy Goals. Fisheries Centre Research Reports 12(1): 158 pp

7. Ainsworth, C.H., Heymans, J.J., Pitcher, T.J. and Vasconcellos, M. (2002) Ecosystem models of Northern British Columbia for the time periods 2000, 1950, 1900 and 1750. Fisheries Centre Research Reports 10(4): 41 pp.
8. Ainsworth, C., Heymans, J.J. and Pitcher, T.J. (2004) Policy Search Methods for Back to the Future. Pages 48–63 in Pitcher, T.J. (ed.) Back to the Future: Advances in Methodology for Modelling and Evaluating Past Ecosystems as Future Policy Goals. Fisheries Centre Research Reports 12(1): 158pp.
9. Ainsworth, C.H., Heymans, J.J, Pitcher, T.J. and Cheung, W.L. (2004) Evaluating Ecosystem Restoration Goals and Sustainable Harvest Policies in Northern British Columbia. (in prep.)
10. Andersen, P.K. (1995) Competition, predation, and the evolution and extinction of Steller's Sea Cow, *Hydrodamalis gigas*. Marine Mammal Science 11 (3): 391-394.
11. Anonymous (1994) Report on the Status of Groundfish Stocks in the Canadian Northwest Atlantic. DFO Atl. Fish. Stocks Status Rep. 94/4.
12. Arcese, P. and Sinclair, A.R.E. (1997) The role of protected areas as ecological baselines. J. Wild. Management61: 587-602.
13. Ardron, J.A., (2002) A Recipe for Determining Benthic Complexity: An Indicator of Species Richness. Chapter 23, Marine Geography: GIS for the Oceans and Seas. In Breman, J. (ed.) ESRI Press, Redlands, CA, USA.
14. Beverton, R.J.H. (1998) Fish, Fact and Fantasy; A Long View. Rev. Fish Biol. Fish. 8(3): 229-249.
15. Buchary, E.A., Cheung, W-L., Sumaila, U.R. and Pitcher, T.J. (2003) Back to the Future: A Paradigm Shift to Restore Hong Kong's Marine Ecosystem. 3rd World Fisheries Congress, Beijing, November 2000. American Fisheries Society Symposium 38: 727-746.
16. Bakun, A. (1996) Patterns in the Ocean: Ocean Processes and Marine Population Dynamics. University of California Sea Grant, San Diego, California, USA. 323 pp.
17. Balmford, A., Bruner, A., Cooper, P., Costanza, R., Farber, S., Green, R.E., Jenkins, M., Jefferiss, R., Jessamy, V., Madden, J., Munro, K., Myers, N., Naeem, S., Paavola, J., Rayment, M., Rosendo, S., Roughgarden, J., Trumper, K. and Turner, R.K. (2002) Economic Reasons for Conserving Wild Nature. Science 297: 950-953.
18. Baum, J.K., Myers, R.A., Kehler, D.G., Worm, B., Harley, S.J. and Doherty, P.A. (2002) Collapse and Conservation of Shark Populations in the Northwest Atlantic. Science 299: 389-392.
19. Blyth, R.E., Kaiser, M.J., Edwards-Jones, G. and Hart, P.J.B. (2002) Voluntary management in an inshore fishery has conservation benefits. Environmental Conservation 29 (4): 493–508.
20. Bradford-Grieve, J., Probert, P., Nodder, S.,Thompson, D., Hall, J., Hanchet, S., Boyd, P., Zeldis, J., Baker, A., Best, H., Broekhuizen, N., Childerhouse, S., Clark, M., Hadfield, M., Safi, K. and Wilkinson, I. (2003) Pilot trophic model for sub-antarctic water over the Southern Plateau, New Zealand: a low biomass, high transfer efficiency system. J. Exp. Mar. Biol. Ecol. 289(2): 223-262.
21. Brothers, N.P., Cooper, J. and Løkkeborg, S. (1999) The incidental catch of seabirds by longline fisheries: worldwide review and technical guidelines for mitigation. FAO Fisheries Circular 937: 100pp.
22. Brown, C.R., Brown, B.E. and Carpenter, C. (1997) Some of the traditional food gathering of the Heiltsuk Nation. British Columbia Ministry of Education, Skills and Training, Victoria, Canada.
23. Brusca, R.C., Campoy Fabela, J., Castillo Sánchez, C, Cudney-Bueno, R., Findley, L.T., Garcia-Hernandez, J., Glenn, E., Granillo, I., Hendricks, M.E., Murrieta, J., Nagel, C., Román, M. and Turk-Boyer, P. (2001) A case of two Mexican biospheres reserves. The Upper Gulf of California/Colorado River Delta and Pinacate/Gran Desierto de Altar Biosphere Reserves. Report for the Unidad CIAD Guaymas, Sonora, Mexico. 96 pp.
24. Caddy, J.F. (2000) Marine catchment basin effects versus impacts of fisheries on semi-enclosed seas. ICES Journal of Marine Science 57(3): 628-640.
25. Callenbach, E. (1995) Bring Back the Buffalo! A sustainable future for America's Great Plains. Island Press, Washington DC, USA. 279 pp.
26. Carraquiry, J. D. and Sánchez, A. (1999) Sedimentation in the Colorado River Delta and Upper Gulf of California after a nearly a century of discharge loss. Marine Geology 158: 125-145.
27. Chavez, F.P., Ryan, J., Lluch-Cota, S.E. and Ñiquen, M.C. (2003) From anchovies to sardines and back: multidecadal change in the Pacific Ocean. Science 299: 217-221.
28. Cheung, W.L. (2001) Changes in Hong Kong's capture fisheries during the 20th century and reconstruction of the marine ecosystem of local inshore waters in the 1950s. M.Phil. thesis, University of Hong Kong, Hong Kong. 186 pp.

29. Cheung, W.L. and Pitcher, T.J. (2003) Designing Fisheries Management Policies that Conserve Marine Species Diversity in the Northern South China Sea. Submitted to Proceedings of the 21st Lowell Wakefield Symposium: Assessment and Management of New and Developed Fisheries in Data-limited Situations. October 22-25, 2003, Alaska Sea Grant, Anchorage, Alaska.
30. Cheung,W.L., Watson, R. and Pitcher, T.J. (2002) Policy Simulation of Fisheries in the Hong Kong Marine Ecosystem. Pages 46-53 in Pitcher, T.J. and Cochrane, K. (eds) The Use of Ecosystem Models to Investigate Multispecies Management Strategies for Capture Fisheries. Fisheries Centre Research Reports Vol. 10(2): 156 pp.
31. Cheung, W. L. and Pitcher, T.J. (2004) An Index Expressing Risk of Local Extinction for Use with Dynamic Ecosystem Simulation Models. Pages 94–102 in Pitcher, T.J. (ed.) Back to the Future: Advances in Methodology for Modelling and Evaluating Past Ecosystems as Future Policy Goals. Fisheries Centre Research Reports 12(1): 158pp.
32. Chichilnisky, G. (1996) An axiomatic approach to sustainable development. Social Choice and Welfare 13: 219-248.
33. Chouinard, G.A., Sinclair, A.F., and Swain, D.P. (2003) Factors implicated in the lack of recovery of southern Gulf of St Lawrence cod since the early 1990s. ICES C.M. 2003/U:04.
34. Christensen, V. (1998) Fishery induced changes in a marine ecosystem: insight from models of the Gulf of Thailand. Journal of Fish Biology 53 (A): 128-142.
35. Christensen, V. (2002) Ecosystems of the past: how can we know since we weren't there? Pages 26-34 in Guénette, S., Christensen, V. and Pauly, D. (eds) Fisheries impacts on North Atlantic ecosystems: models and analyses. Fisheries Centre Research Reports 9(4): 344 pp.
36. Christensen, V. and Pauly, D. (1997) Changes in models of aquatic ecosystems approaching carrying capacity.Ecol. Applic. 8 (1): 104-109.
37. Christensen, V. and Walters, C.J. (2004) *Ecopath* with *Ecosim*: methods, capabilities and limitations. Ecological Modelling 172: 109-139.
38. Christensen, V., Guénette, S., Heymans, J.J., Walters, C.J., Watson, R., Zeller, D. and Pauly, D. (2003) Hundred-year decline of North Atlantic predatory fishes. Fish and Fisheries 4: 1-24.
39. Christensen, V., Walters, C.J. and Pauly, D. (2000) *Ecopath* with *Ecosim*: A User's Guide. Fisheries Centre, University of British Columbia, Vancouver and ICLARM, Penang. 130pp.
40. Clark, C.W. (1973) The economics of overexploitation. Science 181: 630-634.
41. Clarkson, L., Morrissette V. and Régallet, G. (1992) Our Responsibility to The Seventh Generation. International Institute for Sustainable Development, Winnipeg, Canada, 88p.
42. Cisneros-Mata, M.A. and Hamman, M.G. (1995) The rise and fall of the Pacific sardine, Sardinops sagax caeruleus Girad, in the Gulf of California, Mexico. CalCOFI Rep. 36: 136-142.
43. Cochrane, K.L. (2002) The Use of Ecosystem Models to Investigate Ecosystem-Based Management Strategies for Capture Fisheries: Introduction. Pages 5-10 in Pitcher, T.J. and Cochrane, K. (eds) The Use of Ecosystem Models to Investigate Multispecies Management Strategies for Capture Fisheries. Fisheries Centre Research Reports 10(2): 156 pp.
44. Coward, H., Ommer, R. and Pitcher, T.J. (eds) (2000) Just Fish: the Ethics of Canadian Fisheries. Institute of Social and Economic Research Press, St John's, Newfoundland, Canada. 304pp.
45. D'Agrosa, S., Lennert-Cody, C. and Vidal, O. (2000) Vaquita bycatch in Mexico's artisanal gillnets fisheries: driving a small population to extinction. Conservation Biology V14: 110-119.
46. Daws, G., and Fujita, M. (1999) Archipelago: The islands of Indonesia – from the nineteenth-century discoveries of Alfred Russel Wallace to the fate of forests and reefs in the twenty-first century. University of California Press, Berkeley, California, USA. 254pp.
47. Danko, J.P. (1998) Building A Reliable Database from Native Oral Tradition using Fish Related Terms from the Saanich Language. Pages 29-33 in Pauly, D., Pitcher, T.J. and Preikshot, D. (eds) Back to the Future: Reconstructing the Strait of Georgia Ecosystem. Fisheries Centre Research Reports 6(5): 99pp.
48. De Oliveira, J.A.A., Butterworth, D.S., Bole, B.A., Cochrane, K.L. and Brown J.P. (1998) The application of a management procedure to regulate the directed and bycatch fishery of South African sardine, *Sardinops sagax*. S. Afr. J. Mar. Sci. 19: 449-469.
49. Degnbol, P. (2003) Science and the user perspective: the gap co-management must address. Chapter 2 in Wilson, D.C., Nielsen, J.R. and Degnbol, P. (eds) The fisheries co-management experience. Accomplishments, challenges and prospects. Kluwer, Holland.
50. Diamond, J. (1997) Guns, Germs, and Steel: The Fates of Human Societies. Random House, London, UK. 480pp.

51. Dobson, A.P., Bradshaw, A.D. and Baker, A.J.M. (1997) Hopes for the future: restoration ecology and conservation biology. Science 277: 515-522.
52. Dower, J., Leggett, W. and Frank, K. (2000) Improving fisheries oceanography in the future. Pages 263-281 in Harrison, P. and Parsons, T.R. (eds) Fisheries Oceanography: an integrative approach to fisheries ecology and management. Fish and Aquatic Resources Series, Vol. 4. Oxford: Blackwell Science.
53. Doyle, A.C. (1912) The Lost World: being an account of the recent amazing adventures of Professor George E. Challenger, Lord John Roxton, Professor Summerlee, and Mr. E.D. Malone of the 'Daily Gazette'. Hodder and Stoughton, London, New York and Toronto. 140pp.
54. Dulvy, N.K., Sadovy, Y. and Reynolds, J.D. (2003) Extinction vulnerability in marine populations. Fish and Fisheries 4(1): 25-64.
55. Dutil, J.D., Castonguay, M., Gilbert, D. and Gascon, D. (1999) Growth, condition, and environmental relationships in Atlantic cod (*Gadus morhua*) in the northern Gulf of St. Lawrence and implications for the management strategies in the Northwest Atlantic. Can. J. Fish. Aquatic. Sci. 56: 1818-1831.
56. Erfan, A. (2004) The Northern BC historical and interview database for Back-to-the-Future. Fisheries Centre Research Reports (in press).
57. Erdmann, M.V. (2000) Leave the Indonesia's fisheries to Indonesians! Corrupt foreign fishing fleets are depriving locals of food. Inside Indonesia, Issue 63, July-September 2000. (http://www.insideindonesia.org/edit63/erdmann1.htm).
58. Erdmann, M.V. and Pet, J.S. (1999) Krismon and DFP: Some observations on the effects of the Asian financial crisis on destructive fishing practices in Indonesia. SPC Live Reef Fish Information Bulletin 5, March 1999. www.spc.org.nc/coastfish/News/LRF/5/7Krismo.htm.
59. FAO (1995) Code of Conduct for Responsible Fisheries. FAO, Rome, 41pp.
60. Fauzi, A. and E.A. Buchary (2002) A socioeconomic perspective of environmental degradation at Kepulauan Seribu Marine National Park, Indonesia. Coastal Management 30: 167-181.
61. Fearnside, P.M. (2002) Time preference in global warming calculations: a proposal for a unified index. Ecological Economics 41: 21-31.
62. Francis, R.C. and Hare, S.R. (1994) Decadal-scale regime shifts in the large marine ecosystems of the North-east Pacific: a case for historical sciences. Fisheries Oceanography 3: 279-291.
63. Fréchet, A., Gauthier, J., Schwab, P., Bourdages, H., Chabot, D, Collier, F., Grégoire, F., Lambert, Y., Moreault, G., Pageau, L., and Spingle, J. (2003) The status of cod in the Northern Gulf of St. Lawrence (3Pn, 4RS) in 2002. Can. Sci. Adv. Sec. Res. Doc. 2003/65.
64. Froese, R., Luna, S.M. and Capuli, E.C. (1996) Checklist of marine fishes of Indonesia, compiled from published literature. Pages 217-275 in Pauly, D. and Martosubroto, P. (eds) Baseline studies of biodiversity: the fish resources of Western Indonesia. ICLARM Stud. Rev. 23.
65. Fu, C., Mohn, R. and Fanning, L.P. (2001) Why the Atlantic cod (Gadus morhua) stock off eastern Nova Scotia has not recovered. Can. J. Fish. Aquat. Sci. 58(8): 1613-1623.
66. Fulton, E.A. (2001) The effects of model structure and complexity on the behaviour and performance of marine ecosystem models. PhD. Thesis, School of Zoology, University of Tasmania, Hobart, Australia.
67. Fulton, E.A., Smith, A.D.M. and Johnson, C.R. (2003) Effect of complexity on marine ecosystem models. Marine Ecology Progress Series 253: 1-16.
68. Furness, R.W. (2002) Management implications of interactions between fisheries and sandeel-dependent seabirds and seals in the North Sea. ICES Journal of Marine Science, 59: 261–269.
69. Garibaldi, A. and Turner, N. (2004) Cultural keystone species: implications for ecological conservation and restoration. Ecology and Society 9(3): 1.
70. Gibbons, W. and Odum, E.P. (1993) Keeping All the Pieces: Perspectives on Natural History and the Environment. Smithsonian Institution Press, USA, 208pp.
71. Giske, J., Huse, G. and Berntsen, J. (2001) Spatial modelling for marine resource management, with a focus on fish. Sarsia 86: 405-410.
72. Glenn, E.P., Zamora-Arroyo. F., Nagler, P.L., Briggs, M., Shaw, W. and Flessa, K. (2001) Ecology and conservation biology of the Colorado River Delta, Mexico. Journal of Arid Environments 49: 5-15.
73. GOC (2002) Government of Canada Marketable Bonds. Bank of Canada. Department of Monetary and Financial Analysis. www.bankofcanada.ca

74. Gómez-Pompa, A. and Dirzo, R. (1995) Reservas de la biosfera y otras áreas naturales protegidas de México. INE y CONABIO. 67pp. [in Spanish]
75. Goulder, L.H. and Stavins, R.N. (2002) Discounting: an eye on the future. Nature 419: 673-674.
76. Haggan, N. (1998) Reinventing the Tree: reflections on the organic growth and creative pruning of fisheries management structures. Pages 19-30 in Pitcher, T.J., Hart, P.J.B. and Pauly, D. (eds) Reinventing Fisheries Management. Kluwer, Holland. 435pp.
77. Haggan, N. (2000) Back to the Future and Creative Justice: Recalling and Restoring Forgotten Abundance in Canada's Marine Ecosystems. Pages 83-99 in Coward, H., Ommer R. and Pitcher, T.J. (eds) Just Fish: Ethics in the Canadian Coastal Fisheries. ISER Books, St. Johns, Newfoundland. 304pp.
78. Haggan, N. and Beattie, A. (eds) (1999) Back to the Future: Reconstructing the Hecate Strait Ecosystem. Fisheries Centre Research Reports 7(3): 65pp.
79. Haggan, N., Archibald, J. and Salas, S. (1998) Knowledge gains power when shared. Pages 8-13 in Pauly, D., Pitcher, T.J. and Preikshot, D. (eds) Back to the Future: Reconstructing the Strait of Georgia Ecosystem. Fisheries Centre Research Reports 6(5): 99pp.
80. Hall, M. (1988) An ecological view of the tuna-dolphin problem. Rev. Fish Biol. Fish. 8: 1-34.
81. Hall, S.J. (1999) The effects of fishing on marine ecosystems and fisheries. Blackwell, Oxford. 274pp.
82. Hamman, M.G. and Cisneros-Mata, M.A. (1989) Range extension of commercial capture of the northern anchovy, *Engraulis mordax* Girard, in the Gulf of California, Mexico. Calif. Fish and Game 75: 49-53.
83. Harris, C. (1998) Social regime formation and community participation in fisheries management. Pages 261-276 in Pitcher, T.J., Hart, P.J.B. and Pauly, D. (eds) Reinventing Fisheries Management. Kluwer, Holland, 435pp.
84. Harris, D.C. (2001) Fish, Law, and Colonialism: The legal capture of salmon in British Columbia. Univ. Toronto Press, Canada. 306pp.
85. Harris, M. (1998) Lament for an Ocean: The Collapse of the Atlantic Cod Fishery, a True Crime Story. McClellan and Stewart, Toronto, Canada. 342pp.
86. Hart, P.J.B. and Pitcher, T.J. (1998) Conflict, cooperation and consent: the utility of an evolutionary perspective on individual human behaviour in fisheries management. Pages 215-225 in Pitcher, T.J. Hart, P.J.B. and Pauly, D. (eds) Reinventing Fisheries Management. Kluwer, Holland, 435pp.
87. Hays, G.C., Broderick, A.C., Godley, B.J., Luschi., P. and Nichols, W.J. (2003) Satellite telemetry suggests high levels of fishing-induced mortality in marine turtles. Mar. Ecol. Prog. Ser. 262: 305–309.
88. Hayward, T.L. (1997) Pacific oceanic climate change: atmospheric forcing, ocean circulation and ecosystem response. Trends in Ecology and Evolution 12: 150-154.
89. Heymans, J.J. (2003a) Revised model for Newfoundland for the time periods 1985-87 and 1995-97. Pages 40-61 in Heymans, J.J. (ed.) Ecosystem models of Newfoundland and South-eastern Labrador: Additional information and analyses for 'Back to the Future'. Fisheries Centre Research Reports 11(5): 79pp.
90. Heymans, J.J. (2003b) First Nations Impact on the Eastern Newfoundland and Southern Labrador Ecosystem During Pre-Contact Times. Pages 4-11 in Heymans, J.J. (ed.) (2003) Ecosystem Models of Newfoundland and Southeastern Labrador: Additional information and analyses for 'Back to the Future'. Fisheries Centre Research Reports 11(5): 79pp.
91. Heymans, J.J. (2004) Evaluating the Ecological Effects on Exploited Ecosystems using Information Theory. Pages 87–90 in Pitcher, T.J. (ed.) Back to the Future: Advances in Methodology for Modelling and Evaluating Past Ecosystems as Future Policy Goals. Fisheries Centre Research Reports 12(1): 158 pp.
92. Heymans, J.J. and Pitcher, T.J. (2004) Synoptic Methods for Constructing Models of the Past. Pages 11–17 in Pitcher, T.J. (ed.) Back to the Future: Advances in Methodology for Modelling and Evaluating Past Ecosystems as Future Policy Goals. Fisheries Centre Research Reports 12(1): 158 pp
93. Hilborn, R. (1996) The frequency and severity of fish stock collapses and declines. Pages 36-38 in Hancock, D.A., Smith, D.C., Grant, A. and Beumer, J.P. (eds) Developing and Sustaining World Fisheries Resources: The State of Science and Management. Proc. 2nd World Fisheries Congress. CSIRO, Australia.

94. Hilborn, R., Branch, T.A., Ernst, B., Magnusson, A., Minte-Vera, C.V., Scheuerell, M.D. and Valero, J.L. (2003) State Of The World's Fisheries. Annual Review of Environment and Resources 28: 359-399.
95. Hollowed, A.B., Bax, N., Beamish, R., Collie, J., Fogarty, M., Livingston, P., Pope, J. and Rice, J.C. (2000) Are multispecies models an improvement on single-species models for measuring fishing impacts on marine ecosystems? ICES Journal of Marine Science 57: 707-719.
96. Holmgren-Urba, D. and Baumgartner, T. (1993) A 250-year History of pelagic fish in the Central Gulf of California. CalCOFI Rep. 34: 60-67.
97. Hutchings, J.A., and Myers, R.A. (1994) What can be learned from the collapse of a renewable resource? Atlantic cod, Gadus morhua, of Newfoundland and Labrador. Can. J. Fish. Aquat. Sci. 51: 2126-2146.
98. INEGI, 2001. Dirección General de Estadística. Encuesta Nacional de Población y Vivienda 2001. Instituto Nacional de Estadística, Geografía e Informática, México. [In Spanish]
99. Jackson, J.B.C., Kirby, M.X., Berger, W.H., Bjorndal, K.A., Botsford, L.W., Bourque, B.J., Bradbury, R.H., Cooke, R., Erlandson, J., Estes, J.A., Hughes, T.P., Kidwell, S., Lange, S.B., Lenihan, H.S., Pandolfi, J.M., Peterson, C.H., Steneck, R.S., Tegner, M.J. and Warner, R.R. (2001) Historical Overfishing and the Recent Collapse of Coastal Ecosystems. Science 293: 629-637.
100. Johannes, R.E., Freeman, M.M.R. and Hamilton, R.J. (2000) Ignore fishers' knowledge and miss the boat. Fish and Fisheries 1(3): 257-71.
101. Johnson, T.C., Barry, S.L., Chan, Y. and Wilkinson, P. (2001) Decadal record of climate variability spanning the last 700 years in the southern tropics of East Africa. Geology 29(1): 83-86.
102. Jaramillo-Legorreta, C. (1999) Abundance estimated for the vaquita: first step to recovery. Marine Mammal Science. 67(4):967-979 p.
103. Kaiser, M.J., Collie, J.S., Hall, S.J., Jennings, S. and Poiner, I.R. (2002) Modification of marine habitats by trawling activities: prognosis and solutions. Fish and Fisheries 3: 114-136.
104. Kavanagh, P., Newlands, N., Christensen, V. and Pauly, D. (2004) Automated parameter optimization for *Ecopath* ecosystem models. Ecological Modelling 172: 141-149.
105. Kennelly, S.J. and Broadhurst, M.K. (2002) Bycatch begone: changes in the philosophy of fishing technology. Fish and Fisheries 3(4): 340-355.
106. Lavin, M. (1999) On how the Colorado River affected the hydrography of the Upper Gulf of California. Continental Shelf Research 19: 1545-1560.
107. Lavin, M.F., Palacios-Hernández, E. and Cabrera, C. (2003) Sea surface temperature anomalies in the Gulf of California. Geofisica Internacional 42(3): 23-58.
108. Lenihan, H.S. and Peterson, C.H. (1998) How habitat degradation through fishery disturbance enhances impacts of hypoxia on oyster reefs. Ecological Applications 8(1): 128–140.
109. Lichatowich, J. (2001) Salmon Without Rivers: A History of the Pacific Salmon Crisis. Island Press, NY. 333pp.
110. Lin, S.Y. (1949) The fishing industries of Hong Kong. Hong Kong University Fisheries Journal 1: 5-160.
111. Lu, W.H. and Ye, P.R.. (2001) The status of Guangdong bottom trawl fishery resources. Chinese Fisheries 1: 64-66. [Translation, Chinese]
112. Ludwig, D., Hilborn, R. and Walters, C.J. (1993) Uncertainty, resource exploitation, and conservation: Lessons from history. Science 260: 17-18.
113. MacCall, A.D. (1990) The Dynamic Geography of Marine Fish Populations. University of Washington Press, Seattle, USA. 153pp.
114. Mace, P.M. (2001) A new role for MSY in single-species and ecosystem approaches to fisheries stock assessment and management. Fish and Fisheries 2(1): 2-32.
115. Mackinson, S. (2001) Representing trophic interactions in the North Sea in the 1880s, using the *Ecopath* mass-balance approach. Pages 35-98 in Guénette, S., Christensen, V. and Pauly, D. (eds) Fisheries impacts on North Atlantic ecosystems: models and analyses. Fisheries Centre Research Reports 9(4): 344pp.
116. Magna Carta (1215) Reprinted and translated as: A Commentary on the Great Charter of King John, with an Historical Introduction, by William Sharp McKechnie (Glasgow: Maclehose, 1914).
117. Martell, S. and Wallace, S. (1998) Estimating historical lingcod biomass in the Strait of Georgia. Pages 45-48 in Pauly, D., Pitcher, T.J. and Preikshot, D. (eds) Back to the Future: Reconstructing the Strait of Georgia Ecosystem. Fisheries Centre Research Reports 6(5): 99pp.
118. Martell, S.J.D. (2004) Dealing with migratory species in ecosystem models. In Pitcher, T.J. (ed.) Back

to the Future: Advances in Methodology for Modelling and Evaluating Past Ecosystems as Future Policy Goals. Fisheries Centre Research Reports 12 (1): 41-44.
119. McClanahan, T.R. (2002) The near future of coral reefs. Environmental Conservation 29 (4): 460–483.
120. McPhaden, M.J. and Zhang, D. (2002) Slowdown of the meridional overturning circulation in the upper Pacific Ocean. Nature 415: 603-808.
121. McLoughlin, L.C. (2002) Questioning Assumptions and Using Historical Data in Developing an Information Base for Estuarine Management. Coast to Coast 202: 281-285.
122. Meester, G.A., Ault, J.S., Smith, S.G. and Mehrotra, A. (2001) An integrated simulation modeling and operations research approach to spatial management decision making. Sarsia 86(6): 543-558.
123. Mercer, M.C. (1967) Records of the Atlantic Walrus, *Odobenus rosmarus rosmarus*, from Newfoundland. Journal of the Fisheries Research Board of Canada, 24(12): 2631-2635.
124. Merilees, W. (1985) The humpback whales of Strait of Georgia. Waters: Journal of the Vancouver Aquarium 8: 24pp.
125. Miller, A.J. and Schneider, N. (2000) Interdecadal climate regime dynamics in the North Pacific Ocean: theories, observations and ecosystem impacts. Progress in Oceanography 47: 355-379.
126. Montevecchi, W.A., and Kirk, D.A. (1996) Great Auk (*Pinguinus impennis*). Pages 1-20 in: Poole, A. and Gill, F. (eds) The Birds of North America. The American Ornithologists' Union and The Academy of Natural Sciences, Philadelphia, USA. Vol. 260.
127. Morales, A., Rosello, E. and Canas, J.M. (1994) Cueva de Nerja (prov. Malaga): a close look at a twelve thousand year ichthyofaunal sequence from southern Spain. Pages 253- 262 in W. van Neer (ed.) Fish Exploitation in the Past. Ann. Sci. Zool. Mus. Roy. L'Afrique Centrale, Tervuren, Belg. 274.
128. Morissette, L. (2001) Modélisation écosystémique du nord du Golfe du Saint-Laurent. M. Sc. thesis, Université du Québec à Rimouski, Rimouski, QC. [In French]
129. Morissette, L. (2004) Addressing Uncertainty in Marine Ecosystems Modelling. NATO-ASI Strategic Management of Marine Ecosystems 2004.
130. Morissette, L., Savenkoff, C., Castonguay, M., Swain, D.P., Bourdages, H., Chabot, D., and Hammill, M.O. (2004) Contrasting changes between the northern and southern Gulf of St. Lawrence ecosystems associated with the collapse of groundfish stocks. Can. J. Fish. Aquat. Sci. (in press).
131. Morissette, L., Despatie, S.-P., Savenkoff, C., Hammill, M.O., Bourdages, H., and Chabot, D. (2003) Data gathering and input parameters to construct ecosystem models for the northern Gulf of St. Lawrence (mid-1980s). Can. Tech. Rep. Fish. Aquat. Sci. No. 2497.
132. Morrison, M.L. (2002) Wildlife Restoration: Techniques for Habitat Analysis and Animal Monitoring. Island Press, NY. 215pp.
133. Mowat, F. (1984) Sea of Slaughter. Seal Books, McClelland-Bantam Inc., Toronto. 463pp.
134. Myers, R.A. and Worm, B. (2003) Rapid worldwide depletion of predatory fish communities. Nature 423: 280- 283.
135. Newell, D. and Ommer, R.E. (eds) (1999) Fishing Places, Fishing People: Traditions and Issues in Canadian Small-Scale Fisheries. Univ. Toronto Press, Toronto, Canada. 374pp.
136. Orchard, T.J. and Mackie, Q. (2004) Environmental Archaeology: Principles and Case Studies. Pages 64–73 in Pitcher, T.J. (ed.) Back to the Future: Advances in Methodology for Modelling and Evaluating Past Ecosystems as Future Policy Goals. Fisheries Centre Research Reports 12(1): 158 pp.
137. Padilla, E. (2002) Intergenerational Equity and Sustainability. Ecological Economics 41: 69-83.
138. Parsons, T.R. (1996) The impact of industrial fisheries on the trophic structure of marine ecosystems. Pages 352- 357 in Polis, G.A. and Winemiller, K.D. (eds) Food webs: Integration of patterns and dynamics. Chapman and Hall, New York.
139. Pauly, D. (1995) Anecdotes and the shifting baseline syndrome of fisheries. Trends in Ecology and Evolution 10(10): 430.
140. Pauly, D. (1999) Fisheries management: putting our future in places. Pages 355-362 in Newell, D. and Ommer, R. (eds) Fishing Places, Fishing People: Traditions and Issues in Canadian Small-Scale Fisheries. University of Toronto Press, Canada. 374pp.
141. Pauly, D., Christensen, V., Dalsgaaard, J., Froese, R., and Torres, F. Jnr (1998) Fishing down marine food webs. Science 279: 860-863.
142. Pauly, D., Christensen, V., Pitcher, T.J., Sumaila, U.R., Walters, C.J., Watson, R. and Zeller, D. (2002) Diagnosing and overcoming fisheries' lack of sustainability. Nature 418: 689-695.
143. Pauly, D., Pitcher, T.J. and Preikshot, D. (eds) (1998) Back to the Future: Reconstructing the Strait of Georgia Ecosystem. Fisheries Centre Research Reports 6(5): 99pp.

144. Pelletier, D., Mahevas, S., Poussin, B., Bayon, J., Andre, P. and Royer, J.-C. (2001) A conceptual model for evaluating the impact of spatial management measures on the dynamics of a mixed fishery. Pages 53-66 in Kruse, G.H., Bez, N., Booth, A., Dorn, M.W., Hills, S., Lipcius, R.N., Pelletier, D., Roy, C., Smith, S.J. and Witherell, D. (eds) Spatial processes and management of marine populations. Lowell Wakefield Fisheries Symposium Series 17.
145. Pet, J.S. and Mous, P.J. (2002) Marine Protected Areas and benefits to the fishery. Occasional paper from The Nature Conservancy Indonesia Coastal and Marine Program, Sanur, Bali, Indonesia. 15 pp.
146. Pitcher, T.J. (1997) Fish shoaling behaviour as a key factor in the resilience of fisheries: shoaling behaviour alone can generate range collapse in fisheries. Pages 143-148 in Hancock, D.A,, Smith, D.C., Grant, A. and Beumer, J.P. (eds) Developing and Sustaining World Fisheries Resources: the State of Science and Management, Proc. 2nd World Fisheries Congress, CSIRO, Collingwood, Australia. 797pp.
147. Pitcher, T.J. (1998a) 'Back To The Future': a Novel Methodology and Policy Goal in Fisheries. Pages 4-7 in Pauly, D., Pitcher, T.J. and Preikshot, D. (eds) Back to the Future: Reconstructing the Strait of Georgia Ecosystem. Fisheries Centre Research Reports 6(5): 99pp.
148. Pitcher, T.J. (1998b) Pleistocene Pastures: Steller's Sea Cow and Sea Otters in the Strait of Georgia. Pages 49-52 in Pauly, D., Pitcher, T.J. and Preikshot, D. (eds) Back to the Future: Reconstructing the Strait of Georgia Ecosystem. Fisheries Centre Research Reports 6(5): 99pp.
149. Pitcher, T.J. (1999) Rapfish, A Rapid Appraisal Technique For Fisheries, And Its Application To The Code Of Conduct For Responsible Fisheries. FAO Fisheries Circular No. 947: 47pp.
150. Pitcher, T.J. (2000) Fisheries management that aims to rebuild resources can help resolve disputes, reinvigorate fisheries science and encourage public support. Fish and Fisheries 1(1): 99-103.
151. Pitcher, T.J. (2001) Fisheries Managed to Rebuild Ecosystems: Reconstructing the Past to Salvage The Future. Ecological Applications 11(2): 601-617.
152. Pitcher, T.J. (2004a) 'Back To The Future': A Fresh Policy Initiative For Fisheries And A Restoration Ecology For Ocean Ecosystems. Phil. Trans. Roy. Soc. (in press).
153. Pitcher, T.J. (2004b) Introduction to the methodological challenges in 'Back-To-The-Future' research. Pages 4–10 in Pitcher, T.J. (ed.) Back to the Future: Advances in Methodology for Modelling and Evaluating Past Ecosystems as Future Policy Goals. Fisheries Centre Research Reports 12(1): 158pp.
154. Pitcher, T.J. (2004c) Why we have to open the lost valley: criteria and simulations for sustainable fisheries. Pages 78–86 in Pitcher, T.J. (ed.) Back to the Future: Advances in Methodology for Modelling and Evaluating Past Ecosystems as Future Policy Goals. Fisheries Centre Research Reports 12(1): 158 pp.
155. Pitcher, T.J. (2004d) What was the structure of past ecosystems that had many top predators? Pages 18–20 in Pitcher, T.J. (ed.) Back to the Future: Advances in Methodology for Modelling and Evaluating Past Ecosystems as Future Policy Goals. Fisheries Centre Research Reports 12(1): 158pp.
156. Pitcher, T.J. (2004e) The problem of extinctions. Pages 21–28 in Pitcher, T.J. (ed.) Back to the Future: Advances in Methodology for Modelling and Evaluating Past Ecosystems as Future Policy Goals. Fisheries Centre Research Reports 12(1): 158pp.
157. Pitcher, T.J. and Hart, P.J.B. (1982) Fisheries Ecology. Chapman and Hall, London, UK. 414pp.
158. Pitcher, T.J. and Haggan, N. (2003) Cognitive maps: cartography and concepts for ecosystem-based fisheries policy. Pages 456-463 in Haggan, N, Brignall, C. and Wood, L. (eds) Putting Fishers' Knowledge to Work. Fisheries Centre Research Reports 11(1): 540pp.
159. Pitcher, T.J. and Forrest, R. (2004) Challenging ecosystem simulation models with climate change: the 'Perfect Storm'. Pages 29–38 in Pitcher, T.J. (ed.) Back to the Future: Advances in Methodology for Modelling and Evaluating Past Ecosystems as Future Policy Goals. Fisheries Centre Research Reports 12(1): 158pp.
160. Pitcher, T.J., Watson, R., Courtney, A.M. and Pauly D. (1998) Assessment of Hong Kong's inshore fishery resources. Fisheries Centre Research Reports 6 (1): 1-168.
161. Pitcher, T.J., Haggan, N., Preikshot, D. and Pauly, D. (1999) 'Back to the Future': a method employing ecosystem modelling to maximise the sustainable benefits from fisheries. Pages 447-465 in Ecosystem Considerations in Fisheries Management, Alaska Sea Grant, Anchorage, USA. 738pp.
162. Pitcher, T.J., Watson, R., Haggan, N., Guénette, S., Kennish, R., Sumaila, R., Cook, D., Wilson, K. and

Leung, A. (2000) Marine Reserves and the Restoration of Fisheries and Marine Ecosystems in the South China Sea. Bulletin of Marine Science 66(3): 530-566.
163. Pitcher, T.J., Vasconcellos, M., Heymans, S.J.J., Brignall, C. and Haggan, N. (eds) (2002) Information supporting Past and Present Ecosystem Models of Northern British Columbia and the Newfoundland Shelf. Fisheries Centre Research Reports 10(1): 116 pp.
164. Pitcher, T.J., Heymans, J.J. and Vasconcellos, M. (eds) (2002b) Ecosystem models of Newfoundland for the time periods 1995, 1985, 1900 and 1450. Fisheries Centre Research Reports 10(5): 74 pp.
165. Pitcher, T.J., Power, M. and Wood, L. (eds) (2002c) Restoring the Past to Salvage the Future: Report on a Community Participation Workshop in Prince Rupert, BC. Fisheries Centre Research Reports 10(7) : 55 pp.
166. Pitcher, T.J., Heymans, J.J., Ainsworth, C., Buchary, E.A., Sumaila, U.R. and Christensen, V. (2004) Opening The Lost Valley: Implementing A 'Back To Future' Restoration Policy For Marine Ecosystems and Their Fisheries. In Knudsen, E.E., MacDonald, D.D. and Muirhead, J.K. (eds) Sustainable Management of North American Fisheries. American Fisheries Society Symposium 43: 165-193.
167. Polovina, J.J., Mitcum, G.G., Graham, N.E., Craig, M.P., Demartin, E.E. and Flint, E.N. (1994) Physical and Biological consequences of a climate event in the central North Pacific. Fisheries Oceanography 3: 15-21.
168. Power, M.D., Haggan, N. and Pitcher, T.J. (2004) The Community Workshop: how we did it and what we learned from the results. Pages 125–128 in Pitcher, T.J. (ed.) Back to the Future: Advances in Methodology for Modelling and Evaluating Past Ecosystems as Future Policy Goals. Fisheries Centre Research Reports 12(1): 158 pp.
169. Prince, J.D. (2003) The barefoot ecologist goes fishing. Fish and Fisheries 4 (4): 359-371.
170. Reidman, M.L. and Estes, J.A. (1990) The sea otter (*Enhydra lutris*): behavior, ecology and natural history. Biol. Rep. US Fish and Wild. Serv. 90(4).
171. Richardson, A. (1992) The control of productive resources on the Northwest coast of North America. In Williams, N. M. and Hunn, E.S. (eds) Resource managers: North American and Australian hunter-gatherers. Australian Institute of Aboriginal Studies, Canberra, Australia.
172. Roman, J. and Palumbi, S.R. (2003) Whales Before Whaling in the North Atlantic. Science 301: 508-510.
173. Román-Rodríguez, M. and Hamman, G. (1997) Age and Growth of totoaba, *Totoaba macdonaldi*, (Scianidae) in the upper gulf of California. Fishery Bulletin 95: 620-628.
174. Rosello-Izquierdo, E. and Morales-Muniz, A. (2001) A new look at the fish remains from Cueva de Nerja. Fish Remains Working Group, Paihia, New Zealand, October 2001.
175. Rudstam, L., Aneer, G. and Hildren, M. (1994) Top-down control in pelagic Baltic ecosystem. Dana 10: 105-129.
176. Sadovy, Y., and Cheung, W.L. (2003) Near extinction of a highly fecund fish: the one that nearly got away. Fish and Fisheries 4:86-99.
177. Salas, S., Archibald, J. and Haggan, N. (1998) Aboriginal Knowledge and Ecosystem Reconstruction. Pages 22- 28 in Pauly, D., Pitcher, T.J. and Preikshot, D. (eds) Back to the Future: Reconstructing the Strait of Georgia Ecosystem. Fisheries Centre Research Reports 6(5): 103pp.
178. Salomon, A.K., Waller, N., McIlhagga, C., Yung, R. and Walters, C.J. (2002) Modeling the trophic effects of marine protected area zoning policies: a case study. Aquatic Ecology 36: 85-95.
179. Savenkoff, C., Bourdages, H., Castonguay, M., Morissette, L., Chabot, D., and Hammill, M.O. (2004a) Input data and parameter estimates for ecosystem models of the northern Gulf of St. Lawrence (mid-1990s). Can. Tech. Rep. Fish. Aquat. Sci. In press.
180. Savenkoff, C., Bourdages, H., Swain, D.P., Despatie, S.-P., Hanson, J.M., Méthot, R., Morissette, L., and Hammill, M.O. (2004b) Input data and parameter estimates for ecosystem models of the southern Gulf of St. Lawrence (mid-1980s and mid-1990s). Can. Tech. Rep. Fish. Aquat. Sci. In press.
181. Savenkoff, C., Castonguay, M., Chabot, D., Bourdages, H., Morissette, L., and Hammill, M.O. (2004c) Effects of fishing and predation in a heavily exploited ecosystem I: Comparing pre- and post-groundfish collapse periods in the northern Gulf of St. Lawrence (Canada). Can. J. Fish. Aquat. Sci. (in press).
182. Schindler, D.E., Essington, T.E., Kitchell, J.F., Boggs, C., Hilborn, R. (2002) Sharks and tunas: fisheries impacts on predators with contrasting life histories. Ecological Applications 12(3): 735-748.

183. Shindo, S. (1973) General review of the trawl fishery and the demersal fish stocks of the South China Sea. FAO Fish. Tech. Pap. 120: 49pp.
184. Simenstad, C.A., Estes, J.A. and Kenyan, K.W. (1978) Aleuts, sea otters and alternate stable state communities. Science 200: 403-411.
185. Simeone, W. (2004) How traditional knowledge can contribute to environmental research and resource management. Pages 74–77 in Pitcher, T.J. (ed.) Back to the Future: Advances in Methodology for Modelling and Evaluating Past Ecosystems as Future Policy Goals. Fisheries Centre Research Reports 12(1): 158pp.
186. Sinclair, A.R.E., Hik, D.S., Schmitz, O.J., Scudder, G.G.E., Turpin, D.H. and Larter, N.C. (1995) Biodiversity and the need for habitat renewal. Ecological Applications 5: 579-587.
187. Soto-Mardones, L., Marinote, G. and Parés-Sierra, A. (1999) Time and spatial variability of sea surface temperature in the Gulf of California. Ciencias Marinas 25: 1-30.
188. Stanford, R. and Pitcher, T.J. (2004) Ecosystem simulations of the English Channel: climate and trade-offs. Fisheries Centre Research Reports 12(3): 103pp.
189. Steneck, R.S., Graham, M.H., Bourque, B.J., Corbett, D., Erlandson, J.M., Estes, J.A. and Tegner, M.J. (2002) Kelp forest ecosystems: biodiversity, stability, resilience and future. Environmental Conservation 29(4): 436-459.
190. Sumaila, U.R. (2001) Generational Cost Benefit Analysis for the evaluation of marine ecosystem restoration. In Pitcher, T.J. Sumaila, U. and Pauly, D. (eds) Fisheries impacts on North Atlantic Ecosystems: Evaluations and Policy Exploration. Fisheries Centre Research Reports 9(5): 94pp.
191. Sumaila, U.R. and Walters, C.J. (2003) Intergenerational discounting. Pages 19-25 in Sumaila, R. (ed.) Three Essays on the Economics of Fishing. Fisheries Centre Research Reports 11(3): 33pp.
192. Sumaila, U.R. and Walters, C.J. (2004) Intergenerational discounting: a new intuitive approach. Ecological Economics (in press).
193. Sumaila, R.S., Pitcher, T.J., Haggan, N. and Jones, R. (2001) Evaluating the Benefits from Restored Ecosystems: A Back to the Future Approach. Pages 1-7, Chapter 18 in Shriver, A.L. and Johnston, R.S. (eds) Proceedings of the 10th International Conference of the International Institute of Fisheries Economics and Trade, Corvallis, Oregon, USA, July, 2000..
194. Tolman, E.C. (1948) Cognitive Maps in Rats and Men. Psychological Review 55(4): 189-208.
195. Tomascik, T., Mah, A.J., Nontji, A. and Moosa, M.K. (1997a) The ecology of the Indonesian seas: Part One. The Ecology of Indonesia Series. Volume VII. Periplus Editions (HK) Ltd., Hong Kong. 642pp.
196. Tomascik, T., Mah, A.J., Nontji, A. and Moosa, M.K. (1997b) The ecology of the Indonesian seas: Part Two. The Ecology of Indonesia Series. Volume VIII. Periplus Editions (HK) Ltd., Hong Kong. Pages 643-1388.
197. Ulanowicz, R.E. (1999) Life after Newton: an ecological metaphysic. BioSystems 50: 127-142.
198. Van Andel, T.H. (1964) Recent marine sediments of the Gulf of California. In: Van Andel, T.J., Shor, G.G. (eds) Marine Geology of the Gulf of California: A symposium. Am. Assoc. Pet. Geol. Mem. 3: 216-310.
199. Venema S.C. (ed.) (1997) Report on the Indonesia/FAO/DANIDA Workshop on the assessment of the potential of the marine fishery resources of Indonesia. GCP/INT/575/DEN. Report on Activity No. 15. Food and Agricultural Organization of the United Nations, Rome.
200. Wallace, W.S. (1998) Changes in Human Exploitation of Marine Resources in British Columbia Pre-Contact to Present Day. Pages 58-64 in Pauly, D., Pitcher, T.J. and Preikshot, D. (eds) Back to the Future: Reconstructing the Strait of Georgia Ecosystem. Fisheries Centre Research Reports 6(5): 99pp.
201. Walters, C.J. (1986) Adaptive Management of Renewable Resources. MacMillan, New York, USA. 374 pp.
202. Walters, C.J. (1998) Designing fisheries management systems that do not depend on accurate stock assessment. Pages 279-288 in Pitcher, T.J., Hart, P.J.B. and Pauly, D. (eds) Reinventing Fisheries Management. Kluwer, Dordrecht, Holland. 435pp.
203. Walters, C.J. and Maguire, J.J. (1996) Lessons for stock assessment from the northern cod collapse. Rev. Fish Biol. Fish. 6: 125-137.
204. Walters, C.J. and Martell, S.J.D. (2004) Harvest Management for Aquatic Ecosystems. Princeton University Press. Princeton, NJ, USA. (in press).

205. Walters, C.J., Christensen, V. and Pauly, D. (1997) Structuring dynamic models of exploited ecosystems from trophic mass-balance assessment. Reviews in Fish Biology and Fisheries 7: 139-172.
206. Walters, C.J, Pauly, D. and Christensen, V. (1999) *Ecospace*: prediction of mesoscale spatial patterns in trophic relationships of exploited ecosystems, with emphasis on the impacts of marine protected areas. Ecosystems 2: 539-554.
207. Walters, C.J, Pauly, D., Christensen, V. and Kitchell, J.F. (2000) Representing Density Dependent Consequences of Life History Strategies in Aquatic Ecosystems: *Ecosim* II. Ecosystems 3(1): 70-83.
208. Walters, C.J., Christensen, V. and Pauly, D. (2002) Searching for Optimum Fishing Strategies for Fishery Development, Recovery and Sustainability. Pages 11-15 in Pitcher, T.J. and Cochrane, K. (eds) The Use of Ecosystem Models to Investigate Multispecies Management Strategies for Capture Fisheries. Fisheries Centre Research Reports 10(2): 156pp.
209. Walters, C.J., Christensen, V., Martell, S. and Kitchell, J. (2004) Single species versus ecosystem harvest management: Is it true that good single species management would result in good ecosystem management? Oral Presentation. Quantitative Ecosystem Indicators for Fisheries Management. International Symposium. March 31-April 3, 2004. Paris, France.
210. Ward, T., Tarte, D., Hegerl, E. and Short, K, (2002) Policy Proposals and Operational Guidance for Ecosystem-Based Management of Marine Capture Fisheries. World Wide Fund for Nature, Sydney, Australia. 80pp.
211. Watkinson, S. (2001) The importance of salmon carcasses to watershed function. MSc thesis University of British Columbia.Vancouver, 111pp.
212. Watson, J.C., Ellis, G. and Ford, K.B. (1997) Population growth and expansion in the British Columbia sea otter population. Sixth Joint U.S.-Russia sea otter workshop. Forks, Washington,USA.
213. Watson, J.C. (2000) The effects of sea otters (*Enhydra lutris*) on abalone (*Haliotis* spp.) populations. In A. Campbell, A. (ed.) Workshop on rebuilding abalone stocks in British Columbia Can. Spec. Publ. Fish. Aquat. Sci. 130: 123-132.
214. Ward, P. (1997) The call of distant mammoths: why the ice age mammals disappeared. Copernicus Press, NY, USA. 241pp.
215. Weitzman, M.L. (2001) Gamma discounting. American Economic Review 91: 260-271.
216. Welch, D.W, Ishida, Y. and Nagasawa, K. (1998) Thermal limits and ocean migrations of sockeye salmon (*Oncorhynchus nerka*): long-term consequences of global warming. Can. J. Fish. Aquat. Sci. 55: 937-948.
217. Whipple, S.J., Link, J.S., Garrison, L.P. and Fogarty, M.J. (2000) Models of predation and fishing mortality in aquatic ecosystems. Fish and Fisheries 1: 22-40.
218. Widodo, J., Aziz. K.A., Priyono, B.E., Tampubolon, G.H., Naamin, N. and Djamali, A. (eds) (1998) Potential and Distribution of Marine Fisheries Resources in Indonesian Waters. National Stock Assessment Commission, Indonesian Institute of Sciences (LIPI): 251pp. [in Bahasa Indonesian].
219. Williams, J. (1998) Reading rocks: on the possibilities of the use of northwest coast petroglyph and pictograph complexes as source material for ecological modelling of prehistoric sites. Pages 34-37 in Pauly, D., Pitcher, T.J. and Preikshot, D. (eds) Back to the Future: Reconstructing the Strait of Georgia Ecosystem. Fisheries Centre Research Reports 6(5): 99pp.
220. Williams, N.M. and Baines, G. (eds) (1993) Traditional Ecological Knowledge: Wisdom for Sustainable Development. Centre for Resource and Environmental Studies, Australian National University, Canberra, Australia.
221. Winship, A. (1998) Pinnipeds and Cetaceans in the Strait of Georgia. Pages 53-57 in Pauly, D., Pitcher, T.J. and Preikshot, D. (eds) Back to the Future: Reconstructing the Strait of Georgia Ecosystem. Fisheries Centre Research Reports 6(5): 99pp.
222. World Bank (2003) Indonesia at a glance. September 3, 2003. The World Bank Group, website: www.worldbank.org/data/countrydata/aag/idn_aag.pdf
223. World Wide Fund for Nature (1994) Indonesia Programme 1993-1994: Conservation Programme. WWF-Indonesia, Jakarta. 28pp.
224. Worm B., Lotze H.K. and Myers, R.A. (2003) Predator diversity hotspots in the blue ocean. Proceedings of the National Academy of Sciences USA 100: 9884-9888.
225. Zaitsev, Y.P. (1992) Recent changes in the trophic structure of the Black Sea. Fisheries Oceanography 1: 180-189.

226. Zeller, D. and Pauly, D. (2000) How life history patterns and depth zone analysis can help fisheries policy. In Pauly, D. and Pitcher, T.J. (eds) Methods for assessing the impact of fisheries on marine ecosystems of the North Atlantic. Fisheries Centre Research Reports 8(2): 54-63.
227. Zeller, D. and Pauly, D. (2001) Visualisation of standardized life-history patterns. Fish and Fisheries 2: 344-355.

Chapter 4

Management of Contaminated Sediments: Example of Integrated Management Approach

TOWARDS USING COMPARATIVE RISK ASSESSMENT TO MANAGE CONTAMINATED SEDIMENTS

T. BRIDGES, G. KIKER
U.S. Army Engineer Research and Development Center
Environmental Laboratory
3909 Halls Ferry Rd, Vicksburg, MS 39180, USA

J. CURA
Menzie-Cura & Associates, Inc.
8 Winchester Place
Winchester, MA 01890, USA

D. APUL
Department of Civil Engineering
The University of Toledo
2801W Bancroft St., Toledo, OH 43606, USA

I. LINKOV
Cambridge Environmental Inc.
58 Charles Street, Cambridge, MA 02141, USA

Abstract

Comparative risk assessment (CRA) has been used as an environmental decision making tool at a range of regulatory levels in the past two decades. Contaminated and uncontaminated sediments are currently managed using a range of approaches and technologies; however, a method for conducting a comprehensive, multidimensional assessment of the risks, costs and benefits associated with each option has yet to be developed. The development and application of CRA to sediment management problems will provide for a more comprehensive characterization and analysis of the risks posed by potential management alternatives. The need for a formal CRA framework and the potential benefits and key elements of such a framework are discussed.

1. CRA – Background and Applications

Comparative risk analysis has been most commonly applied within the realm of policy analysis, in that it supports tradeoff decisions with broad implications. Andrews et al. (2004) distinguish CRA uses at macro and micro scales.

At the macro scale, programmatic CRA has helped to characterize environmental priorities on regional and national levels by comparing the multi-dimensional risks associated with policy alternatives. U.S. government agencies at various levels have logged significant experience with policy-oriented, macro-level CRA. Gutenson (1997) suggests that the starting point was a series of Integrated Environmental Management Projects performed during the 1980s. These took place in Santa Clara, CA, Philadelphia, PA, Baltimore, MD, Denver, CO, and the Kanwha Valley, WV. Their common goal was to improve local environmental decision-making by supporting it with quantitative risk analysis. However, it was the national-level "Unfinished Business" report (US EPA, 1987) that created sustained momentum for the use of CRA. During this time period the USEPA offered grants to encourage U.S. regions, states, and localities to undertake similar projects. From 1988 to 1998, some twenty-four states and more than a dozen localities undertook comparative risk projects. See Andrews et al., 2004, Andrews, 2002, and Jones, 1997 for detailed descriptions of the nature of these projects. International CRA applications are reviewed in Tal and Linkov, 2004 and in Linkov and Ramadan, 2004.

At smaller scales, so-called micro studies in CRA have been used to compare interrelated risks involved in a specific policy choice (e.g. drinking water safety: chemical versus microbial disease risks). In these micro applications, the CRAs often had more focused objectives within the general goal of evaluating and comparing possible alternatives and their risks in solving problems.

The number and varied nature of CRA applications suggest that there is a measurable degree of acceptance for CRA as a decision-making tool even though no clear or specific guidance exists on how to conduct a CRA or use the information the analysis produces. This paper expands the discussion on the micro applications of CRA by identifying the benefits and key elements of an effective CRA framework that can be used for managing contaminated sediments.

2. CRA and Management of Sediments

Several different management alternatives and technologies exist for contaminated and uncontaminated sediments. While the number of management options is constrained, the relevant exposure pathways, receptors at potential risk, and implementation costs and benefits vary broadly. Example sediment management alternatives include unrestricted open-water placement, confined aquatic disposal (i.e., capping), use in constructed wetlands projects, disposal in lined landfills, chemical or physical stabilization technologies, cement manufacture, lightweight aggregate production, topsoil production, and in situ treatment (Seager and Gardner; 2005). Selection of the 'best' option is complicated by the varied nature of the risks, costs and benefits associated with the options. A "mind-opening" approach for analyzing the risks, costs and benefits for each potential management alternative will provide for more informed and credible decision making.

Cura et al. (2004) reviewed CRA in an attempt to rectify or at least demonstrate the differences among the varied definitions offered in the literature, explain the use of CRA at various regulatory levels, and search for an application of CRA at operational, as opposed to policy, levels. They reviewed the status of CRA within the context of environmental decision-making, evaluated its potential application as a decision-making framework for selecting alternative technologies for managing dredged material and made recommendations for implementing such a framework. Cura et al. (2004) emphasize in their review that CRA, however conducted, is an inherently subjective, value-laden process. They found that while there was some objection to this lack of total scientific objectivity ('hard version" of CRA), that the 'hard versions" provided little help in suggesting a method that surmounted the psychology of choice in decision-making schemes. The application of CRA in the decision making process at dredged material management facilities will involve the use of value-based professional judgments. The literature suggested that the best way to incorporate this subjectivity and still maintain a defensible comparative framework was to develop a method that was logically consistent and addressed this issue of uncertainty by comparing risks on the basis of more than one set of criteria, more than one set of categories, and more than one set of experts.

3. CRA Outcomes and Framework

There is no single, precise definition for CRA. In the context of sediment management problems, it can be viewed as a part of a decision making process that relies on estimated relative risks or impacts associated with each management alternative under consideration. These risks can be expressed in terms of their relationship to the natural environment, human health, the legal or regulatory context, and socioeconomics. Estimation and interpretation (possibly weighing or ranking) of relative risks is also an integral part of the CRA process. From this broad definition, it is clear that application of CRA to sediment management problems inherently avoids an overly narrow and isolated consideration of alternatives as CRA begins with the notion that various options are available and need to be evaluated with respect to the differences that exist among them.

An effective CRA of sediment management alternatives can provide a range of benefits to the decision-making process. A CRA can be used to promote a structured, fair, and open exchange of ideas among scientists, citizens, and government officials on a broad range of issues related to characterizing the risks, costs and benefits of management options. The comprehensive and comparative nature of a CRA will lead to assumptions being more transparent and reduce the hidden influence of undeclared biases. For these reasons CRA should result in more consistent and reproducible decision making.

While the benefits of CRA as a decision tool are well recognized, a formal, accepted procedure for conducting a CRA does not exist. We propose that CRA can provide the necessary data inputs concerning risks, costs and benefits that pertain to identifying the 'best" sediment management alternative; however, a guidance framework for applying

CRA to sediment problems must be developed in order to realize the benefits of the approach. A process that provides for incorporating specific stakeholder concerns and perspectives about the attributes of the problem at hand will make for a transparent and comprehensive analysis of risks, costs and benefits.

To be effective, a CRA framework must guide the analysis and comparison of disparate endpoints (e.g., human health hazard indices and cancer risks, ecological toxicity quotients, management costs, habitat loss/creation etc.) that are expressed in different units and scales. To address the needs of decision makers a CRA should identify the management alternative, or set of alternatives, that maximize risk reduction and minimize the risk of not achieving the risk reduction objective per unit cost (e.g., dollars per m^3 of sediment managed or remediated). A comprehensive CRA framework should lead to providing the answers to such questions as:

- How will the analysis address the decision drivers
- How will the sufficiency of the analysis be determined?
- What are the bases of comparison among the alternatives?
- What are the analytical components/causal pathways in the analysis?
- Which alternative maximizes the risk reduction per unit cost?
- Where is the best place to put this material?
- Which alternative minimizes risk?
- Which alternative minimizes cost?
- Which alternative minimizes uncertainty?
- What are the sources of uncertainty and variability in the assessment?
- Which alternative reflects stakeholder preferences?
- Does the analysis address the decision drivers?

Implementing a CRA framework to achieve specific programmatic or regulatory objectives raises several additional questions including:

- What, if any, programmatic or regulatory constraints exist for using CRA to inform decision making?
- How will the CRA incorporate stakeholder concerns and preferences?
- How will uncertainties in the analysis be accounted for in the decision making process?
- How will the results of the analysis be communicated to stakeholders and decision-makers ?

4. CRA Methods

Cura et al. (2004) identified and compared nine important analytical elements in nine different CRAs and found that most or all of the studies categorized elements to be compared, developed criteria as the basis of comparison, and scored the categories. Many of the studies did not incorporate public advice and only a few convened a Delphi panel, employed iteration, weighed the criteria, and defined the sources of uncertainty and quantified it. Their detailed analysis of these various assessments resulted in eight recommendations (Table 1) regarding the development and application of a CRA method for the management of dredged material. Expanding on the work of Cura et al. (2004), the key elements of an ideal framework are presented in Figure 1. The outer circle represents the logical implementation of activities starting with determining the alternatives and progressing towards making a decision. A circular presentation of activities was selected to convey the iterative nature of an ideal framework. The steps in the process will be revisited and refinements made as information and data are collected and judgments are made regarding the sufficiency of the information for decision making.

Other activities and essential elements of an effective framework are included in the center of the figure. An explicit understanding of the linkage among these elements of the process has yet to be developed by risk practitioners; however, these particular activities cut across many of the activities that are central to assessing and comparing risks. Their inclusion is intended to provide some direction for future development of CRA as a practical and operational tool.

4.1 A PROPOSED LOGICAL PROGRESSION OF ACTIVITIES IN CRA

Step A. Determine Alternatives: The purpose of a CRA is to compare the risks, costs and benefits associated with more than one alternative management option. Comparing a number of alternatives requires that a list of alternatives be prepared prior to beginning the CRA. Depending on the context for the CRA the amount effort and potential contention associated with selecting the alternatives to be compared could vary considerably among sites. A list of alternatives could be developed through a formal regulatory process (e.g., a feasibility study) or through close coordination with the stakeholder community at the site. Ensuring that the product of the CRA is responsive to the needs of the decision-making process requires that the list of alternatives must be sufficiently complete to represent the full suite of feasible options.

The list of alternatives will form the basis for the Problem Formulation of the CRA, including the development of conceptual models for each of the alternatives.

Step B. Identify Categories of Comparison We propose five broad categories of risks for evaluation at a sediment site: human health, ecological, socioeconomic, legal/regulatory, and engineering reliability. The extent to which risks within these broad categories would be evaluated for a specific site will depend on the scope of decision making and objectives defined by the regulatory context Widely accessible guidance for assessing risks to human health and the environment are available for application to the problem of contaminated sediment and for relevant subcategories for these risks. Example subcategories that may be identified for purposes of making comparisons among alternatives include risks from various exposure pathways, risks from specific contaminants, or cancer and non-cancer risks. Framing the risk questions

Table 1 Recommendations for development and application of a CRA framework (adapted from Cura et al., 2004)

Recommendation	Rationale	Implementation
The approach for dredged material management should reflect and use realistically estimated risks	Use realistic estimates as much as possible because the management decision that uses the comparison results in tangible actions and effect.	The tiered approach for conducting environmental evaluations of dredged material (USEPA/USACE 1992) results in a robust site or technology specific database for those sites that reach Tier IV in the process.
Develop and specify the CRA model and its specific elements in advance of the analysis	There is no general model for CRA especially at the operational level. Therefore the USACE will have to develop a CRA model and its elements to assure that there is an even application of CRA.	Recommend that the USACE incorporate Lash's (1996) suggestion that a CRA should involve at least five areas: defining what is meant by risk; selecting endpoints to consider; categorizing the risks for comparison; selecting a time frame for evaluating adverse effects; and gauging the seriousness of the consequences.
Identify and convene a Delphi panel that represents the goals and interests of the USACE to define the categories of risks for ranking and the attributes against which to judge them	The literature recognizes the importance of such a panel in a system that is so value laden.	The panel should represent the diverse nation-wide interests of the USAE. This will avoid the potential problem that Lash (1996) recognizes, narrow questions that focus on narrow standards (technology, cancer risk, cost) in the decision making process can be so simple as to produce that answers that are wrong.
Define the risk categories	All assessments develop or assume categories for ranking and criteria for judging them. A clear definition of risk categories for comparison will assure a more even application of a comparative risk.	Morgan et al. (2000) recognize that categorization of risks will define the course of the analysis, and they provide explicit guidance for their development including; grouping risks into a manageable number of categories; choosing a set of attributes to evaluate each category, and developing descriptions of each category in terms of the attributes.
Develop criteria for ranking the categories	The criteria provide the bases for comparing specific categories among different alternatives or technologies.	Specific criteria should be either numeric (e.g. human cancer endpoint or narrative (e.g. comparison of ecological properties to a reference area). Obviously, the criteria will have to match the categories.
Apply the categories and criteria in the planning and problem formulation stage of the ecological and human health risk assessment	Assures that the output of the assessments will be compatible with the CRA framework and consistent with recent SAB recommendations (USEPA, 2000). This should avoid the need for second order assumptions or "retrofitting".	Implementation of this recommendation requires that the engineering, business, regulatory, and societal concerns be represented in the planning of the risk assessment.
Score the categories against the criteria	If the CRA is to move beyond the "comparison of risks" it must integrate or rank the categories. Some method of scoring or ranking is an obvious solution.	The literature indicates that one way to generate some confidence in the systems is to have more than one set of categories against which to make the comparisons or have more than one panel of experts making the scores.
Identify sources of uncertainty and present them probabilistically where possible	This broadens the field of knowledge open to the decision maker and informs the decision in a quantitative manner.	The USACE has already proceeded to identify sources of uncertainty and even prioritize them for a subset of categories (Vorhees et al., 2002)

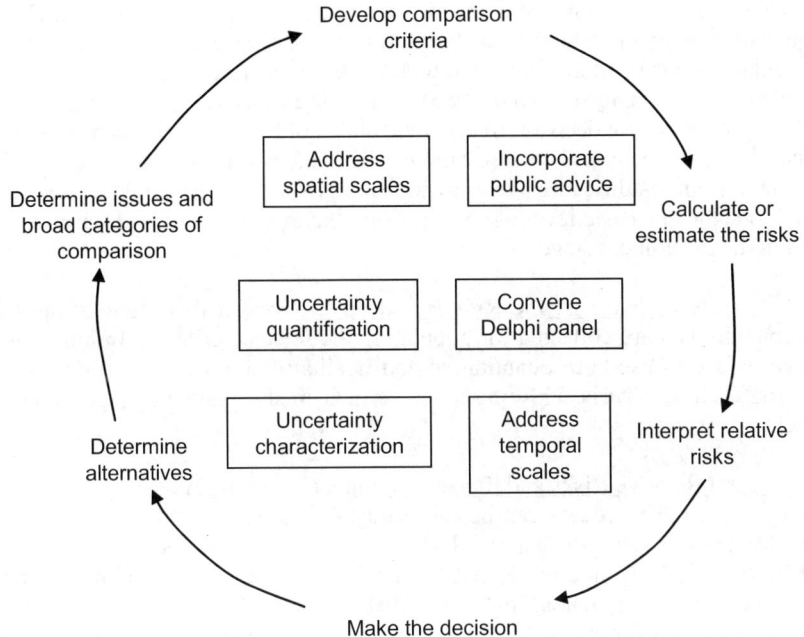

Figure 1 Schematic of relevant CRA activities.

or hypotheses for the other categories of risk will be more complicated given the lack of specific, accepted guidance for evaluating these risks. For example, in the case of engineering reliability, evaluating the risks that alternatives will be unable to attain the desired risk reduction over a specific time horizon will involve use of standardized environmental risk assessment methods as well as engineering methods associated with conducting a failure analysis for designed structures. Assessing the socioeconomic risks associated with a range of alternative management options poses particular challenges given the broad array of subcategories of risks involved, including cultural impacts, lost use of natural resources, issues related to environmental justice, and the difficult in describing the uncertainties associated with those risks.

The degree to which the categories and subcategories of risks are aggregated or disaggregated will have significant consequences for how the risk information is interpreted and used in decision making. For example, the aggregated risk to receptors in contact with surface water may be estimated as high; however, when disaggregated into contaminant classes actual risks may vary to a considerable degree over a range of difference temporal and spatial scales.

Step C. Develop Criteria for Basis of Comparison Once the relevant risk categories have been identified, comparison criteria and sub-criteria can be developed. In this step, endpoints, or criteria, are selected for each category and subcategory of risk that will form the basis for comparison among the selected alternatives. The criteria may be a combination of general and site-specific features. Factors such as spatial-scale, temporal variation or uncertainty can be incorporated as specific criteria, if desired. Criteria should be assessed and modified to achieve an exhaustive and consistent set of measures for further analysis (Roy, 1985). The development and acceptance of the criteria/sub-criteria set by decision-makers and stakeholders is a significant step before costs and effort are expended in assessing criteria values for the various alternatives. While some criteria modifications or clarification may be necessary in later stages, it is beneficial to achieve a basic level of consensus on the overall criteria set for clarity and direction in the assessment stage.

Step D. Calculate or Estimate the Risk Based on Criteria At this stage of the CRA, information and data are collected for populating the selected criteria. In most cases it will be necessary to use both quantitative and qualitative information to describe the relevant risks. The criteria/sub-criteria set defined in the previous stage guides the assessment effort.

Step E. Interpret Relative Risks and Potential Tradeoffs Among Alternatives
At this stage, the CRA process can be efficiently linked with a multi-criteria decision analysis (MCDA) process so that risk-based criteria values can be interpreted along with other non-risk decision criteria. MCDA methods are most useful in integrating the various criteria to show potential tradeoffs between alternatives in terms of the various decision criteria. Sensitivity of the alternative ranking to variations in criteria values through spatial and temporal variation or uncertainty can be explored at this stage. In addition, MCDA allows decision-makers to explore the variation of alternative rankings to criteria weights and stakeholder values. Within some MCDA processes, risk neutral, risk averse and risk tolerant perspectives can be explored with respect to the decision criteria. The analytical methods used for integrating the criteria are described in more detail in the MCDA review chapter.

4.2 SUMMARY

CRA has provided a useful framework for organizing the various risks inherent in environmental decisions into a logical and transparent format for decision-makers and stakeholders. Its cross-cutting perspective provides a valuable complement to indicators and benchmarking efforts, and issue-specific regulatory programs. Its broadly scoped analysis approach complements the narrower, more detailed risk assessments performed in support of specific regulatory proposals. In private industry, many product and service design choices involve environmental tradeoffs, and comparative risk analysis helps inform those tradeoffs. The role of CRA in systematically gathering risk-related information links efficiently with frameworks such as MCDA that can integrate a broader scope of risk and non-risk-related criteria into structured and transparent decisions.

5. References

1. Andrew, C.J., Apul, D.S., and Linkov, I. (2004) Comparative risk assessment: past experience, current trends and future directions. In Linkov, I. And Ramadan, A. eds. 'Comparative Risk Assessment and Environmental Decision Making"Kluwer, 2004
2. Andrews, C.J. (2002) *Humble Analysis: The Practice of Joint Fact-Finding*,Praeger, Westport, CT.
3. Beierle, T.C. (2002). The quality of stakeholder-based decisions. Risk Analysis 22(4):739-749
4. Cura, J.J., Bridges, T.S., McAdrle, M.E. (2004) Comparative risk assessment methods and their applicability to dredged material management decision-making. Human and Ecological Risk Assessment, 10:485-503
5. Gutenson, D. (1997) "Comparative risk: what makes a successful project?" Duke Environmental Law and Policy Forum, Vol VII, No1,pp 69-80.
6. Jones, K. (1997) Can comparative risk be used to develop better environmental decisions? Duek Environmental Law and Policy Forum, Vol VII, No1, pp 33-46
7. Lash, J. (1996) Integrating science, values, and democracy through comparative risk assessment. In Finkel, A.M., and Golding, D. (eds), Worst Things First? The Debate Over Risk-Based National Environmental Priorities, pp. 69-86, Resources for the Future, Washington, DC
8. Linkov, I., Ramadan, A., eds 'Comparative Risk Assessment and Environmental Decision Making,"Kluwer, Amsterdam 2004.
9. Morgan, M.G., Florig, H.K., DeKay, M.L. (2000) Categorizing risks for risk ranking. Risk Analysis 20(1):49-58
10. Roy, B. (1985, translation 1996). Multicriteria methodology for decision aiding. Kluwer, Boston.
11. Seager, T., and Gardner, K.H. (2005) Barriers to adoption of novel environmental technologies: contaminated sediment. In J.M. Proth, E. Levner, I. Linkov, eds. *'Strategic Management of Marine Ecosystems"* Kluewer (in press)
12. Tal, A., Linkov, I. (2004, in press). The Role of Comparative Risk Assessment in Addressing Environmental Security in the Middle East. Risk Analysis.
13. U.S. Environmental Protection Agency (USEPA), Office of Policy Analysis, Office of Policy, Planning, and Evaluation (1987) *Unfinished Business: A Comparative Assessment of Environmental Problems*, Overview report and technical appendices, USEPA, Washington, DC.
14. USEPA (U.S. Environmental Protection Agency)/USACE (US Army Corpts of Engineers) (1992) Evaluating environmental effects of dredged material management alternatives – A technical framework. EPA 842-B-92-008. Department of the Army, Washington, DC.
15. USEPA (U.S. Environmental Protection Agency). (2000). Toward integrated environmental decision-making. EPA-SAB-EC-00-011. Science advisory board, Washington, DC.
16. Vorhees, D.J., Driscoll, S.B., von Stackelberg, K. (2002) An evaluation of sources of uncertainty in a dredged material assessment. Human Ecological Risk Assessment. 8(2):369-89

MULTI-CRITERIA DECISION ANALYSIS: A FRAMEWORK FOR MANAGING CONTAMINATED SEDIMENTS

I. LINKOV and S. SAHAY
Cambridge Environmental Inc.
58 Charles Street, Cambridge, MA 02141, USA
Linkov@CambridgeEnvironmental.com.

G. KIKER and T. BRIDGES
U.S. Army Engineer Research and Development Center
Environmental Laboratory
3909 Halls Ferry Rd, Vicksburg, MS 39180, USA

T.P. SEAGER
Center for Contaminated Sediments Research
University of New Hampshire
Durham, NH 03824, USA

Abstract

Decision-making in environmental projects can be complex and seemingly intractable, principally due to the inherent existence of tradeoffs between sociopolitical, environmental, and economic factors. One tool that has been used to support environmental decision-making is comparative risk assessment (CRA). Central to CRA is the construction of a two-dimensional decision matrix that contains project alternatives' scores on various criteria. The projects are then evaluated by either qualitatively comparing the projects' scores on the different criteria or by somehow quantitatively aggregating the criterion scores for each project and comparing the aggregate scores. Although CRA is laudable in its attempts to evaluate projects using multiple criteria, it has at least one significant drawback. That drawback is the unclear or unsupported way in which it combines performance on criteria to arrive at an optimal project alternative. In the case of qualitative comparison of project scores using CRA, it can be unclear why an alternative is chosen if it performs better only on some criteria compared to another alternative. Quantitative CRAs are often unsupported in how they determine the relative importance of each criterion in determining an aggregate score for each alternative.

Multicriteria decision analysis (MCDA) not only provides better-supported techniques for the comparison of project alternatives based on decision matrices but also has the added ability of being able to provide structured methods for the incorporation of project stakeholders' opinions into the ranking of alternatives. In this paper, we provide a brief overview of common MCDA techniques and their use in regulatory agencies in the USA and EU. Then, we discuss existing literature in which MCDA techniques have been applied to decision-making involving aquatic ecosystems including decisions related to the remediation of contaminated sediments. Finally, we develop a straw man decision analytic framework specifically tailored to deal with decision-making related to contaminated sediments.

1. Current and Evolving Decision Analysis Methodologies

Environmental decisions are often multi-faceted, involving many different stakeholders with different priorities and objectives. These decisions present exactly the type of problem that behavioral decision research has shown humans are poorly equipped to solve unaided. Most people, when confronted with such problems, will attempt to use intuitive or heuristic approaches to simplify the complexity until the problem seems more manageable. In the process, important information may be lost, opposing points of view may be discarded, and elements of uncertainty may be ignored. In short, there are many reasons to expect that, on their own, individuals — including experts — will often experience difficulty making informed, thoughtful choices in a complex decision-making environment involving value tradeoffs and uncertainty (McDaniels et al., 1999).

Moreover, environmental decisions typically draw upon multidisciplinary knowledge incorporating the natural, physical, and social sciences, medicine, politics, and ethics. This fact and the tendency of environmental issues to involve shared resources and broad constituencies mean that *group* decision processes are often necessary. These may have some advantages over individual processes: more perspectives may be put forward for consideration, the probability of benefiting from the presence of natural systematic thinkers is higher, and groups often learn to rely upon more deliberative, well-informed members. However, groups are also susceptible to the tendency to establish entrenched positions (defeating compromise solutions) or to prematurely adopt a common perspective that excludes contrary information, a tendency termed "group think" (McDaniels et al., 1999).

For environmental management projects, decision makers may receive input classifiable into four broad categories: 1) the results of modeling/monitoring studies, 2) risk analysis, 3) cost or cost-benefit analysis, and 4) stakeholder preferences (Figure 1a). However, decision techniques currently in use typically offer little guidance on how to integrate or judge the relative importance of information from the categories. Some types of information — modeling and monitoring results — do not depend on much qualitative judgement, others — risk assessment and cost-benefit analyses — may incorporate a higher degree of qualitative judgment, while others — stakeholder opinions or concerns — may be presented in solely qualitative terms. Structured information about stakeholder preferences may not be presented to the decision maker at all. In cases where the decision maker does receive information on stakeholder preferences, the information may be handled in an *ad hoc* or subjective manner that exacerbates the difficulty of defending the decision process as reliable and fair.

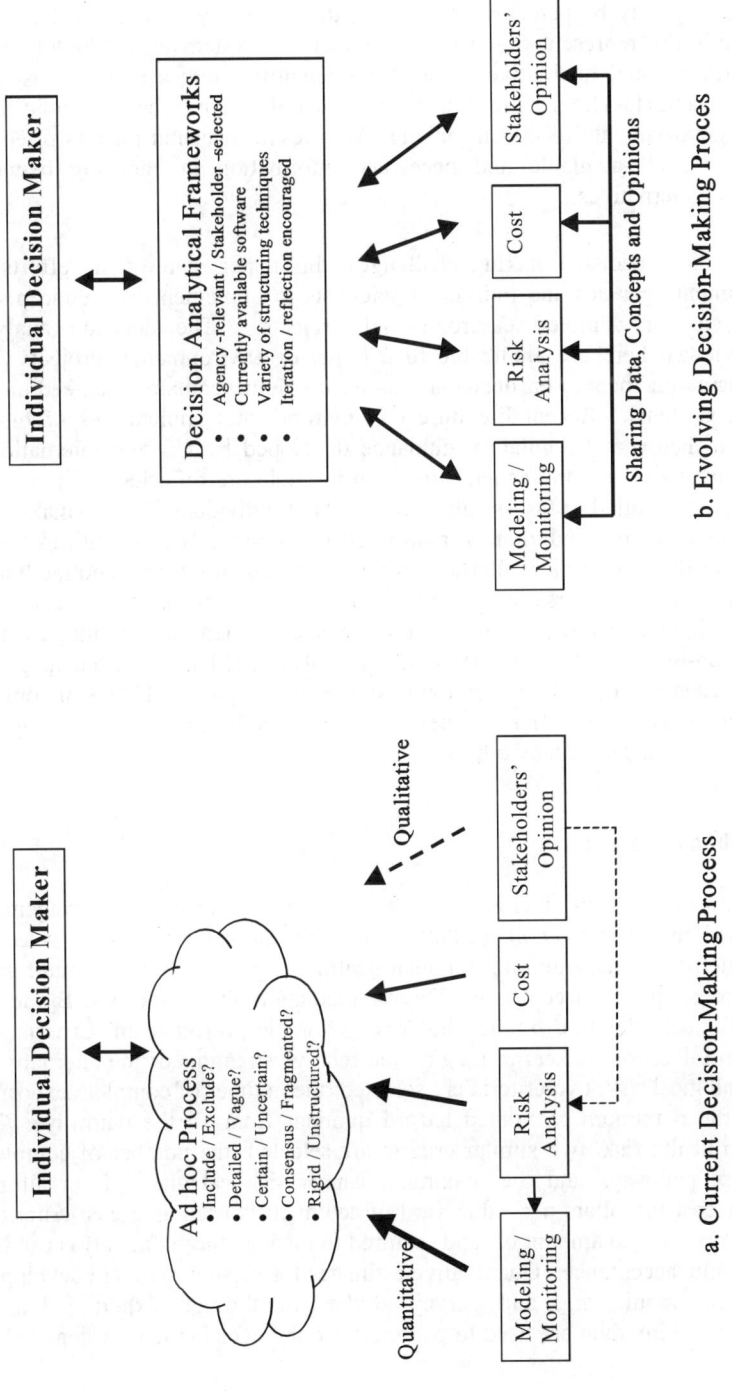

Figure 1: Current and evolving decision-making processes for contaminated sediment management.

Moreover, where structured approaches to combining the four categories of information *are* employed, they may be perceived as lacking the flexibility to adapt to localized concerns or faithfully represent minority viewpoints. A systematic methodology to combine quantitative and qualitative inputs from scientific studies of risk, cost and benefit analyses, and stakeholder views to rank project alternatives has yet to be fully developed for environmental decision making. As a result, decision makers often do not optimally use all available and necessary information in choosing between identified project alternatives.

In response to these decision-making challenges, this paper reviews the efforts of several government agencies and individual scientists to implement new concepts in decision analysis for complex environmental projects. The decision analytic approaches reviewed here are applicable to a range of environmental projects, but subsequent discussion focuses on decision making involving contaminated sediments and aquatic ecosystems. Recent literature on environmental applications of multi-criteria decision theory and regulatory guidance developed by US and international agencies is summarized, and the general trends in the field are reflected in Figure 1b. MCDA tools can be applied to assess value judgments of individual decision makers or multiple stakeholders. For individuals, risk-based decision analysis quantifies value judgments, scores different project alternatives on the criteria of interest, and facilitates selection of a preferred course of action. For group problems, the process of quantifying stakeholder preferences may be more intensive, often incorporating aspects of group decision-making. One of the advantages of an MCDA approach in group decisions is the capacity for calling attention to similarities or potential areas of conflict between stakeholders with different views, which results in a more complete understanding of the values held by others.

2. MCDA Methods and Tools

Figure 2 illustrates decision dilemmas for a contaminated sediment management project discussed in Driscoll *et al.* (2002). The decision-makers seek to select a management alternative that minimizes human health and ecological risks, minimizes cost, and maximizes public acceptance. Three remediation alternatives (A, B and C) are identified for consideration by stakeholders and/or the project team. Criteria are established to aid decision-makers in judging the relative strengths of the alternatives. To evaluate ecological risk, two criteria are selected: the number of complete exposure pathways and the maximum calculated hazard quotient from all the pathways. To evaluate human health risk, two similar criteria are selected: the number of complete human exposure pathways and the maximum cancer risk calculated from all the pathways. The cost in dollars per cubic yard of sediment is used as a cost criterion. The impacted area (*i.e.* the amount of land required to manage the sediment) is used as a measure of public acceptance. Quantitative estimates for these criteria are developed through research, monitoring, and survey studies or through expert judgment elicitation. The resulting data are used to parameterize the decision matrix depicted in Figure 2.

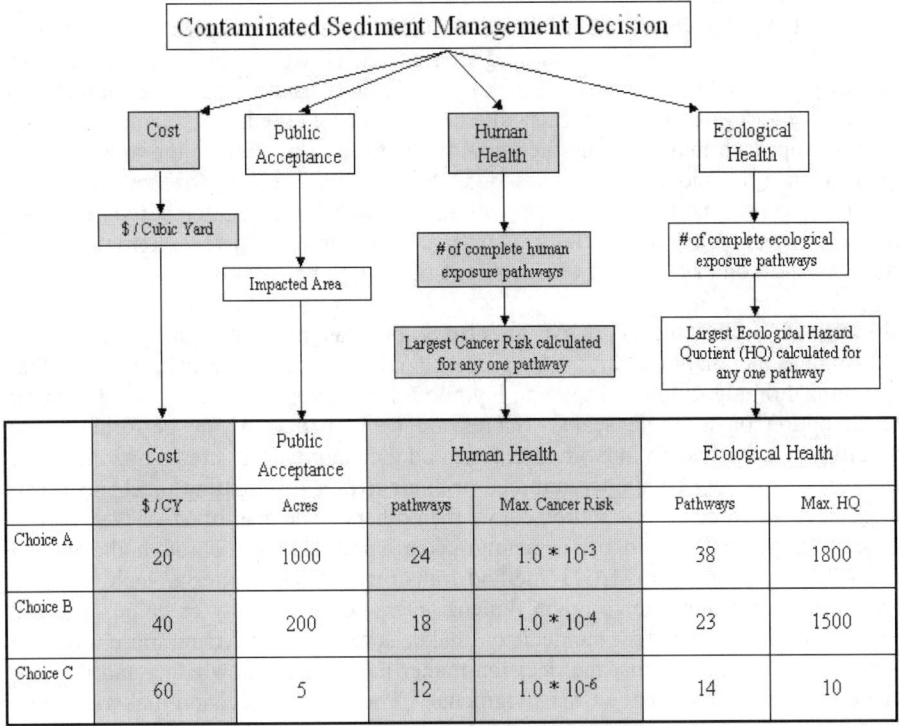

Figure 2: Example decision criteria and matrix.

A decision matrix in a form similar to Figure 2 is usually the final product of feasibility studies for Superfund projects or similar investigations. Decisions are typically based on an informal, *ad hoc* comparison of the alternatives. MCDA methods have evolved as a response to the observed inability of people to effectively analyze multiple streams of dissimilar information. There are many different MCDA methods and a detailed analysis of the theoretical foundations of these methods and their comparative strengths and weaknesses is presented in Belton and Stewart (2002). The common purpose of MCDA methods is to evaluate and choose among alternatives based on multiple criteria using systematic analysis that overcomes the limitations of unstructured individual or group decision-making.

Almost all decision analysis methodologies share similar steps of organization in the construction of the decision matrix. Each MCDA methodology synthesizes the matrix information and ranks the alternatives by different means (Yoe, 2002). Different methods require diverse types of value information and follow various optimization algorithms. Some techniques rank options, some identify a single optimal alternative, some provide an incomplete ranking, and others differentiate between acceptable and unacceptable alternatives.

Within MCDA, elementary methods can be used to reduce complex problems to a singular basis for selection of a preferred alternative. However, these methods do not necessarily to weight the relative importance of criteria and combine the criteria to produce an aggregate score for each alternative. For example, an elementary goal aspiration approach may rank the dredging alternatives in relation to the total number of performance thresholds met or exceeded. While elementary approaches are simple and can, in most cases, be executed without the help of computer software, these methods are best suited for single-decision maker problems with few alternatives and criteria, a condition that is rarely characteristic of environmental projects.

Multi-attribute utility theory (MAUT), also known as multi-attribute value theory (MAVT), and the analytical hierarchy process (AHP) are more complex methods that use optimization algorithms. They employ numerical scores to communicate the merit of each option on a single scale. Scores are developed from the performance of alternatives with respect to individual criteria and then aggregated into an overall score. Individual scores may be simply summed or averaged, or a weighting mechanism can be used to favor some criteria more heavily than others. The goal of MAUT/MAVT is to find a simple expression for decision-makers' preferences. Through the use of utility/value functions, the MAUT method transforms diverse criteria, such as those shown in Figure 2 into one common dimensionless scale of utility or value. MAUT relies on the assumptions that the decision-maker is rational (preferring more utility to less utility, for example), that the decision-maker has perfect knowledge, and that the decision-maker is consistent in his judgments. The goal of decision-makers in this process is to maximize utility/value. Because poor scores on criteria can be compensated for by high scores on other criteria, MAUT is part of a group of MCDA techniques known as "compensatory" methods.

Similar to MAUT, AHP completely aggregates various facets of the decision problem into a function which determines how good a solution is(objective function). The goal is to select the alternative that results in the greatest value of the objective function. Like MAUT, AHP is a compensatory optimization approach. However, AHP uses a quantitative comparison method that is based on pair-wise comparisons of decision criteria, rather than utility and weighting functions. All individual criteria must be paired against all others and the results compiled in matrix form. For example, in examining the choices in the remediation of contaminated sediments, the AHP method would require the decision-maker to answer questions such as, "With respect to the selection of a sediment alternative, which is more important, public acceptability or cost?" The user uses a numerical scale to compare the choices and the AHP method moves systematically through all pair-wise comparisons of criteria and alternatives. The AHP technique thus relies on the supposition that humans are more capable of making relative judgments than absolute judgments. Consequently, the rationality assumption in AHP is more relaxed than in MAUT.

Unlike MAUT and AHP, outranking is based on the principle that one alternative may have a degree of *dominance* over another (Kangas et al., 2001). Dominance occurs when one option performs better than another on at least one criterion and no worse than the other on all criteria (ODPM, 2004). However, outranking techniques do not presuppose that a single best alternative can be identified. Outranking models compare the performance of two (or more) alternatives at a time, initially in terms of each criterion, to identify the extent to which a preference for one over the other can be asserted. The criteria are not necessarily compared on a single scale. Outranking techniques then aggregate the preference information across all relevant criteria and seek to establish the strength of evidence favoring selection of one alternative over another. For example, an outranking technique may entail favoring the alternative that performs the best on the greatest number of criteria. Thus, outranking techniques allow inferior performance on some criteria to be compensated for by superior performance on others. They do not necessarily, however, take into account the magnitude of relative underperformance in a criterion versus the magnitude of over-performance in another criterion. Therefore, outranking models are known as "partially compensatory." Outranking techniques are most appropriate when criteria metrics are not easily aggregated, measurement scales vary over wide ranges, and units are incommensurate or incomparable.

3. Governmental/Regulatory Uses of MCDA

Decision process implementation is often based on the results of physical modeling and engineering optimization schemes. Even though federal agencies are required to consider social and political factors, the typical decision analysis process does not provide specifically for explicit consideration of such issues. Comparatively little effort is applied to engaging and understanding stakeholder perspectives or to providing for potential learning among stakeholders. One result of this weakness in current and common decision models is that the process tends to quickly become adversarial where there is little incentive to understand multiple perspectives or to share information. However, our review of regulatory and guidance documents revealed several programs where agencies are beginning to implement formal decision analytical tools (such as multi-criteria decision analysis) in environmental decision-making.

3.1. U.S. ARMY CORPS OF ENGINEERS

Historically, the U.S. Army Corps of Engineers has used essentially a single measure approach to civil-works, planning decisions through its Principles and Guidelines (P&G) framework (USACE, 1983). The Corps has primarily used net National Economic Development (NED) benefits as the single measure to choose among different alternatives. The P&G method makes use of a complex analysis of each alternative to determine the benefits and costs in terms of dollars and other non-dollar measures (environmental quality, safety, *etc.*); the alternative with the highest net NED benefit (with no environmental degradation) is usually selected. The USACE uses a variety of mechanistic/deterministic fate and transport models to provide information in quantifying the various economic development and ecological restoration accounting requirements as dictated by P&G procedures. The level of complexity and scope addressed by these models is determined at the project level by a planning team. Issues such as uncertainty and risk are also addressed through formulation at the individual project management level.

While the P&G method is not specifically required for planning efforts related to military installation operation and maintenance, regulatory actions, or operational and maintenance dredging, it is a decision approach that influences many USACE decisions. The USACE planning approach is essentially a mono-criterion approach, where a decision is based on a comparison of alternatives using one or two factors (Cost Benefit Analysis, as commonly used, is an example of a mono-criterion approach). The P&G approach has its challenges in that knowledge of the costs, benefits, impacts, and interactions is rarely precisely known. This approach is limiting and may not always lead to an alternative or decision process satisfactory to key stakeholders.

In response to a USACE request for a review of P&G planning procedures, the National Research Council (1999) provided recommendations for streamlining planning processes, revising P&G guidelines, analyzing cost-sharing requirements, and estimating the effects of risk and uncertainty integration in the planning process. As an integration mechanism, the National Research Council (1999) review recommended that further decision analysis tools be implemented to aid in the comparison and quantification of environmental benefits from restoration, flood damage reduction, and navigation projects. In addition, new USACE initiatives such as the Environmental Operating Principles within USACE civil works planning have dictated that projects adhere to a concept of environmental sustainability that is defined as "a synergistic process whereby environmental and economic considerations are effectively balanced through the life of project planning, design, construction, operation, and maintenance to improve the quality of life for present and future generations" (USACE 2003a, p. 5). In addition, revised planning procedures have been proposed to formulate more sustainable options through "combined" economic development/ecosystem restoration plans (USACE, 2003b). While still adhering to the overall P&G methodology, USACE (2003b) advises project delivery teams to formulate acceptable, combined economic development and ecosystem restoration alternatives through a multi-criteria/trade-off methodology (Males, 2002). Despite the existence of new guidance and revisions on the application of MCDA techniques to environmental projects, there remains a need for a systematic strategy to implement these methods within specific USACE mission areas (navigation, restoration) as well as linkage with existing risk analysis and adaptive management procedures.

3.2. U.S. ENVIRONMENTAL PROTECTION AGENCY

Stahl (2002, 2003) has recently reviewed the decision analysis process in U.S. Environmental Protection Agency (EPA) and observed that EPA could improve its decision processes to more effectively encourage stakeholder participation, integration of perspectives, learning about new alternatives, and consensus building. According to Stahl, the decision-framing process usually conforms to EPA's mission but does not always recognize different stakeholder perspectives. The problem formulation process may be influenced explicitly and implicitly by political factors which create a barrier to the integration of physical science concerns and relevant social science concerns. Stahl concludes that this approach can compromise the cohesive analysis of human and ecological impacts of a project and may result in decisions unfairly supportive of the interests of some stakeholders at the expense of others.

Similar to the USACE, the EPA uses a variety of modeling tools to support its current decision-making processes. The majority of these tools are "quantitative multimedia systems that assess benefits and risks associated with each proposed alternative with the objective of selecting the 'best option'" (Stahl, 2003). Our review has identified several EPA guidance documents that introduce decision-analytical tools and recommend their use. Multi-criteria Integrated Resource Assessment (MIRA) is being proposed as an alternative framework to existing decision analytic approaches at the U.S. EPA (Stahl et al., 2002, Stahl, 2003, USEPA, 2002). MIRA is a process that directs stakeholders to organize scientific data and establishes links between the results produced by the research community and applications in the regulatory community. MIRA also encompasses a tool that utilizes AHP-based tradeoff analysis to determine the relative importance of decision criteria. MIRA was developed by EPA Region 3's Air Protection Division as an effort to link its decisions to environmental impacts.

Multi-attribute product evaluation is inherent in the nature of Life Cycle Assessment (LCA) that has been rapidly emerging as a tool to analyze and assess the environmental impacts associated with a product, process, or service (Miettinen and Hamalainen, 1997; Seppala *et al.* 2002). The EPA developed the *Framework for Responsible Environmental Decision-Making* (FRED) to assist the Agency's Office of Pollution Prevention and Toxics in their development of guidelines for promoting the use of environmentally preferable products and services (USEPA, 2000). The FRED decision-making method provides a foundation for linking life cycle indicator results with technical and economic factors for decision-makers when quantifying the environmental performance of competing products.

3.3. U.S. DEPARTMENT OF ENERGY

Similar to the USACE and USEPA, the U.S. Department of Energy (DOE) uses a variety of models to support its decision-making process. A recent review (Corporate Project 7 Team, 2003) concluded that even though there are a significant number of guidance documents, systems, and processes in use within the DOE to determine, manage, and communicate risk, there is a great need for comparative risk assessment tools, risk management decision trees, and risk communication tools that allow site managers to reach agreement with their regulators and other stakeholders while achieving mutual understanding of the relationship between risk parameters, regulatory constraints, and cleanup. Because of DOE mandates, many DOE models are developed specifically for dealing with radiologically contaminated sites and sites with dual (chemical and radiological) contamination. Many models are deterministic, although probabilistic models are also used (U.S. DOE, 2003).

Our review has identified several guidance documents produced by DOE that introduce decision-analytical tools and recommend their use. Generic guidance developed for a wide variety of DOE decision needs (Baker *et al.*, 2001), breaks the decision process into eight sequential steps: 1) defining the problem, 2) determining the requirements, 3) establishing the goals of the project, 4) identifying alternative methods or products, 5) defining the criteria of concern, 6) selecting an appropriate decision-making tool for the particular situation, 7) evaluating the alternatives against the criteria, and 8) finally validating the solution or solutions against the problem statement. This guidance then focuses on how to select a decision-making tool — it recommends five evaluation methods and analyzes them. These methods are: 1) pros and cons analysis, 2) Kepner-Tregoe (K-T) decision analysis, 3) analytical hierarchy process, 4) multi-attribute utility theory, and 5) cost-benefit analysis.

The DOE produced a standard for selecting or developing a risk-based prioritization (RBP) system, entitled "Guidelines for Risk-Based Prioritization of DOE Activities", in April 1998. The standard describes issues that should be considered when comparing, selecting, or implementing RBP systems. It also discusses characteristics that should be used in evaluating the quality of an RBP system and its associated results. DOE (1998) recommends the use of MAUT as an RBP model since it is a flexible, quantitative decision analysis technique and management tool for clearly documenting the advantages and disadvantages of policy choices in a structured framework. MAUT merits special consideration because it provides sound ways to combine quantitatively dissimilar measures of costs, risks, and benefits along with decision-maker preferences, into high-level, aggregated measures that can be used to evaluate alternatives. MAUT allows full aggregation of performance measures into one single measure of value that can be used for ranking alternatives. However, DOE (1998) cautions that the results of MAUT analysis should not normally be used as the principal basis for decision-making, as decision making will generally require taking into account factors that cannot be readily quantified, e.g. equity. Furthermore, the guidance states that no technique can eliminate the need to rely heavily on sound knowledge, data, and judgments or the need for a critical appraisal of results.

The DOE used a multi-attribute model as the core of its Environmental Restoration Priority System (ERPS) for prioritizing restoration projects developed in the late 1980's (Jenni et al., 1995). Although ERPS was designed to operate with any specified set of values and trade-offs, its use was limited to values that were elicited from DOE managers, including values based on risk analysis. DOE headquarters decided not to apply ERPS because of stakeholder opposition, although similar decision support systems have since been adopted for use at various DOE sites (CRESP, 1999). DOE has also attempted to use simple weighting to aid program planning and budget formulation processes (CRESP, 1999).

3.4. EUROPEAN UNION

A detailed review of the regulatory background and use of decision analytical tools in the European Union was recently conducted within the EU-sponsored Contaminated Land Rehabilitation Network for Environmental Technologies project (CLARINET) (Bardos et al., 2002). The review found that environmental risk assessment, cost-benefit analysis, life cycle assessment, and multi-criteria decision analysis were the principal analytical tools used to support environmental decision-making for contaminated land management in sixteen EU countries (Austria, Belgium, Denmark, Finland, France, Germany, Greece, Ireland, Italy, Netherlands, Norway, Portugal, Spain, Sweden, Switzerland, and the United Kingdom). Similar to the U.S., quantitative methods like ERA and CBA are presently the dominant decision support approaches in use while MCDA and explicit tradeoffs are used less frequently.

Pereira and Quintana (2002) reviewed the evolution of decision support systems for environmental applications developed by the EU Joint Research Center (JRC). The concept of environmental decision support has evolved from highly technocratic systems aimed at improving understanding of technical issues by individual decision makers to a platform for helping all parties involved in a decision process engage in meaningful debate. Applications developed in the group include water resources management, siting of waste disposal plants, hazardous substance transportation, urban transportation, environmental management, and groundwater management

4. MCDA Applications in the Management of Contaminated Sediments and Related Areas

Our non-exhaustive review of recent literature shows that MCDA has been used to support decision-making related to contaminated sediment management and related applications in aquatic ecosystem management. We summarize in this section decision analysis applications published in English language journals over the last 10 years that were located through Internet and library database searches. Each identified article is classified based on whether it depends solely on technical data and expert evaluations or whether it incorporates stakeholder preferences. The articles are summarized in Tables 1 and 2.

4.1. APPLICATIONS OF MCDA BASED SOLELY ON TECHNICAL CRITERIA AND EXPERT EVALUTATIONS

MCDA techniques based solely on technical criteria and expert evaluations have been applied to optimize policy selection in the remediation of contaminated sediments and aquatic ecosystems, the reduction of contaminants entering ecosystems, the optimization of water and coastal resource management, and the management of fisheries (Table 1).

Table 1 MCDA applications based solely on technical criteria and expert evaluations

Application Area	Method	Decision Context	Funding Agency	Citation
Remediation of contaminated sediments and aquatic ecosystems	Risk-cost trade-off analysis, fuzzy set theory, composite programming	Disposal of dredged materials	USACE and University of Nebraska	Stansbury et al., 1999
	Risk-cost trade-off analysis	Disposal of dredged materials	URS Greiner Inc.; University of Nebraska-Lincoln	Pavlou and Stansbury, 1998
	SMART	Choosing a remedial action alternative at Superfund Site	USACE	Wakeman, 2003
	MAUT	Remediation of aquatic ecosystems contaminated by radionuclides using MOIRA	EC projects	Rios-Insua et al., 2002; Gallego, 2004
	MAUT	Remediation of mixed-waste subsurface disposal site	DOE	Grelk, 1997; Grelk, 1998; Parnell et al., 2001
Reduction of contaminants introduced into aquatic ecosystems	Cost-effectiveness analysis	Optimizing method to reduce nitrogen discharge to the Potomac River by 40%	SAIC	Doley et al., 2001
	Cost-benefit analysis	Protection of groundwater through choosing from among various alternatives for reducing sulfur dioxide, nitrogen oxides, and ammonia airborne emissions	Environment and Climate Program, European Union	Wladis et al., 1999
	MAUT	Wastewater planning management.	Agricultural University of Tehran, Iran	Kholgi, 2001
	Outranking (ELECTRE), distance (compromise programming)	Wastewater recycling and reuse in the Mediterranean	Aristotle University, Greece	Ganoulis, 2003
Optimization of water and coastal resources	Outranking (PROMETHEE-I, II; GAIA; MCQA-I, II, III), distance (compromise programming; cooperative game theory)	Pick optimal use of Danube region between Vienna and Slovakian border from choices like hydroelectric station and a national park	NSF and USACE	Ozelkan and Duckstein, 1996

Application Area	Method	Decision Context	Funding Agency	Citation
	Distance (compromise programming) and outranking (ELECTRE III)	Water allocation in the Upper Rio Grande	USACE, NSF, US-Hungarian Joint Research and Technology Fund	Bella et al., 1996
	Distance	Allocating waters of Jordan River basin to bordering nations	Birzeit University, Palestine	Mimi and Sawalhi, 2003
	MAUT	Consideration expansion of water supply to Cape Town, South Africa, at the expense of regional mountain flora	University of Cape Town	Joubert et al., 1997
	MAUT	Selection of management alternative Missouri River	University of Missouri-Columbia	Prato, 2003
	AHP, sensitivity analysis, MAUT	Optimizing the extent and location of a reclaimed coastline	Chinese government, John Swire and Sons, University College Oxford	Ni et al., 2002; Qin et al., 2002
	MAUT	Designing a water quality monitoring network for a river system	National Cheng-Kung University, Taiwan	Ning and Chang, 2002
Fishery management	AHP	Determining how to allocate funds for research into fisheries	Alaska Department of Fish and Game	Merritt, 2001
	MAUT	Fisheries management	Fisheries and Oceans Canada	McDaniels, 1995
	Fuzzy set theory and if-then rules	Analyzing plan to increase salmon population in Columbia River	Washington State University	Gurocak and Whittlesey, 1998
	MAUT	Estimating fishery fleet size for the North Sea	EU	Mardle and Pascoe, 2002

Table 2   MCDA applications with stakeholder involvement

Application Area	Method	Application Format	Decision Context	Funding Agency	Citation
Remediation of contaminated sites	Outranking (PROMETHEE)	Interviews and surveys	Selecting novel technological alternatives for sediment management	Dartmouth College and the University of New Hampshire	Rogers et al., 2004
	MAUT	Individual surveys	Identifying radioactive waste cleanup priorities at DOE sites	DOE/NSF	Arvai and Gregory, 2003
	AHP, MAUT	Questionnaires	Ranking of remedial alternatives at hazardous waste sites	DOE	Apostolakis, 2001; Bonano, 2000; Accorsi et al., 1999a&b
Reduction of contaminants introduced into aquatic ecosystems	Fuzzy outranking (NAIADE)	Interest groups	Choosing a sustainable wastewater treatment system in Surahammar, Sweden	Swedish Foundation for Strategic Environmental Research	van Moeffaert, 2003
	Outranking (PROMETHEE)	Workshop	Prioritization of wastewater projects in Jordan	Staffordshire University, UK	Al Rashdan et al., 1999
	Elicitation of criteria from stakeholders	Surveys, meetings, interviews	Determining the effects of a proposed 30% reduction in nitrogen loading to the Neuse Estuary in North Carolina	University of North Carolina	Borsuk, 2001
Optimization of water resources	Outranking (PROMETHEE)	Interviews, discussions, committees	Choosing the extent of groundwater protection versus economic development in an area of Elbe River in Germany	UFZ Center for Environmental Research, Germany	Klauer et al., 2002
	MAUT	Individual surveys	Water use planning	University of British Columbia, Compass Resource Management	Gregory and Failing, 2002
	MAUT+AHP	Questionnaires, interviews, surveys	Regulation of water flow in a Lake-River system	Academy of Finland	Hamalainen et al., 2001

Application Area	Method	Application Format	Decision Context	Funding Agency	Citation
	AHP & MAUT/SMART	Interview	Environmental Impact Assessment of 2 water development projects on a Finnish river	Finnish Environmental Agency, Helsinki University of Technology	Marttunen and Hamalainen, 1995
	MAUT	Small-group sessions	Consensus building for water resource management in Oregon	NSF, EPA, Carnegie Mellon University	Gregory and Wellman, 2001
	Committee consensus	Stakeholder committee	Water management in British Columbia	B.C. Hydro, Social Sciences and Humanities Research Council of Canada, NSF	McDaniels et al., 1999; Gregory et al., 2001
	Mental modeling	Individual surveys, workshop	Watershed management	EPA	Whitaker and Focht, 2001; Focht et al., 1999
Management of other resources	AHP	Interviews	Developing better management strategies for the Wonga Wetlands on the Murray River in Australia	La Trobe University, Australia	Herath, 2004
	AHP	Survey	Managing a coral reef	East West Center and WWF The Netherlands	Fernandes et al., 1999
	Trade-off analysis	Focus groups, surveys, interviews	Choosing among four development scenarios in for the Buccoo Reef Marine Park in Tobago	UK Deparment for International Development	Brown et al., 2001
	AHP	Completed by individuals representing stakeholder groups	Analyzing priorities in fishery management	European Commission	Mardle et al., 2004
	AHP	Completed by individuals representing stakeholder groups	Fishery management in Trinidad and Tobago	Food and Agriculture Organization of the United Nations	Soma, 2003

4.1.1 Remediation of Contaminated Sediments and Aquatic Ecosystems Only a few papers have been written that directly apply MCDA techniques to the remediation of aquatic systems. In a series of papers (Gallego et al., 2004; Rios-Insua et al., 2002), Gallego, Rios-Insura, and colleagues describe and apply the MOIRA system for the analysis of remedial alternatives for lakes contaminated by radionuclides. MOIRA is a MAUT model tailored to take into consideration criteria — environmental, economic, and social — associated with radiological contamination. Wakeman (2003) uses the simple multiattribute rating technique (SMART) to analyze alternatives for dredging contaminated sediments at a Superfund site in Montana. Factors considered in the study include the availability of materials and services, the ability to construct alternatives, and reliability. Pavlou and Stansbury (1998) apply a formal analysis of the tradeoff between environmental risk reduction and cost to contaminated sediment disposal. They evaluate cost, risk reduction, and potential beneficial uses of fill materials associated with three alternative methods of sediment remediation. Stansbury et al. (1999) augment the use of risk-cost tradeoff analysis with fuzzy set theory and composite programming in another paper examining contaminated sediment management. The use of fuzzy set theory formalizes the treatment of uncertainty in the analysis, while composite programming is used to find the optimal remediation strategy.

Many contaminated aquatic sites are on the EPA National Priorities List and thus go through the Superfund cleanup process. Grelk (1997), Grelk et al. (1998), and Parnell et al. (2001) have developed a CERCLA-based decision analysis value model. The model incorporates five criteria — implementability; short-term effectiveness; long-term effectiveness; reduction of toxicity, mobility, or volume through treatment; and cost — that are further subdivided into a set of 21 measures. MAUT was used to determine weights associated with each individual measure. The model was used to perform analysis of remedial alternatives for a mixed-waste subsurface disposal site at Idaho National Environmental Engineering Laboratory (INEEL).

4.1.2 Reduction of Contaminants Introduced into Aquatic Ecosystems In addition to being used in the remediation of aquatic ecosystems, MCDA techniques have been used in attempts to reduce of the amount of pollution entering those ecosystems. Doley et al. (2001) use cost-effectiveness analysis to find an optimal way to reduce nitrogen discharge into the Potomac River. They couple a water quality model with an optimization model to assess the best way to reduce nitrogen discharges from various land use types. Wladis et al. (1999) evaluate alternative emission control scenarios for NO_x, SO_2, and NH_3, considering how these pollutants effect groundwater. Specifically, they use cost-benefit analysis to evaluate two emission control scenarios and their effects on aluminum and nitrate levels in groundwater. Kholgi (2001) and Ganoulis (2003) apply MCDA to decide how to manage waste water in North America and the Mediterranean, respectively. Kholgi uses MAUT to decide among alternatives, while Ganoulis illustrates the use of a distance technique through a case study.

4.1.3 Optimization of Water and Coastal Resources MCDA techniques have also been used to help balance the sometimes conflicting demands of environmental conservation and business development with regards to water allocation and coastal development. Analyses of water bodies in the United States (Bella et al., 1996; Prato, 2003), Europe (Ozelkan and Duckstein, 1996), and South Africa (Joubert et al., 1997) have examined various uses for water bodies such as consumption, recreation, conservation, and power generation.

A MAUT-based method was applied to compare current and alternative water control plans in the Missouri River (Prato, 2003). Structural modifications to the river have significantly altered its fish and wildlife habitat and thus have resulted in the need for careful ecosystem management. The following criteria were considered: flood control, hydropower, recreation, navigation, water supply, fish and wildlife, interior drainage, groundwater, and preservation of historic properties. The analysis supported the implementation of a modified plan that incorporates adaptive management, increased drought conservation measures, and changes in dam releases. Ni, Borthwick, and Qin in two papers (Ni et al., 2002; Qin et al., 2002) describe their use of AHP in determining the optimal length and location for a coastline reclamation project considering both developmental and environmental factors. In one of their studies, AHP is used to determine preference weights, while in the other study a specially developed questionnaire is used. The objectives are then optimized using the preference weights. Ning and Chang (2002) use MAUT to optimize the location of water quality monitors in a water quality monitoring system in Taiwan. Other MCDA methods (such as distance techniques like compromise programming and game theory) have also been used. For example, a study of the Jordan River (Mimi and Sawalhi, 2003) attempts to optimize the allocation of water from the river to countries that border it using a distance technique.

4.1.4 Fishery management
Many studies have been completed using MCDA techniques to optimize fishery management. Most studies attempt to find an optimal level of fish use versus conservation. For example, McDaniels (1995) uses a MAUT approach to select among alternatives for a commercial fishery in the context of conflicting long-term objectives for salmon management. Similarly, Mardle and Pascoe (2002) use MAUT in fishery management while Gurocak and Whittlesey (1998) use a combination of fuzzy set theory and if-then rules. Merritt (2001) uses AHP to optimally allocate funds for research into fish stocks.

4.2 APPLICATIONS OF MCDA THAT INCORPORATE STAKEHOLDER VALUE JUDGEMENT

MCDA tools have been used to explicitly incorporate and sometimes quantify stakeholder values in deciding among cleanup and management alternatives (Table 2). These analyses have used a variety of techniques to elicit stakeholder opinion including focus groups, surveys, meetings, interviews, discussions, workshops, and questionnaires. The stakeholder opinions have then been integrated into numerous MCDA methods. Stakeholder values are often considered as one attribute among many such as costs and risk reduction. In addition to having the advantage of providing decision-makers with stakeholder input, MCDA can also have the benefit of providing a framework that permits stakeholders to structure their thoughts about the pros and cons of different remedial and environmental management options. Often, MCDA applications incorporating stakeholder opinions focus on the same issues addressed in the MCDAs reviewed in section. MCDA applications for group decision-making in other areas were also reviewed by Bose et al. (1997) and Matsatsinis and Samaras (2001).

4.2.1 Remediation of Contaminated Sites Our review has identified only one study dealing with the application of decision-analytical tools to include stakeholder involvement at contaminated sediment sites. However, we have identified several studies dealing with stakeholder involvement for contaminated terrestrial sites (Table 2). Rogers et al. (2004) employ a PROMETHEE outranking method to incorporate stakeholder values into the process of selecting one of a group of novel technological alternatives for sediment management. The authors found systematic outranking analysis to be effective at sorting out complex trade-offs. They identified dominated alternatives and studied the sensitivity of second-best alternatives to preference weightings. The stakeholders involved were eager to have their values heard and incorporated into the management decision process.

Arvai and Gregory (2003) compared two approaches for involving stakeholders in identifying radioactive waste cleanup priorities at DOE sites: 1) a traditional approach that involved communication of scientific information that is currently in use in many DOE, EPA, and other federal programs and 2) a values-oriented communication approach that helped stakeholders in making difficult tradeoffs across technical and social concerns. The second approach has strong affinity to the MAUT-based tradeoffs discussed earlier in this chapter. The authors concluded that the incorporation of value-based tradeoff information leads stakeholders to making more informed choices.

Apostolakis and his colleagues (Apostolakis, 2001; Bonano et al., 2000; Accorsi et al., 1999a&b) developed a methodology that uses AHP, influence diagrams, MAUT, and risk assessment techniques to integrate the results of advanced impact evaluation techniques with stakeholder preferences. In this approach, AHP is used to construct utility functions encompassing all the performance criteria. Once the utility functions have been constructed, MAUT is applied to compute expected utilities for alternatives. The authors used this approach to elicited stakeholder input and select a suitable technology for the cleanup of a contaminated terrestrial site.

4.2.2 Reduction of Contaminants Introduced into Aquatic Ecosystems We located multiple MCDAs involving stakeholders that analyze ways to reduce contaminants entering aquatic ecosystems. van Moeffaert (2003) attempts to find the optimal wastewater treatment system among alternatives considered in Surahammar, Sweden. He uses a fuzzy outranking technique and combines the rankings with the opinions of various interest groups to choose " 'the best defendable' alternative." Al Rashdan et al. (1999) use outranking to prioritize wastewater projects in Jordan. They select criteria to judge the projects with the help of stakeholders through a brainstorming session. The methodology was found to be very useful in solving problems with conflicting criteria.

Borsuk et al. (2001) examine the effects of a proposed 30% reduction in nitrogen loading on the Neuse River estuary in North Carolina. They elicit stakeholder opinion to determine which criteria should be examined in analyzing the effects of the reduction.

4.2.3 Optimization of Water Resources Many MCDAs involving stakeholder opinion seek to improve resource allocation and management. Klauer et al. (2002) attempt to use outranking to optimize groundwater protection strategies in an area of the Elbe River in Germany. Through interviews, discussions, and committees, Klauer uses stakeholder opinion to develop alternatives and criteria to rank them with. Unfortunately, the decision-making body in Germany decided to withdraw from Klauer's MCDA process and make a decision without considering its results. A number of other analyses (Gregory and Failing, 2002; Hamalainen et al., 2001; Marttunen and Hamalainen, 1995; Gregory and Wellman, 2001; McDaniels et al., 1999; Gregory et al., 2001; Whitaker and Focht, 2001) seek to optimize water use planning using MAUT, AHP, and other MCDA techniques eliciting user opinions to determine alternatives, criteria, and criteria values. In the management of the Illinois River basin in eastern Oklahoma, a novel technique called Mental Modeling was used (Focht et al., 1999, Whitaker and Focht, 2001). Mental Modeling (Morgan et al., 2002) is a promising tool for assessing individual judgments. It involves individual, one-on-one interviews leading participants through a jointly determined agenda of topics.

4.2.4 Management of Other Resources MCDAs involving stakeholder involvement are also used to manage wetlands, coral reefs, and fisheries. Herath (2004) uses AHP to incorporate stakeholders' opinions in deciding how much a wetland in Australia should be developed to increase nature-based tourism. When faced with the same choices as Herath (2004) except regarding coral reefs, Fernandes et al. (1999) also use AHP to incorporate stakeholder opinions while Brown et al. (2001) use stakeholder workshops to elicit stakeholder opinions and a less-quantitative tradeoff analysis to select a management option for Buccoo Reef Marine Park in Tobago; criteria evaluated included ecological, social and economic factors. In two papers (Mardle et al., 2004; Soma, 2003), MCDA analysis involving stakeholder opinion is applied to fishery management. In both of these analyses, stakeholders value the importance of criteria through AHP.

5. Straw Man Application Framework of MCDA Methods and Tools for Sediment Management

Successful environmental decision-making in complex settings will depend on the extent to which three key ingredients are integrated within the process: people, process and tools. Based on our review of MCDA concepts and applications, we have synthesized our understanding into a systematic decision framework (Figure 3). This framework is intended to provide a generalized road map to the environmental decision-making process

290

People:

Policy Decision Maker(s)

Scientists and Engineers

Stakeholders (Public, Business, Interest groups)

Process:

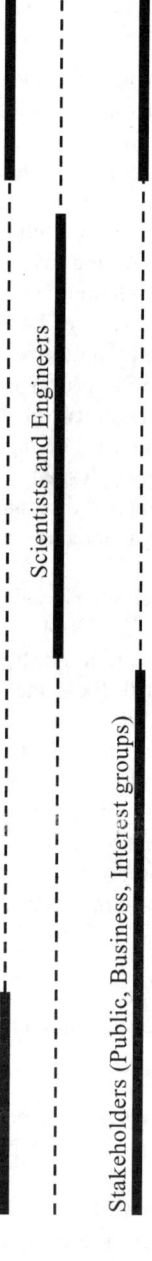

Define Problem & Generate Alternatives

Identify criteria to compare alternatives

Gather value judgments on relative importance of the criteria

Screen/eliminate clearly inferior alternatives

Determine performance of alternatives for criteria

Rank/Select final alternative(s)

Tools:

Environmental Assessment/Modeling (Risk/Ecological/Environmental Assessment and Simulation Models)

Decision Analysis (Group Decision Making Techniques/Decision Methodologies and Software)

Figure 3: Straw man framework

Having the right combination of *people* is the first essential element in the decision process. The activity and involvement levels of three basic groups of people (decision-makers, scientists and engineers, and stakeholders) are symbolized by dark lines for direct involvement and dotted lines for less direct involvement. While the actual membership and the function of these three groups may overlap or vary, the roles of each are essential in gathering the most utility from human input to the decision process. Each group has its own way of viewing the world, its own method of envisioning solutions, and its own societal responsibility. Policy- and decision-makers spend most of their effort defining the problem context and the overall constraints on the decision. In addition, they may have responsibility for the selection of the final decision and its implementation. Stakeholders may provide input in defining the problem but contribute the most input into helping to formulate performance criteria and contributing value judgments for weighting the various success criteria. Depending on the problem and regulatory context, stakeholders may have some responsibility in ranking and selecting the final option. Scientists and engineers have the most focused role in that they provide the measurements or estimations of the desired criteria that determine the success of various alternatives. While they may take a secondary role as stakeholders or decision-makers, their primary role is to provide the technical input as necessary in the decision process.

The framework (Figure 3) places *process* in the center of the overall decision process. While it is reasonable to expect that the decision-making process may vary in specific details among regulatory programs and project types, emphasis should be given to designing an adaptable structure so that participants can modify aspects of the project to suit local concerns, while still producing a structure that provides the required outputs. The process depicted in Figure 3 follows two basic themes: 1) generating management alternatives, success criteria, and value judgments and 2) ranking the alternatives by applying the value weights. The first part of the process generates and defines choices, performance levels, and preferences. The latter section methodically prunes non-feasible alternatives by first applying screening mechanisms (for example, overall cost, technical feasibility, general societal acceptance) followed by a more detailed ranking of the remaining options by decision analytical techniques (AHP, MAUT, Outranking) that utilize the various criteria levels generated by environmental tools, monitoring, or stake-holder surveys.

As shown in Figure 3, the *tools* used within group decision-making and scientific research are essential elements of the overall decision process. As with *people*, the applicability of the tools is symbolized by solid lines (direct or high utility) and dotted lines (indirect or lower utility). Decision analysis tools help to generate and map preferences of stakeholder groups as well as individual value judgments into organized structures that can be linked with the other technical tools from risk analysis, modeling and monitoring, and cost estimations. Decision analysis software can also provide useful graphical techniques and visualization methods to express the gathered information in understandable formats. When changes occur in the requirements or decision process, decision analysis tools can respond efficiently to reprocess and iterate with the new inputs. The framework depicted in Figure 3 provides a focused role for the detailed scientific and engineering efforts invested in experimentation, environmental monitoring, and modeling that provide the rigorous and defendable details for evaluating criteria performance under various alternatives. This integration of decision and scientific and engineering tools allows each to have a unique and valuable role in the decision process without attempting to apply either tool beyond its intended scope.

As with most other decision processes reviewed, it is assumed that the framework in Figure 3 is iterative at each phase and can be cycled through many times in the course of complex decision-making. The same basic *process* is used initially with rough estimates to sketch out the basic elements and challenges in the decision process with a few initial stakeholders and screening-level analysis or models. A first-pass effort may efficiently point out challenges that may occur, key stakeholders to be included or modeling/analysis studies that should be initiated. As these challenges become more apparent one iterates again through the framework to explore and adapt the process to address the more subtle aspects of the decision with each iteration giving an indication of additional details would benefit the overall decision.

6. Conclusion

Effective environmental decision-making requires an explicit structure for coordinating joint consideration of the environmental, ecological, technological, economic, and socio-political factors relevant to evaluating and selecting among management alternatives. Each of these factors includes multiple sub-criteria, which makes the process inherently multi-objective. Integrating this heterogeneous information with respect to human aspirations and technical applications demands a systematic and understandable framework to organize the people, processes, and tools for making a structured and defensible decision.

Stakeholder involvement is increasingly recognized as being an essential element of successful environmental decision making. The challenge of capturing and organizing that involvement as structured inputs to decision-making alongside the results of scientific and engineering studies and cost analyses can be met through application of the tools reviewed in this paper. The current environmental decision-making context limits stakeholder participation within the "decide and defend" paradigm that positions stakeholders as constraints to be tested, rather than the source of core values that should drive the decision-making process. Consequently, potentially controversial alternatives are eliminated early and little effort is devoted to maximizing stakeholder satisfaction with either the decision process or outcome. Instead, the final decision may be something to which no one objects too strenuously. Ultimately, this process does little to serve the needs or interests of the people who must live with the consequences of an environmental decision: the public.

The increasing volume of complex and often controversial information being generated to support environmental decisions and the limited capacity of any one individual decision maker to integrate and process that information emphasize the need for developing tractable methods for aggregating the information in a manner consistent with decision makers' values. The field of MCDA has developed methods that can help in developing a decision analytical framework useful for environmental management, including the management of contaminated sites. The purpose of MCDA is not always to single out the "correct" decision, but to help improve understanding in a way that facilitates a decision-making process involving risk, multiple criteria, and conflicting interests. MCDA visualizes tradeoffs among multiple, conflicting criteria and quantifies the uncertainties necessary for comparison of available remedial and abatement alternatives. This process helps technical project personnel as well as decision makers and stakeholders systematically consider and apply value judgments to derive a favorable management alternative. MCDA also provides methods for participatory decision-making where stakeholder values are elicited and explicitly incorporated into the decision process.

Different MCDA methods have their associated strengths and limitations. No matter which analytical decision tool is selected, implementation requires complex tradeoffs. This complexity is probably one of the main reasons why MCDA is still not widely used in practical applications. However, explicit, structured approaches will often result in a more efficient and effective decision process compared with the often intuition- and bias-driven decision processes that are currently used.

Formal applications of MCDA in management of contaminated sites are still rare. Applications in related areas are more numerous, but to date they have remained largely academic exercises with some exception in the use of AHP-based methods in natural resources planning. Nevertheless, the positive results reported in the studies reviewed in this paper as well as the availability of recently developed software tools provides more than an adequate basis for recommending the use of MCDA in contaminated site management.

Environmental decision-making involves complex trade-offs between divergent criteria. The traditional approach to environmental decision-making involves valuing these multiple criteria in a common unit, usually money, and thereafter performing standard mathematical optimization procedures. Extensive scientific research in the area of decision analysis has exposed many weaknesses in the cost-benefit analysis (Belton and Steward, 2002). At the same time, new methods that facilitate a more rigorous analysis of multiple criteria have been developed. These methods, collectively known as MCDA methods, are increasingly being adopted in environmental decision-making. This paper surveyed the principal MCDA methods currently in use and cited numerous environmental applications of these methods. While MCDA offers demonstrable advantages, choosing among MCDA methods is a complex task. Each method has strengths and weaknesses; while some methods are better grounded in mathematical theory, others may be easier to implement. Data availability may also act as a constraint on applicable methods. It is therefore unavoidable that the decision-maker will have to choose, on a case-by-case basis, the most suitable MCDA technique applicable to each situation. This paper has set out a decision analytic framework to facilitate such a selection process and thereafter provides guidance on the implementation of the principal MCDA methods within a larger context of the people, processes, and tools used in decision-making. The extensive growth over the last 30 years in the amount and diversity of information required for environmental decision-making has exceeded the capacity of common, unstructured decision models. Focused effort directed at integrating MCDA principles and tools with existing approaches, including the use of risk and cost/benefit analysis, will lead to more effective, efficient, and credible decision making.

7. Acknowledgements

The authors are grateful to Drs. Ganoulis, Cooke, Small, Valverde, Gloria, Yoe, Florig, Sullivan, Dortch, and Gardner for useful suggestions. Special thanks to Dr. Stahl for making her dissertation available. Support for this study was provided by the USACE Hazard/Risk Focus Area of the Environmental Quality Technology Program (GAK, IL, SS) and the Center for Contaminated Sediments (TSB). Additional support was provided by NOAA through the Cooperative Institute for Coastal and Estuarine Environmental Technology (TPS). Permission was granted by the Chief of Engineers to publish this material.

8. References

1. Accorsi, R., Apostolakis, G. E., and Zio, E. (1999a). Prioritizing stakeholder concerns in environmental risk management. *Journal of Risk Research 2*(1):11-29.
2. Accorsi, R., Zio, E., and Apostolakis, G.E. (1999b). Developing utility functions for environmental decision-making. *Progress in Nuclear Energy 34*(4):387-411.
3. Al-Rashdan, D., Al-Kloub, B., Dean, A. and Al-Shemmeri, T.T. (1999). Environmental impact assessment and ranking the environmental projects in Jordan. *European Journal of Operational Research 18*:30-45.
4. Apostolakis, G.E. (2001). *Assessment and Management of Environmental Risks.* I. Linkov and J. Palma-Oliveira eds. Kluwer Academic Publishers, p. 211-220.
5. Arvai, J. and Gregory, R. (2003). Testing alternative decision approaches for identifying cleanup priorities at contaminated sites. *Environmental Science & Technology 37*(8):1469-1476.
6. Baker, D., Bridges, D., Hunter, R., Johnson, G., Krupa, J., Murphy, J., and Sorenson, K. (2001). Guidebook to decision-making methods. Developed for the Department of Energy. WSRC-IM-2002-00002.
7. Bardos, P., Lewis, A., Nortcliff, S., Matiotti, C., Marot, F., and Sullivan, T. (2002). CLARINET report: review of decision support tools for contaminated land management, and their use in Europe. Published by Austrian Federal Environment Agency, on behalf of CLARINET.
8. Bella, A., Duckstein, L., and Szidarovszky, F. (1996). A multicriterion analysis of the water allocation conflict in the upper Rio Grande River basin. *Applied Mathematics and Computation 77*:245-265.
9. Belton, V. and Steward, T. (2002). *Multiple Criteria Decision Analysis: An Integrated Approach.* Kluwer Academic Publishers: Boston, MA.
10. Bonano, E.J., Apostolakis, G.E., Salter, P.F., Ghassemi, A., and Jennings, S. (2000). Application of risk assessment and decision analysis to the evaluation, ranking and selection of environmental remediation alternatives. *Journal of Hazardous Materials 71*:35-57.
11. Borsuk, M., Clemen, R.L., Maguire, L.A., and Reckhow, K. (2001). Stakeholder values and scientific modeling in the Neuse River watershed. *Group Decision and Negotiation 10*:355-373.

12. Bose, U., Davey, A.M., and Olson, D.L. (1997). Multi-attribute utility methods in group decision making: past applications and potential for inclusion in GDSS. *International Journal of Management Sciences 25*(6):691-706.
13. Brown, K., Adger, N.W., Tompkins, E., Bacon, P., Shim, D., and Young, K. (2001). Trade-off analysis for marine protected area management. *Ecological Economics 37*:417-434.
14. Corporate Project 7 Team (2003). Assessment report. Corporate Project 7: A cleanup program driven by risk-based end states. U.S. Department of Education.
15. Peer Review Committee of the Consortium for Risk Evaluation with Stakeholder Participation (CRESP) (1999). Peer review of the U.S. Department of Energy's use of risk in its prioritization process.
16. Doley, T.M., Benelmoouffok, D., and Deschaine, L.M. (2001). Decision support for optimal watershed load allocation. Paper presented at the Society for Modeling and Simulation International Conference, Seattle, 22-26 April.
17. Driscoll, S.B.K., Wickwire, T.W., Cura, J.J., Vorhees, D.J., Butler, C.L., Moore, D.W., and Bridges, T.S. (2002). A comparative screening-level ecological and human health risk assessment for dredged material management alternatives in New York/New Jersey harbor. *Human and Ecological Risk Assessment 8*:603-626.
18. Fernandes, L., Ridgley, M., and van't Hof, T. (1999). Multiple criteria analysis integrates economic, ecological and social objectives for coral reef managers. *Coral Reefs 18(4)*:393-402.
19. Focht, W., DeShong, T., Wood, J., and Whitaker, K. (1999). A protocol for the elicitation of stakeholders' concerns and preferences for incorporation into policy dialogue. Proceedings of the Third Workshop in the Environmental Policy and Economics Workshop Series: Economic Research and Policy Concerning Water Use and Watershed Management, Washington. 1-24.
20. Gallego, E., Jiménez, A., Mateos, A., Sazykina, T., and Ríos-Insua, S. (2004). Application of Multiattribute Analysis (MAA) to search for optimum remedial strategies for contaminated lakes with the MOIRA system. Paper presented at the 11th annual meeting of the International Radiation Protection Association, Madrid, 23-28 May.
21. Ganoulis, J. (2003). Evaluating alternative strategies for wastewater recycling and reuse in the Mediterranean area. *Water Science and Technology:Water Supply 3(4)*:11-19.
22. Gregory, R., McDaniels, T., and Fields, D. (2001). Decision aiding, not dispute resolution: creating insights through structured environmental decisions. *Journal of Policy Analysis and Management 20(3)*:415-432.
23. Gregory, R. and Wellman, K. (2001). Bringing stakeholder values into environmental policy choices: a community-based estuary case study. *Ecological Economics 39*:37-52.
24. Gregory, R. and Failing, L. (2002). Using decision analysis to encourage sound deliberation: water use planning in British Columbia, Canada. *Professional Practice*:492-499.
25. Grelk, B.J. (1997). A CERCLA-based decision support system for environmental remediation strategy selection. Department of the Air Force, Air Force Institute of Technology, Thesis.
26. Grelk, B., Kloeber, J.M., Jackson, J.A., Deckro, R.F., and Parnell, G.S. (1998). Quantifying CERCLA using site decision maker values. *Remediation 8*(2):87-105.
27. Gurocak, E.R. and Whittlesey, N.K. (1998). Multiple criteria decision making: a case study of the Columbia River salmon recovery plan. *Environmental and Resource Economics 12*(4):479-495.
28. Hämäläinen, R.P., Kettunen, E., Ehtamo, H., and Marttunen, M. (2001). Evaluating a framework for multi-stakeholder decision support in water resources management. *Group Decision and Negotiation 10*(4):331-353.
29. Herath, G. (2004). Incorporating community objectives in improved wetland management: the use of the analytic hierarchy process. *Journal of Environmental Management 70*(3):263-73.
30. Jenni, K.E., Merkhofer, M.W., and Williams, C. (1995). The rise and fall of a risk-based priority system: lessons from DOE's Environmental Restoration Priority System. *Risk Analysis 15*(3):397-410.
31. Joubert, A.R., Leiman, A., de Klerk, H.M., Katua, S., and Aggenbach, J.C. (1997). Fynbos (fine bush) vegetation and the supply of water: a comparison of multi-criteria decision analysis and cost-benefit analysis. *Ecological Economics 22*:123-140.
32. Kangas, J., Kangas, A., Leskinen, P., and Pykalainen, J. (2001). MCDM methods in strategic planning of forestry on state-owned lands in Finland: applications and experiences. *Journal of Multi-Criteria Decision Analysis 10*:257-271.
33. Kholghi, M. (2001). Multi-criterion decision-making tools for wastewater planning management. *Journal of Agricultural Science and Technology 3*:281-286.

34. Klauer, B., Drechsler, M., and Messner, F. (2002). Multi-criteria analysis more under uncertainty with IANUS - method and empirical results. UFZ discussion GmbH, Leipzig.
35. Males, R.M. (2002). Beyond expected value: making decisions under risk and uncertainty. RMM Technical Services, under contract to Planning and Management Consultants, Ltd. Prepared for U.S. Army Corps of Engineers, Institute for Water Resources. IWR Report.
36. Mardle, S. and Pascoe, S. (2002). Modeling the effects of trade-offs between long and short-term objectives in fisheries management. *Journal of Environmental Management 65*:49-62.
37. Mardle, S., Pascoe, S., and Herrero, I. (2004). Management objective importance in fisheries: an evaluation using the analytic hierarchy process (AHP). *Environmental Management 33*(1):1-11.
38. Marttunen, M. and Hämäläinen, R.P. (1995). Decision analysis interviews in environmental impact assessment. *European Journal of Operational Research 87*(3):551-563.
39. Matsatsinis, N.F. and Samaras, A.P. (2001). MCDA and preference disaggregation in group decision support systems. *European Journal of Operational Research 130*:414-429.
40. McDaniels, T. (1999). A decision analysis of the Tatshenshini-Alsek Wilderness Preservation alternatives. *Journal of Environmental Management 19*(3):498-510.
41. McDaniels, T.L. (1995). Using judgment in resource management: an analysis of a fisheries management decision. *Operations Research 43(3)*:415-426.
42. Merritt, M. (2001). Strategic plan for Salmon Research in the Kuskokwim River Drainage. Alaska Department of Fish and Game. Division of Sport Fish. Special Publication No. 01-07.
43. Miettinen, P. and Hamalainen, R.P. (1997). How to benefit from decision analysis in environmental Life Cycle Assessment (LCA). *European Journal of Operational Research 102*(2):279-294.
44. Mimi, Z.A. and Sawalhi, B.I. (2003). A decision tool for allocating the waters of the Jordan River Basin between all riparian parties. *Water Resources Management 17*:447-461.
45. Morgan, M.G., Fischhoff, B., Bostrom, A., and Atman, C.J. (2002). *Risk Communication*. Cambridge University Press.
46. National Research Council (1999). *New Directions in Water Resources Planning for the U.S. Army Corps of Engineers*. National Academy Press. Washington DC.
47. Ni, J.R., Borthwick, A.G.L., and Qin, H.P. (2002). Integrated approach to determining postreclamation coastlines. *Journal of Environmental Engineering 128*(6):543-551.
48. Ning, S.K. and Chang, N. (2002). Multi-objective, decision-based assessment of a water quality monitoring network in a river system. *Journal of Environmental Monitoring 4*:121-126.
49. Office of the Deputy Prime Minister (ODPM) (2004). *DLTR Multi-Criteria Decision Analysis Manual*. Available at http://www.odpm.gov.uk/stellent/groups/odpm_about/documents/page/odpm_about_608524-02.hcsp. Downloaded June 16, 2004.
50. Özelkan, E.C. and Duckstein, L. (1996). Analyzing water resources alternatives and handling criteria by multi criterion decision techniques. *Journal of Environmental Management 48*(1):69-96.
51. Parnell, G.S., Frimpon, M., Barnes, J., Kloeber, Jr., J.M., Deckro, R.F., and Jackson, J.A. (2001). Safety risk analysis of an innovative environmental technology. *Risk Analysis 21*(1):143-155.
52. Pavlou, S.P. and Stansbury, J.S. (1998). Risk-cost trade off considerations for contaminated sediment disposal. *Human and Ecological Risk Assessment 4*(4):991-1002.
53. Pereira, A.G. and Quintana, S.C. (2002). From technocratic to participatory decision support systems: responding to the new governance initiatives. *Journal of Geographic Information and Decision Analysis 6*(2):95-107.
54. Prato, T. (2003). Multiple attribute evaluation of ecosystem management for the Missouri River. *Ecological Economics 45*:297-309.
55. Qin, H.P., Ni, J.R., and Borthwick, A.G.L. (2002). Harmonized optimal postreclamation coastline for Deep Bay, China. *Journal of Environmental Engineering 128*(6):552-561.
56. Rogers, S.H., Seager, T.P., and Gardner, K.H. (2004). Combining expert judgement and stakeholder values with Promethee: a case study in contaminated sediments management. In:*Comparative Risk Assessment and Environmental Decision Making*. I. Linkov and A. Bakr Ramadan eds. Kluwer Academic Publishers, p. 305-322.
57. Ríos-Insua, S., Gallego, E., Mateos, A., and Jiménez, A. (2002). A decision support system for ranking countermeasures for radionuclide contaminated aquatic ecosystems: the MOIRA project, p. 283-297.
58. Seppala, J., Basson, L., and Norris, G.A. (2002). Decision analysis frameworks for life-cycle impact assessment. *Journal of Industrial Ecology 5*(4):45-68.

59. Soma, K. (2003). How to involve stakeholders in fisheries management--a country case study in Trinidad and Tobago. *Marine Policy 27*:47-58.
60. Stahl, C.H., Cimorelli, A.J., and Chow, A.H. (2002). A new approach to Environmental Decision Analysis: Multi-Criteria Integrated Resource Assessment (MIRA). *Bulletin of Science, Technology and Society 22*(6):443-459.
61. Stahl, C.H. (2003). Multi-Criteria Integrated Resource Assessment (MIRA): a new decision analytic approach to inform environmental policy analysis. Vol 1. Dissertation submitted to Faculty of the University of Delaware in partial fulfillment of the requirements for the degree of Doctor of Philosophy in Urban Affairs and Public Policy.
62. Stansbury, J., Member, P.E., Bogardi, I., and Stakhiv, E.Z. (1999). Risk-cost optimization under uncertainty for dredged material disposal. *Journal of Water Resources Planning and Management 125*(6):342-351.
63. U.S. Army Corps of Engineers (USACE) (1983). *Economic and environmental principles and guidelines for water and related land resources implementation studies.*
64. U.S. Army Corps of Engineers (USACE) (2003a). USACE environmental operating principles and implementation guidance. Available at http://www.hq.usace.army.mil/CEPA/7%20Environ%20Prin%20web%20site/Page1.html.
65. U.S. Army Corps of Engineers (USACE) (2003b). Planning civil works projects under the environmental operating principles. Circular 1105-2-404. Available at http://www.usace.army.mil/inet/usace-docs/eng-circulars/ec1105-2-404/entire.pdf.
66. U.S. Department of Energy (U.S. DOE) (1998). Guidelines for risk-based prioritization of DOE activities. U.S. Department of Energy. DOE-DP-STD-3023-98.
67. U.S. Department of Energy (U.S. DOE) (2003). Policy DOE P 455.1 – Subject: use of risk-based end states. Initiated by Office of Environmental Management.
68. U.S. Environmental Protection Agency (U.S. EPA) (2000). Framework for Responsible Environmental Decision-making (FRED): using life cycle assessment to evaluate preferability of products. Prepared by Science Applications International Corporation, Research Triangle Institute, EcoSense Inc., Roy F. Weston Inc., Five Winds International. EPA/600/R-00/095.
69. U.S. Environmental Protection Agency (USEPA) (2002). Consistency and transparency in determination of EPA's anticipated ozone designations. Report no. 2002-S-00016. Office of Inspector General, Special Review.
70. van Moeffaert, D. (2002). *Multi-Criteria Decision Aid in Sustainable Urban Water Management.* Masters thesis. Royal Institute of Technology. Stockholm, Sweden.
71. Wakeman, J.B. (2003). "RE: decision analysis." Email to Igor Linkov, dated April 4, 2003. Includes Attachment A: descision analysis based upon implementability of the action alternatives at Milltown Dam.
72. Whitaker, K. and Focht, W. (2001). Expert modeling of environmental impacts. *Oklahoma Politics 10*:179-186.
73. Wladis, D., Rosén, L., and Kros, H. (1999). Risk-based decision analysis of atmospheric emission alternatives to reduce ground water degradation on the European scale. *Ground Water 37*(6):818-826.
74. Yoe, C. (2002). *Trade-Off Analysis Planning and Procedures Guidebook.* Prepared for Institute for Water Resources, U.S. Army Corps of Engineers. April.

BARRIERS TO ADOPTION OF NOVEL ENVIRONMENTAL TECHNOLOGIES: CONTAMINATED SEDIMENTS

T.P. SEAGER, K.H. GARDNER
Center for Contaminated Sediments Research, Gregg Hall
University of New Hampshire
Durham, NH 03824, USA

Abstract

New technologies face high barriers to adoption compared to existing technologies for several reasons including a perceived sense of increased risk, a lack of experience with the new technologies among managers and/or regulators, or simply the fact that decision makers are not aware of the availability of the technology. Environmental technologies, however, may be especially difficult to move from innovation to commercialization. Partly this may be because environmental resources exist largely in the public domain where private industry may be unable to fully capture the economic benefits of novel technologies. But also, it may be because environmental projects often involve multiple stakeholder groups with competing or mutually exclusive interests. No single technology is likely to emerge which is perceived by all stakeholders as superior to all competing alternatives on all decision criteria. Therefore, novel technologies are likely to involve tradeoffs that engender both support and objections. This chapter provides a review of some of the difficulties in implementing novel contaminated sediments technologies in the marketplace. Brief descriptions of several technologies are provided, and the contrasting objectives and perspectives of different groups essential to environmental innovation are discussed.

1. Introduction

The prohibition on ocean dumping of contaminated sediments has created an acute need for new technological alternatives for management of contaminated dredged materials. A wide range of possibilities are available, including highly engineered approaches such as hazardous waste landfills, to "softer" solutions such as artificial wetlands construction or restoration. No single alternative is best for all sediments or all stakeholders. Priorities can (and should) change from site to site, depending upon the level and nature of contamination, the objectives of the involved stakeholder groups, and site or time specific opportunities. However, the range of possibilities may be depicted in a continuum of highly structured to highly unstructured technologies (Figure 1).

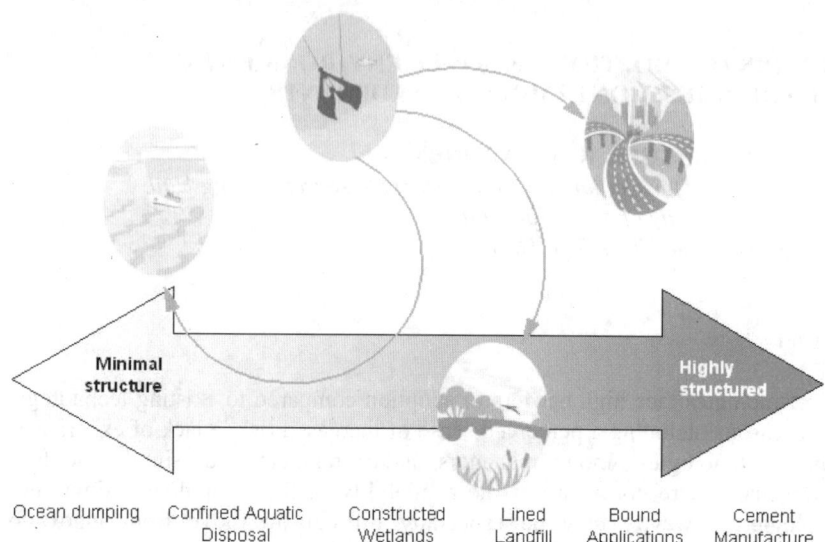

Figure 1: The Industrial Ecology of Contaminated Sediments Management.

On the left edge of the scale, representing minimal structure and control, is ocean dumping. The primary decision variables are the location and timing of disposal. However, once the sediments are discharged to the ocean, control over their fate or the disposition of contaminants is abandoned. Confined aquatic disposal (CAD) represents a slightly more highly engineered solution. In addition to timing and location, additional design variables such as thickness and composition of a clean cap afford an additional element of protection. Nonetheless, unexpected events such as sediment scouring, groundwater upwelling or seismic events may disrupt the stability of contaminants in CAD cells. Wetlands construction and restoration further separate contaminated sediments from the aquatic environment and may have the advantage of restoring lost ecological services (such as wildlife habitat or stormwater attenuation). Additional design parameters may include the hydrology of the wetland areas, such as controlled flooding, type of vegetative cover, and capping or surrounding of contaminated sediments with clean materials. Upland disposal provides an additional element of control over hydrologic parameters by removing the sediments from the aquatic environment altogether. Depending upon the level of protective measures taken, leachate from upland disposal sites may be captured and treated prior to release. Alternatively, sediments may be stabilized (e.g., by adding cement) prior to disposal in order to further minimize migration of contaminants from the disposal area.

In general, cost is lowest at the left edge (fewest design parameters, least intervention and control) and increases as the alternatives under consideration move towards the middle. However, as we near the right edge of the scale (greatest design parameters, greatest intervention and control), a new type of highly engineered alternative may emerge in which the sediments are employed as raw materials for industrial or manufacturing processes. In these alternatives, the contaminants may be destroyed, treated, or removed by pollution control equipment. The design and testing efforts may be considerably larger to ensure that the sediments can meet required engineering performance standards. However, overall net costs may be lower if existing capital equipment can be employed (lower capital costs), or the materials manufactured from contaminated sediments have a significant market value that offsets treatment or environmental control expenses. Therefore, the potential exists at the right edge of the spectrum to find management alternatives that are as cost-effective as those on the left, provided an initially expensive period of research, development, and trial-and-error experience can be overcome.

Nevertheless, industrial innovation and environmental protection have historically been viewed as antagonistic objectives. Only recently has a new view emerged that places technological advancement at the center of a multifaceted strategy for achieving eco-efficiency, economic development, and environmental protection. In the new paradigm, the potential exists for creating economic value, reducing dependence upon virgin raw materials or fossil fuels, and enhancing environmental protection simultaneously – in short, for moving towards *sustainability* (Pastakia 1998, Crittenden 2002, Anderson 1998). Whereas the old view is exemplified by the left end of the continuum representing dumping or dilution of contaminated materials, the right end is consistent with the new environmental paradigm of industrial ecology, in which the waste materials generated by one process (e.g., dredging) become the raw materials from other processes (e.g., cement manufacture, construction, or soil amendment).

Environmental innovation is therefore an area in which the potential benefits of collaborative partnerships among academe, industry, and government policy-makers or regulators could not be more obvious (Heaton and Banks 1998). However, the shift in thinking can both create new opportunities and new problems. One of the advantages of the old paradigm is its simplicity. There are few decision variables and few criteria (e.g., cost) by which alternatives should assessed. In exploring the new paradigm, it becomes apparent that an expanding number of decision variables leads to a more complicated decision process. Moreover, the criteria by which alternatives might be judged expands, as do the number of people vested in the outcome and the priorities or values they hold. A lack of prior collaborative experience between the essential parties may present a serious impediment to technological innovation -- despite the potential for newfound feelings of mutual interest.

Moreover, the increased complexity may be resisted by decision makers accustomed to simpler ways. A framework for structuring the assessment and decision process regarding new sediments management technologies is badly needed to facilitate adoption of technologies that may face multiple barriers to adoption. This paper briefly summarizes a number of new types of technologies, discusses some potential barriers to new technology adoption, and lays out a brief agenda for research and further discussion.

2. Environmental Innovation

Commercial exploitation of many innovative environmental technologies often stalls at nascent developmental stages. This is due to a number of factors including: the novelty of the technologies, the need for pilot scale (compared with bench scale) demonstration projects, potential regulatory or socio-political obstacles, the reluctance of conventional commercial enterprises to handle contaminated materials, or a lack of information about innovative technologies (Eggers et al. 2000, Hostager & Neil 1998, Krueger 1998). Overcoming these obstacles will likely require a multidisciplinary team of research, industrial and government partners. The fundamental needs are:

- To further basic and applied research fostering the commercial development of environmental technologies.
- To assess the life cycle environmental profiles, economic potential, and regulatory status of these technologies in joint industrial, government, and academic pilot-scale demonstration and development projects.
- To create multidisciplinary undergraduate, Master's and Doctoral education and training opportunities with the potential to support a broad range of technology-related entrepreneurial ventures.
- To strengthen the existing network of technology transfer and collaborative commercialization activities.

Significant challenges to the collaborative commercialization of novel technologies may exist, including a lack understanding between university and industry cultures, inconsistent or insufficient incentive structures, and a perceived lack of flexibility among the involved parties (Siegel et al. 2003). Moreover, the absence of key stakeholders from the R&D process may undermine some innovation efforts (Douthwaite et al. 2001). To overcome these potential obstacles, several activities may be required to support any particular innovative venture including: team-building kickoff meetings intended to span cultural barriers, early involvement of key stakeholders, entrepreneurial internship experiences for researchers at the graduate and faculty levels, multidisciplinary education programs, and close involvement with centers of innovation support such as university offices of intellectual property management and other state, national, or research organizations.

3. Novel Technologies for Contaminated Sediments Management

There are many environmental areas in which partnerships might be productive from an entrepreneurial, *intra*preneurial and social standpoint (Lober 1998, Keough & Polonsky 1998). However, regarding contaminated sediments management there are essentially two pathways by which innovative technologies may create commercial and environmental appeal: in-situ remediation and beneficial reuse. The latter may be especially appropriate for management of contaminated sediments dredged for navigational purposes whereas the former may be most appropriate for sediments that require management due to elevated ecological or human health risks. Below are listed brief descriptions of novel technologies under development by the University of New Hampshire Center for Contaminated Sediments Research and other places. While far from exhaustive, this list provides a basis for studying examples from which a more general discussion may be formed.

3.1 THERMAL TREATMENT TECHNOLOGIES

One class of technologies for management of contaminated dredged material relies on thermal treatment of the sediment for decontamination. In the paradigm of beneficial use, high temperature processes requiring a mineral source similar in composition to dredged materials are sought such that dredged material can be considered an alternative feedstock material. The premise is to capitalize on the inherent oxide composition of dredged material to produce valuable and marketable commodities such as portland cement, bricks or tile, or glass. While organic contaminants are destroyed by thermal oxidation, metals may be bound up in the final product and thereby environmentally unavailable. In the case of portland cement manufacture, existing cement production infrastructure may be readily employed (Dalton et al., 2004). One drawback may be the need to add mercury stack emission controls, which may increase expenses.

3.1.1 Cement Manufacture. The principle oxides present in dredged sediments applicable to cement manufacture include Al_2O_3, SiO_2, and Fe_2O_3, all of which are key ingredients in portland cement production (Figure 2). The manufacturing process consists of grinding raw materials, mixing them intimately in certain proportions, and burning them in a large rotary kiln at a temperature of ~1450°C – at which point the material sinters and fuses into balls known as clinker. The clinker is then cooled, ground to a fine powder with some gypsum added to yield the commercial product. After mixing with water, portland cement is the glue that binds sand and gravel together into the rock-like mass known as concrete.

Presently, portland cement is manufactured from natural materials quarried in close proximity to the kiln, such as limestone and silica, or in the case of alumina and iron, brought to the kiln in bulk from other natural deposits. While lime (i.e. calcium oxide) is not found in nature, it is produced during cement manufacture by heating calcium carbonate, which is found in raw materials such as limestone or calcite. In conventional cement manufacture, raw materials such as clay or shale are used as sources of the silica, alumina and ferric oxides. Contaminated dredged material is comprised primarily of clays and silts (sandy fractions are typically not contaminated and are suitable for either ocean disposal or other beneficial uses such as beach nourishment), which can substitute for traditional sources of clays and shales due to a very similar mineralogy.

Preliminary economic analysis has indicated that the potential for substantial cost savings may exist for both port/facility operators and the cement producers. Savings would result from reduced tipping fees, whereas for the cement company, savings would be realized from reduced raw material costs. Cement has a market value of

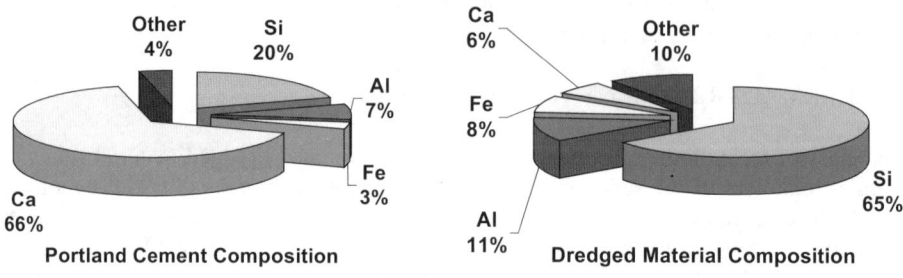

Figure 2: Typical mineral compositions of sediments and cement.

approximately $75/ton, which has a substantial impact on the economic viability of dredged material use in cement manufacture.

A proprietary process called Cement-Lock™ has been developed that is similar to the description above. Cement-Lock embodies the description above, although it requires a dedicated kiln (which creates the potential for incorporating more sophisticated emissions controls) rather than existing infrastructure and it does not produce portland cement but rather a blended, construction-grade cement.

3.1.2 Lightweight aggregate production. Another valuable product for which dredged material is a suitable feedstock substitute is lightweight aggregate, with a market value of approximately $50-$75/ton. Lightweight aggregate is used to make lightweight concrete, which has many applications from production of concrete blocks to use in roofing and flooring in multistory buildings. Lightweight aggregate is also used as geotechnical fill material where constructed works are underlain by weak soils. Lightweight aggregate is typically produced from clays, slates, and shales that are fed to a rotary kiln that reaches temperatures in excess of 1,200 °C. As the temperature increases the minerals become plastic and the gases inside the materials expand producing disconnected pore spaces. When the material cools the pores remain, creating an aggregate material that is significantly lighter than traditional crushed rock aggregates. Pilot scale research has shown that fine-grained dredged material from NY/NJ harbor is a suitable material for lightweight aggregate production (JCI/UPCYCLE Associates, 2002). There are additional steps required, most notably dewatering to specific water content and extrusion of dredged material to produce pellets to feed to the rotary kiln. The cost to manage material via lightweight aggregate production was estimated from the pilot scale work to be approximately $42 per cubic yard (estimate assumed dewatering facility in NY/NJ harbor and transport via rail to an existing kiln/LWA production facility).

3.2 CEMENT STABILIZATION (FLOWABLE FILL)

Flowable fill is a cementitious material used in general construction. It's most common use is to backfill around pipes, in trenches cut through roads, and similar applications. Its flowable nature makes it ideal for these types of applications, so void spaces are not left that can lead to future failure, and its high early strength allows traffic to resume after a short period of time. Flowable fill is traditionally produced similar to concrete: cement, water, and a fine aggregate (sand) are mixed to make very "runny" concrete. It is also designed to provide the appropriate strength for the particular application: it most typically is a relatively weak concrete, having a strength similar to soil such that it can be removed easily with conventional earth moving equipment.

Preliminary research has demonstrated that dredged material can be used in an as-dredged condition (i.e. with very high water content) to produce an acceptable flowable fill material with the addition of cement and coal fly ash (Melton, 2004). A similar approach was used to produce a fill material for a Brownfield's site in New Jersey and a demonstration of mine reclamation in Pennsylvania (PADEP, 2003; Sadat, 2001), both of which used dredged material from NY/NJ harbor with the addition of cement and fly ash to reach acceptable strengths.

A significant question that is particularly germane to the production of flowable fill using contaminated sediments is whether such flowable fill is acceptable for general use, or whether such use would represent an unacceptable risk (i.e. a transfer of risk from the aquatic to the terrestrial environment).

3.3 SOIL WASHING / TOPSOIL PRODUCTION

A soil washing process has been developed and marketed as the BioGenesis^{SM} Sediment Washing Technology, which has been pilot tested and has proven to remove organic and inorganic contaminants from a range of sediment particle sizes. The pilot-scale process tested included seven steps. At the beginning of treatment, the process uses proprietary surfactants, specialty chemicals, and chelators for metal separations. Floatable organics are removed in an aeration step, and collision impact forces are used to strip the sorbed contaminants and organic coatings from the solid particles in the sediment washing step. Destruction of the organic material which has been removed from the sediment particles is accomplished in a cavitation and oxidation step. The decontaminated sediment particles are removed from the water phase in the liquid/solid separation step using mechanical methods (centrifuge), with the resulting cake (post-treated material) containing cleaned sand, silt and clay particles at approximately 30% moisture. In the beneficial use step, post-treated material is mixed with amendments to create soil products suitable for beneficial use. Biogenesis has estimated costs for this process at a full-scale facility to be $29-$35 per cubic yard.

3.4 IN-SITU TREATMENT TECHNOLOGIES

The treatment of contaminated sediments in-situ has lagged significantly behind technological advances in ex-situ treatment, in part due to the difficulty of treating subaqueous sediments and in part due to historically relatively less focus on cleaning up contaminated sediments sites. Traditionally, there has been a single dominant option for treatment, which is dredging. More recently sand caps (i.e. non-reactive caps) have become a viable option and have been used at many sites (e.g. Palos Verdes shelf in California, Grasse River in New York).

3.4.1 Reactive Capping. Reactive caps are a logical extension of non-reactive caps, providing increased resistance to contaminant flux to the water column. Reactive caps can consist of reagents that degrade contaminants as they are transported through the cap or can sequester contaminants via sorption or precipitation mechanisms. Caps that sequester contaminants may face greater scrutiny because contaminants remain in place and may be concentrated within the cap material. A recent demonstration project has placed two reactive materials in the Anacostia river, a coke material for sorption of organic contaminants retained in a geotextile mat, and an apatite mineral for sorption and surface precipitation of metals which was broadcast as a loose granular material (c.f. Melton et al., 2003).

3.4.2 In-situ sequestration. Similar to reactive caps that sequester contaminants, sequestration technologies seek to reduce the bioavailability of contaminants and the flux of contaminants to the water column. These technologies developed from the recognition that bioavailability and flux to the water column depend on pore water aqueous phase concentration and to the easily exchangeable fraction of sorbed contaminants (c.f. Ghosh et al., 2003). Addition of organic carbon-based sorbents (such as granular or powdered activated carbon and coke), has been demonstrated to reduce bioavailability and release to overlying water. Similar to reactive (or non-reactive) capping, the technology leaves contamination in place or concentrates the contamination on certain types of particles (e.g. granular activated carbon particles).

4. Metrics for Technology Assessment

New technologies are only attractive if they can offer quantifiable benefits to one or more stakeholder groups vested in a problem. The benefits might be tangible, such as reduced cost or reduced toxicological risk, or they may be intangible (but still quantifiable), such as enhanced corporate or community image. In any case, one of obstacles to adoption of new technologies is making the potential benefits readily apparent to potential early adopters, and demonstrating that the benefits are distributed fairly or equitably. To do this, and to be able to set goals and gauge progress, it is necessary to be able to *measure* the benefit in some kind of quantifiable metric. A broad discussion of multi-criteria metrics for oil spill response are discussed in a separate chapter (Seager et al 2004). Much of the discussion is applicable to novel contaminated sediment technology assessment as well, although the specific measure should always be site specific and sensitive to stakeholder values.

5. Barriers to Adoption

To understand the economic and environmental potential of the novel technologies, and to prove the final products and processes as a commercially viable, three critical areas require further analysis:

5.1 ENGINEERING

Laboratory experiments do not effectively simulate in situ conditions. Material requirements, costs, environmental effectiveness, and ecological impacts are often dependent upon the scale of technology deployment. Pilot scale experiments are required to prove the concept under field conditions, and to investigate the critical factors that will determine the commercial viability of the technology. Moreover, experience at the pilot scale with novel technologies may result in new knowledge or improvements that improve overall efficiencies and reduce costs.

5.2 ENVIRONMENTAL REGULATION

Contaminated sediments management represents a significant regulatory challenge in balancing social objectives (such as human and ecological health protection) with economic considerations. Public policies undergo constant reevaluation. Although the risk of new regulatory action may be perceived as an impediment to environmental technology investment, these challenges can also create entrepreneurial opportunities for ventures that are positioned to bridge public and private interests (Mohr 2001). New research efforts must be designed exactly for this purpose – to fill the gap between non-profit environmental research activities with a primarily social agenda and for-profit commercial ventures motivated primarily by private interests (Figure 3). One of the unique research needs is for partnership with federal and state agencies that are potential users and regulators of the technologies developed, which should aid in early identification and management of potential regulatory obstacles, and mitigate the risk of commercialization failure due to lack of user input.

5.3 INDUSTRIAL ECOLOGY

Characterization of the potential life cycle environmental and economic benefits of environmental technologies depends upon the environmental profile of the current alternatives displaced and the life-cycle demands of the new technologies developed. In the case of in situ technologies, dredging and material handling savings must be weighed against the resource requirements of manufacturing and deploying amendments or reactive barriers.

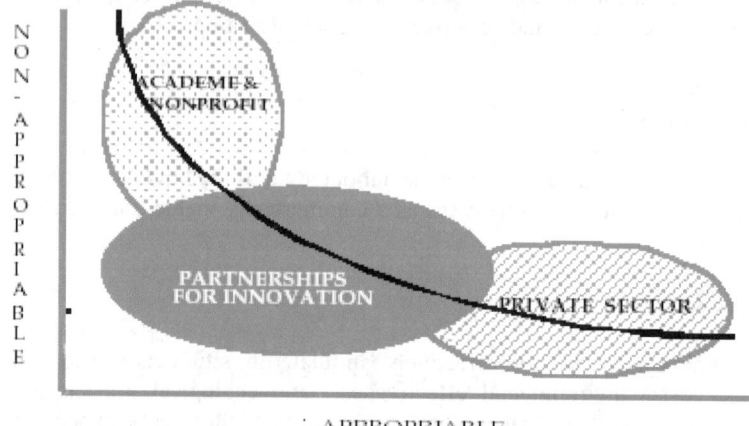

Figure 3: Environmental technology development typically falls in the gap between public and private goods (Larson and Brahmakulam 2001).

6. Education and Entrepreneurial Incubation

One approach to furthering novel technologies is to adopt a philosophy of conducting specialized research and commercial development in the broader, interdisciplinary context of economic and environmental sustainability and in cooperation with industrial and government cohorts that are positioned to contribute to and benefit from the research. Current literature demonstrates that social support and cultural norms are essential factors in cultivating successful entrepreneurs (Kassicieh et al. 1998, Jelinek 1996, Greve & Salaff 2003). A team-based approach to building partnerships between academe, industry, government, and undergraduate and graduate students could foster a supportive climate for innovation. However, it is critical to the success of the technologies to recognize that team members may have different motives, perspectives, motivations, or characteristics, as summarized in Table 1.

There may be multiple methods of fostering interdisciplinary and interagency collaboration, enhancing graduate, undergraduate and employee education and training, and fostering development of *social capital* – that is, the network of relationships, cultural norms, and trust essential to maintain a culture of innovation (Fountain 1998, Gittell and Thompson 2002). However, there are several steps or milestones that must be achieved on the path to this goal (Vohora et al. 2004). A combination of process-focused metrics, such as meetings, internships or research products, and outcome-focused metrics, such as patents, licensing or royalty agreements, and pilot projects can be employed to measure "innovative performance" (Hagedoorn and Cloodt 2003 -- see Table 2).

TABLE 1: Characteristics of Innovation Stakeholders

Participant	Actions	Primary Motives	Secondary Motives	Culture
University scientist	Discovery and education	Recognition	Financial gain, secure addn. Resources	Scientific
University Technology Transfer Office	Support faculty and entrepreneurs in partnership deals	Protect and market university's intellectual property	Facilitate technology diffusion; secure addn. research funding	Bureaucratic
Industry	Commercialize new technologies.	Financial gain	Maintain monopoly on proprietary technology	Entrepreneurial
Government	Regulate and provide public resources.	Social benefits: jobs, environmental protection, education	Regional or international competitiveness	Bureaucratic

(adapted from Siegel et a. 2003)

TABLE 2: Innovation Metrics

Process-based Metrics	Outcome-based Metrics
Partner meetings.	Invention disclosures.
Joint research papers published.	Patent filings and awards.
Joint R&D funding proposals written.	Intellectual property and royalty agreements.
New pilot-scale funding commitments secured.	Technology-related revenues & new hires.
Students enrolled in new courses and internships; research assistantships.	Mass (volume) of remediated sediment.
Engineering Systems Design doctorates & MBAs awarded.	Mass (volume) of natural raw material resources saved.

(adapted from Hagedoorn and Cloodt 2003)

In any case, the most significant obstacles to adoption of novel environmental technologies (such as contaminated sediments management) are likely to be found beyond the laboratory walls in which the technologies are customarily developed. The fact that no single organization is in a position to advocate alone for any particular technology suggests that advancements must be made through cooperative arrangements, with innovation "champions" within each organization willing to assume the leadership roles required. For these champions to be able to effectively work together, they will have to trust each other, understand one another's perspectives, recognize the importance of multiple organizational objectives, and seek strategies for commercialization that can be fairly be described as win-win.

7. Acknowledgements

The authors are grateful to Drs. Ross Gittell and Jeffrey Sohl in the Whittemore School of Business at the University of New Hampshire for their useful discussions and suggestions regarding entrepreneurship. Funding for this work was provided by NOAA through the Cooperative Institute for Coastal and Estuarine Environmental Technology.

8. References

1. Anderson AR. 1998. Cultivating the Garden of Eden: Environmental entrepreneuring. *J Organization Change Management.* 11(2):135-144.
2. Biogenesis Enterprises, Inc. and Roy F. Weston, Inc.. 1999. *BioGenesis^{SM} Sediment Washing Technology Full-Scale, 40 Cy/Hr, Sediment Decontamination Facility For The NY/NJ Harbor.* Final Report On The Pilot Demonstration Project.
3. Crittenden J. 2002. Engineering the quality of life. *Clean Tech Env Policy.* 4:6-7.
4. Dalton JL, Gardner KH, Seager TP, Weimer ML, Spear JCM, Magee BJ. 2004. Properties of Portland cement made from contaminated sediments. *Resources, Conservation and Recycling.* 41:227-241.
5. Douthwaite B, Keatinge JDH, Park JR, Why Promising Technologies Fail: The Neglected Role of User Innovation During Adoption, *Research Policy,* 2001, 30(5):819-836.
6. Eggers DM, Villani J, Andres R. 2000. Third party information provider and innovative environmental technology adoption. *Amer. Behavioral Scientist.* 44(2):265-276.
7. Fountain J. 1998. Social capital: A key enabler of innovation. In *Investing for Innovation: Creating a Research and Innovation Policy that Works* edited by Branscomb LM and Keller JH. MIT Press: Cambridge MA.
8. Greve A, Salaff, JW. 2003. Social networks and entrepreneurship. *Entrepreneurship-Theory And Practice.* 28(1):1-22.
9. Gittell R, Thompson P. 2002. Making social capital work: Social capital and community economic development in *Social Capital and Poor Communities,* edited by Saegert S, Thompson P, Warren M. Russell Sage Foundation Press: New York NY.
10. Hagedoorn J, Cloodt M, Measuring innovative performance: Is there an advantage in using multiple indicators?, *Research Policy,* 32(8):1365-1379.
11. Heaton GR, Banks RD. 1998. Toward a new generation of environmental technology. In *Investing for Innovation: Creating a Research and Innovation Policy that Works* edited by Branscomb LM and Keller JH. MIT Press: Cambridge MA.
12. Hostager TJ, Neil TC. 1998. Seeing environmental opportunities: Effects of intrapreneurial ability, efficacy, motivation, and desirability. *J Organizational Change Management.* 11(1):11-25.
13. JCI/UPCYCLE Associates, LLC. 2002. *Final Summary Report: Sediment Decontamination and Beneficial Use Pilot Project.* Prepared for New Jersey Department of Transportation (Project

AO# 9350203), Office of Maritime Resources and USEPA Region 2 through Brookhaven National Laboratory (Contract No. 48172).
14. Jelinek M. 1996. 'Thinking technology' in mature industry firms: Understanding technology entrepreneurship. *Int. J Technology Management*, 11(7-8):799-813.
15. Kassicieh SK, Radosevich HR, Banbury, CM. 1998. Using attitudinal, situational, and personal characteristics variables to predict future entrepreneurs from national laboratory inventors. *IEEE Transactions On Engineering Management*, 44(3):248-257.
16. Keough PD, Polonsky MJ. 1998. Environmental commitment: A basis for environmental entrepreneurship? *J Organizational Change Management*. 11(1):38-49.
17. Krueger N. 1998. Encouraging the identification of environmental opportunities. *J Organization Change Management*. 11(2):174-183.
18. Larson EV, Brahmakulam IT. 2001. *Partnerships: Building a New Foundation for Innovation: Results of a Workshop for the National Science Foundation*. Rand Corporation: Arlington VA. http://www.rand.org/publications/DRU/DRU2651/
19. Lober DJ. 1998. Pollution prevention as corporate entrepreneurship. *J Organizational Change Management*. 11(1):26-37.
20. Melton J. 2004. Personal communication, 2004.
21. Melton JS, Crannell BS, Eighmy TT, Wilson C, Reible DD, 2003. Field Trial of the UNH Phosphate-Based Reactive Barrier Capping System for the Anacostia River, EPA Grant Number: R819165-01-0
22. Mohr R. 2001. Environmental Policy and the Adoption of Technology. Ph.D. Dissertation. University of Texas at Austin: Austin TX.
23. Morgan MG, Fischhoff B, Bostrom A, Atman CJ. 2002. *Risk Communication: A Mental Models Approach*. Cambridge University Press: Cambridge UK.
24. Murray F. 2004. The role of academic inventors in entrepreneurial firms: Sharing the laboratory life. *Research Policy*. Article in press.
25. Pastakia A. 1998. Grassroots ecopreneurs: Change agents for a sustainable society. *J Organization Change Management*. 11(2):157-173.
26. Pennsylvania Department of Environmental Protection and New York / New Jersey Clean Ocean and Shore Trust. 2003. *The Use Of Dredged Material In Abandoned Mine Reclamation*. Final report for the Bark Camp demonstration project.
27. Sadat Associates, Inc. 2001. *Use of Dredged Materials for the Construction of Roadway Embankments*. submitted to New Jersey Maritime Resources.
28. Siegel DS, Waldman D, Link A, Assessing the impact of organizational practices on the relative productivity of university technology transfer offices: An exploratory study, *Research Policy*, 2003, 32:27-48.
29. Vohora A, Wright M, Lockett A. 2004. Critical junctures in the development of university high-tech spinout companies, *Research Policy*, 33:147–175.

Index of authors

AINSWORTH	199
ANGEL	53
APUL	261
ATOYEV	179
BLACK	53
BOUDOURESQUE	29
BOURIAKO	159
BRIDGES	261, 271
BUCHARY	199
CADIOU	29
CHEUNG	199
COOPER	169
CURA	261
EDEN	53
FORREST	199
GARDNER	299
HAGGAN	199
KATZ	53
KIKER	261, 271
LE DIRÉAC'H	29
LEVNER	95
LINKOV	169, 261, 271
LOZANO	199
MORATO	199
MORISSETTE	127, 199
NAIDENKO	159
PARSHIKOVA	81
PITCHER	199
PROTH	95, 159
SAHAY	271
SEAGER	169, 271, 299
SHERMAN	65
SILVERT	109
SPANIER	53
TAPIERO	143
YABLOKOV	11